OCR A level

Chemistry A

Oxford Cambridge and RSA

is is an OCR endorsed resource

Second Edition

2

Sam Holyman
David Scott
Victoria Stutt

PEARSON

Published by Pearson Education Limited, 80 Strand, London, WC2R 0RL.

www.pearsonschoolsandfecolleges.co.uk

Text © Pearson Education Limited 2015
Edited by Sue Gardner and Sharon Thorn
Designed by Elizabeth Arnoux and James Handlon for Pearson Education Limited
Typeset by Tech Set Ltd, Gateshead
Original illustrations © Pearson Education Limited 2015
Illustrated by Tech Set Ltd, Gateshead
Cover design by Juice Creative
Picture research by Chrissie Martin
Cover photo/illustration © **Science Photo Library Ltd:** Eye of Science

The rights of Samantha Holyman, Victoria Stutt and Dave Scott to be identified as authors of this
work have been asserted by them in accordance with the Copyright, Designs and Patents Act 1988.

First edition published 2008
This edition published 2015

2022
12

British Library Cataloguing in Publication Data
A catalogue record for this book is available from the British Library

ISBN 978 1 447 99081 9

Websites
Pearson Education Limited is not responsible for the content of any external internet sites. It is essential
for tutors to preview each website before using it in class so as to ensure that the URL is still accurate,
relevant and appropriate. We suggest that tutors bookmark useful websites and consider enabling
students to access them through the school/college intranet.

Printed in Great Britain by Ashford Colour Press Ltd.

In order to gain OCR endorsement this resource has undergone an independent quality check. OCR
has not paid for the production of this resource, nor does OCR receive any royalties from its sale. For
more information about the endorsement process please visit the OCR website www.ocr.org.uk

Acknowledgements

The publisher would like to thank Adelene Cogill, Chris Curtis and Chris Ryan for their contributions to the Maths skills and Preparing for your exams sections of this book.

The authors and publisher would like to thank the following individuals and organisations for permission to reproduce photographs:

(Key: b-bottom; c-centre; l-left; r-right; t-top)

Alamy Images: Justin Kase z08z 112br, Michael Neelon 111, Simon Dack 138; **Fotolia.com**: Alex Yeung 93 (2-L), faula 52–3, Philippe Devanne 48; **Getty Images**: David Becker 84, Stockbyte / Comstock 44; **Martyn F. Chillmaid**: 65, 74, 93 (3), 94, 105, 107, 108, 109, 110, 164; **Pearson Education Ltd**: Trevor Clifford 67br, 184; **Science Photo Library Ltd**: 88–9, 101, Andrew Brookes, National Physical Laboratory 182–3, Andrew Lambert Photography 8–9, 73bl, 73br, 77, 112tl, 115, 132, 134, 136, 137tr, 148–9, 172cl, 172bl, 173tr, Astrid & Hanns-Frieder Michler 95t, Biophoto Associates 144, British Antarctic Survey 116, Charles D. Winters 29, 38, 96, Clive Freeman, The Royal Institution 170, Colin Cuthbert 188, Dennis Schroeder, NREL / US Department of Energy 187, Dick Luria 95b, Food & Drug Administration 185, GIPhotoStock 67cl, J.C. Revy, ISM 106b, Jerry Mason 104, Laguna Design 120–1, Martyn F. Chillmaid 16, 24, 34, 137bl, 173cr, 173br, Patrice Latron / Look at Science 123, Paul Rapson 165, Power and Syred 106t, Sebastian Kaulitzki 154; **SciLabware Ltd**: 172tl; **Shutterstock.com**: Martin Kemp 93 (2-R), Ron Ellis 93 (2-C), S.Borisov 93 (1), Syda Productions 70

All other images © Pearson Education

We are grateful to the following for permission to reproduce copyright material.

Article on page 48 from *Mussels don't stick around in acidic ocean water*, http://www.dailyclimate. org/tdc-newsroom/2014/09/acidification-mussels, Sep 9, 2014. By Miguel Llanos and The Daily Climate; Article on page 84 adapted from Hydrogen gets onboard, *Chemistry World*, March (Gutowski M, Autrey T 2006), Royal Society of Chemistry http://www.rsc.org/chemistryworld/ Issues/2006/March/HydrogenOnBoard.asp; Article on page 116 adapted from *Survive Brutally Cold Temperatures* Posted by Stefan Sirucek in *Weird & Wild*, http://newswatch.nationalgeographic. com/2013/07/10/blue-blood-helps-octopus-, *National Geographic*; Article on page 144 adapted from Pesticides linked to vitamin D deficiency, *Chemistry World*, January (King, A 2012), Royal Society of Chemistry http://www.rsc.org/chemistryworld/News/2012/January/pesticides-ddt-vitamin- deficiency.asp; Article on page 178 adapted from The sweet scent of success, *Chemistry World*, February (Davies, E 2009), Royal Society of Chemistry http://www.rsc.org/chemistryworld/ Issues/2009/February/TheSweetScentOfSuccess.asp; Article on page 202 adapted from Making pain history, *Chemistry World*, March (Sutton, M), Royal Society of Chemistry http://www.rsc.org/ chemistryworld/2012/12/aspirin-history

Contents

Module 6
Organic chemistry and analysis

How to use this book

Welcome to your OCR A level Chemistry A student book. In this book you will find a number of features designed to support your learning.

Chapter openers

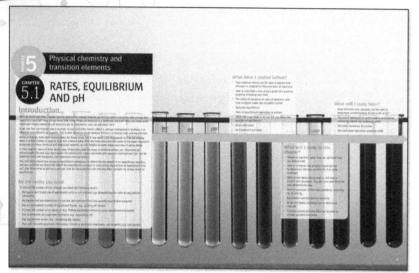

Each chapter starts by setting the context for that chapter's learning.

- Links to other areas of Chemistry are shown, including previous knowledge that is built on in the chapter and future learning that you will cover later in your course.
- The **All the maths you need** checklist helps you to know what maths skills will be required.

Main content

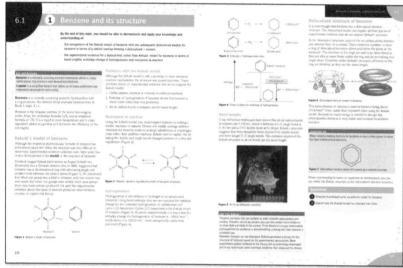

The main part of the chapter covers all of the points from the specification you need to learn. The text is supported by diagrams and photos that will help you understand the concepts.

Within each topic, you will find the following features:

- **Learning objectives** at the beginning of each topic highlight what you need to know and understand.
- **Key terms** are shown in bold and defined within the relevant topic for easy reference.
- **Worked examples** show you how to work through questions, and how your calculations should be set out.
- **Investigations** provide a summary of practical experiments that explore key concepts.
- **Learning tips** help you focus your learning and avoid common errors.
- **Did you know?** boxes feature interesting facts to help you remember the key concepts.

At the end of each topic, you will find **questions** that cover what you have just learned. You can use these questions to help you check whether you have understood what you have just read, and to identify anything that you need to look at again. Answers to all questions in this student book are available at http://www.pearsonschoolsandfecolleges.co.uk/Secondary/Science/16Biology/OCR-A-level-Science-2015/FreeResources/FreeResources.aspx.

Thinking Bigger

At the end of each chapter there is an opportunity to read and work with real-life research and writing about science. These sections will help you to expand your knowledge and develop your own research and writing techniques. The questions and tasks will help you to apply your knowledge to new contexts and to bring together different aspects of your learning from across the whole course. The timeline at the bottom of the spread highlights which other chapters of your book the material relates to.

These spreads will give you opportunities to:

* read real-life material that's relevant to your course
* analyse how scientists write
* think critically and consider relevant issues
* develop your own writing
* understand how different aspects of your learning piece together.

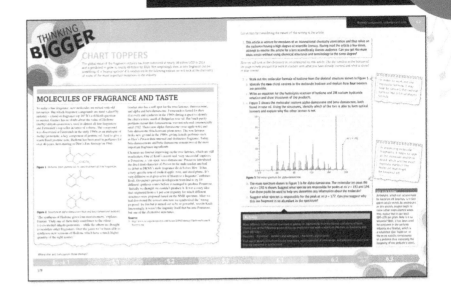

Practice questions

At the end of each chapter, there are practice questions to test how fully you have understood the learning.

Answers to all questions in this student book are available at http://www.pearsonschoolsandfecolleges. co.uk/Secondary/Science/16Biology/OCR-A-level-Science-2015/FreeResources/FreeResources.aspx.

Maths Skills

At the end of the book there is a **Maths Skills** section that focuses on key mathematical concepts to provide greater depth of explanation and enhance your understanding through worked examples.

Preparing for your exams

The book concludes with a section that offers some practical advice about preparing for your exams, including sample questions and answers that allow you to see where common mistakes are made and how you can improve your responses.

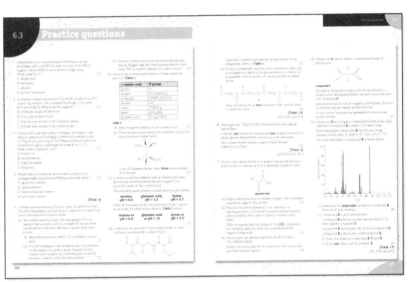

Getting the most from your ActiveBook

Your ActiveBook is the perfect way to personalise your learning as you progress through your OCR A level Chemistry course. You can:

* access your content online, anytime, anywhere
* use the inbuilt highlighting and annotation tools to personalise the content and make it really relevant to you
* search the content quickly.

Highlight tool

Use this to pick out key terms or topics so you are ready and prepared for revision.

Annotations tool

Use this to add your own notes, for example, links to your wider reading, such as websites or other files. Or make a note to remind yourself about work that you need to do.

MODULE **5**

Physical chemistry and transition elements

CHAPTER **5.1**

RATES, EQUILIBRIUM AND pH

Introduction

Why do some reactions happen quickly and others slowly? How do we find out which reactants will change the speed of a reaction? How do we know how many steps are involved in a chemical reaction? Why are some acids weak and others strong? Do all reactions go to completion, and can we affect this?

If we can find out exactly how a reaction occurs and what factors affect it, we can manipulate it to make it as effective and efficient as possible. This is why chemists study reaction kinetics, including rates and equilibrium.

Acids and bases have been known about for many years, but it has taken a lot of research to find out exactly what it is that makes an acid an acid and a base a base. Acid and base reactions are some of the most important processes in many chemical and biological systems, so it is helpful to study these reactions in some detail.

In this chapter, you will learn about rates of reactions and the ways in which reactions are influenced by temperature. You will also learn about the information (about reactants and reaction mechanisms) that can be obtained from rate equations, rate constants and rate graphs.

You will learn about how to use an equilibrium expression to determine the extent of an equilibrium reaction, and you will find out about the role of the equilibrium constant in controlling the position of equilibrium. You will also learn what acids and bases are, how to calculate their pH and why their strength all comes down to equilibrium.

All the maths you need

To unlock the puzzles of this chapter you need the following maths:

- Recognise and make use of appropriate units in calculations (*e.g. determining the units for equilibrium constants*)
- Recognise and use expressions in decimal and ordinary form (*e.g. quoting equilibrium amounts*)
- Use an appropriate number of significant figures (*e.g. quoting pH values*)
- Change the subject of an equation (*e.g. finding equilibrium amounts using equilibrium constants*)
- Use exponential and logarithm functions (*e.g. calculating pH*)
- Use logarithmic scales (*e.g. calculating pK_a values*)
- Plot and translate graphical information, including gradients, intercepts, and tangents (*e.g. rate graphs*)

What have I studied before?

- How collision theory can be used to predict how changes in conditions influence rates of reactions
- How to calculate a rate using a graph of a physical quantity changing over time
- The effect of catalysts on rates of reactions and how catalysts lower the activation barrier
- Dynamic equilibrium
- How an equilibrium expression is written
- What the magnitude of K_c can tell you about the position of equilibrium
- Acids and bases
- Le Chatelier's principle

What will I study later?

- How titrations and indicators can be used to determine concentrations of ions such as Fe^{2+}
- The acidic behaviour of some organic compounds, including phenol and carboxylic acids
- The basic behaviour of amines
- The acid–base behaviour of amino acids

What will I study in this chapter?

- Orders of reactions: what they are and how they are determined
- How to interpret rate graphs, including how to determine the rate constant for first order reactions
- What a rate-determining step is, and how to predict rate equations that are consistent with the rate-determining step
- How to calculate further rate constants, including K_p, K_a and K_w
- Equilibrium and equilibrium constants
- Acids and bases, including how to determine their pH
- Titration curves and how they can be used to choose suitable indicators

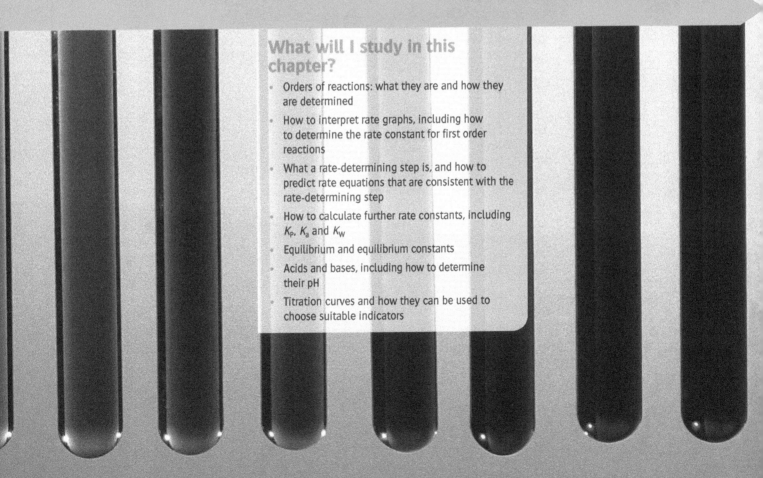

Orders, rate equations and rate constants

By the end of this topic, you should be able to demonstrate and apply your knowledge and understanding of:

* use of the terms: *rate of reaction, order, overall order, rate constant*

* explanation and deduction of:
  (i) orders from experimental data
  (ii) a rate equation from orders of the form: rate = $k[A]^m[B]^n$, where m and n are 0, 1 or 2

* calculation of the rate constant, k, and related quantities, from a rate equation including determination of units

---

KEY DEFINITIONS

Rate of reaction is the change in concentration of a reactant or a product per unit time.
The **order** with respect to a reactant is the power to which the concentration of the reactant is raised in the rate equation.
The **rate constant**, k, is the constant that links the rate of reaction with the concentrations of the reactants raised to the powers of their orders in the rate equation.
The **rate equation** for a reaction $A + B \rightarrow C$ is given by: rate = $k[A]^m[B]^n$
where m is the order of reaction with respect to A and n is the order of reaction with respect to B.
The **overall order** of a reaction is the sum of the individual orders, $m + n$

You learnt about collision theory in Book 1, topic 3.2.7. You will recall that reactions only proceed when successful collisions occur – that is, particles must collide with the correct orientation and with enough energy to overcome the activation barrier (i.e. the activation energy). How frequently these successful collisions occur will determine the rate of a reaction.

Rate of reaction

The amount of reactant that is used up, or the amount of product that is made, over a given time is known as the **rate of reaction**.

$$\text{rate of reaction} = \frac{\text{change in concentration of reactant or product}}{\text{time}}$$

Rates are usually measured in $\text{mol dm}^{-3}\,\text{s}^{-1}$ (mol per dm^3 per s), but other units may sometimes be more appropriate. If a reaction is very slow, a larger timescale (such as minutes) may be used. In this case, the units would be $\text{mol dm}^{-3}\,\text{min}^{-1}$. If it is difficult to measure the concentration, you may use another measurement which allows you to monitor the amount of a product or reactant. For example, if a gas is produced in a reaction, you could measure the volume of gas produced over a time period. In this case, the units would be $\text{cm}^3\,\text{s}^{-1}$.

Orders of reactions

If more than one reactant is involved in a reaction, each reactant can affect the rate of the reaction differently. The effect of the individual reactants is described by stating an **order** with respect to each reactant.

Consider a reactant, A. Its concentration affects the rate of a reaction. This can be expressed mathematically as:

$$\text{rate} \propto [A]^x$$

The order is always specific to each of the reactants present. There are three main types of order.

Zero order

If the order is 0 with respect to a reactant A, then: rate $\propto$ [A]^0

- The rate is unaffected by changing the concentration of A.
- Note that any number to the power 0 is equal to 1.

First order

If the order is 1 with respect to a reactant B, then: rate $\propto$ [B]^1

The rate is directly proportional to the concentration.

- If [B] increases by 2 times, the rate also increases by 2 times.
- If [B] increases by 3 times, the rate also increases by 3 times.

Second order

If the order is 2 with respect to a reactant C, then: rate $\propto$ [C]^2

The change in rate will be equal to the change in concentration squared.

- If [C] increases by 2 times, the rate increases by $2^2 = 4$ times.
- If [C] increases by 3 times, the rate increases by $3^2 = 9$ times.

Rate reactions and overall orders

Chemists use rate equations to mathematically express the influence each reactant has on a reaction.

Take the reaction A + B + C $\rightarrow$ products

If the orders for A, B and C were x, y and z respectively, you could write the following expression for the rate:

rate $\propto$ [A]^x[B]^y[C]^z

The sign for proportional can be removed if a constant is added into the equation. Chemists use the **rate constant, k**. The rate constant links the concentrations and orders of reactants to the rate. The expression then becomes a **rate equation**:

rate $= k$[A]^x[B]^y[C]^z

If any reactant is zero order then it will not appear in the rate equation. This is because it does not affect the rate. For instance, in our example above, where the orders are 0, 1 and 2 with respect to A, B and C, the rate equation would be:

rate $= k$[A]^0[B]^1[C]^2

However, because any number raised to the power zero = 1, that reactant can be removed. Powers of 1 can also be omitted from outside the brackets. Do not remove the reactant from the rate equation though! Therefore this rate equation would become:

$$\text{rate} = k[B][C]^2$$

Overall order

The **overall order** of a reaction is the sum of the individual orders.

In the example above, rate = $k[B]^1[C]^2$ and the overall order is $1 + 2$, which is 3.

The rate equation can be determined only from experimental results. Note that the orders are not the same as the numbers used to balance an equation.

> **LEARNING TIP**
>
> Remember, $dm^3\,mol^{-1}\,s^{-1}$ is the same as $dm^3/mol/s$ or dm^3 per mol per second. This is because if a quantity x is divided by a quantity y, the division x/y, can be replaced by multiplying x by y raised to the power -1: $x\,y^{-1}$.

Calculating the value and units for rate constants

The rate constant is calculated by substituting values for concentration and rates into the rate equation and rearranging to find k.

The units of k depend on the overall order of the rate reaction. The units of k are determined by substituting units for rate and concentration into the rate equation.

Zero order, rate = $k[A]^0 = k$ $\quad k = \dfrac{\text{rate}}{1}$ $\quad$ units of $k = mol\,dm^{-3}\,s^{-1}$ $\quad = mol\,dm^{-3}\,s^{-1}$

First order, rate = $k[A]$ $\quad k = \dfrac{\text{rate}}{[A]}$ $\quad$ units of $k = \dfrac{(\cancel{mol\,dm^{-3}}\,s^{-1})}{(\cancel{mol\,dm^{-3}})}$ $\quad = s^{-1}$

Second order, rate = $k[A]^2$ $\quad k = \dfrac{\text{rate}}{[A]^2}$ $\quad$ units of $k = \dfrac{(\cancel{mol\,dm^{-3}}\,s^{-1})}{(mol\,dm^{-3})^{\cancel{2}}}$ $\quad = dm^3\,mol^{-1}\,s^{-1}$

Third order, rate = $k[A]^2[B]$ $\quad k = \dfrac{\text{rate}}{[A]^2\,[B]}$ $\quad$ units of $k = \dfrac{(\cancel{mol\,dm^{-3}}\,s^{-1})}{(mol\,dm^{-3})^2\,(\cancel{mol\,dm^{-3}})}$ $\quad = dm^6\,mol^{-2}\,s^{-1}$

> **WORKED EXAMPLE 1**
>
> The following experimental results were obtained for the reaction A + B + C → products
>
> | Experiment number | Concentration of reactants/mol dm^{-3} | | | Rate/mol dm^{-3} s^{-1} |
> |---|---|---|---|---|
> | | A | B | C | |
> | 1 | 1.0×10^2 | 1.0×10^2 | 1.0×10^2 | 3.0×10^4 |
> | 2 | 2.0×10^2 | 1.0×10^2 | 1.0×10^2 | 6.0×10^4 |
> | 3 | 1.0×10^2 | 2.0×10^2 | 1.0×10^2 | 3.0×10^4 |
> | 4 | 1.0×10^2 | 1.0×10^2 | 2.0×10^2 | 12.0×10^4 |
>
> Use these results to determine the order with respect to each reactant and the overall order. Write a rate equation for the reaction.
> - Between experiment 1 and experiment 2, the concentration of A has doubled. The rate has also doubled. The order with respect to reactant A is **first order**.
> - Between experiment 1 and experiment 3, the concentration of B has doubled. The rate has remained unchanged. The order with respect to reactant B is **zero order**.
> - Between experiment 1 and experiment 4, the concentration of C has doubled. The rate has increased by 4, i.e. 2^2. This is equal to the change in concentration squared. The order with respect to C is **second order**.
>
> The rate equation can be written as:
> $$\text{rate} = k\,[A]^1[B]^0[C]^2$$
> This can be simplified to:
> $$\text{rate} = k[A][C]^2$$
> The overall order of the reaction is **third order** as $1 + 2 = 3$.

Remember that any number raised to a power of 0 is 1. Also remember that the power 1 does not need to be written.
You can check your conclusion by comparing other pairs of experiments.

WORKED EXAMPLE 2

The reaction between hydrogen, H_2, and nitrogen monoxide, NO, has the following rate equation:

rate $= k[H_2(g)][NO(g)]^2$

In a $1\,dm^3$ reaction vessel, 6.0×10^{-3} mol of $H_2(g)$ and 3.0×10^{-3} mol NO(g) were reacted together.
The initial rate of this reaction was $4.5 \times 10^{-3}\,mol\,dm^{-3}\,s^{-1}$.
Calculate the rate constant, k, for this reaction and state its units.

Answer

- Rearrange the equation to make k the subject: $k = \dfrac{\text{rate}}{[H_2(g)]\,[NO(g)]^2}$

- Substitute values and calculate k: $k = \dfrac{4.5 \times 10^{-3}}{6.0 \times 10^{-3} \times (3.0 \times 10^{-3})^2} = 8.3 \times 10^4$

- Units of $k = \dfrac{(mol\,dm^{-3}\,s^{-1})}{(mol\,dm^{-3})\,(mol\,dm^{-3})^2} = dm^6\,mol^{-2}\,s^{-1}$

- $k = 8.3 \times 10^4\,dm^6\,mol^{-2}\,s^{-1}$

LEARNING TIP

It is a mathematical convention that positive indices are written first. In this example, the units were determined to be $s^{-1}\,mol^{-2}\,dm^6$ but, following the convention, this is written as $dm^6\,mol^{-2}\,s^{-1}$. Where there is more than one unit with the same sign, they are written alphabetically, so mol^{-2} comes before s^{-1}.

Questions

1. In a reaction between R, S and T:
   - when the concentration of R is doubled, the rate is unchanged
   - when the concentration of S is quadrupled, the rate increases by 16 times
   - when the concentration of T is halved, the rate is halved.
   (a) Deduce the orders with respect to R, S and T.
   (b) Hence, write down the rate equation for the reaction.

2. The rate equation for a reaction between P and Q is rate $= k[P]^2[Q]$. How will the rate change if:
   (a) the concentration of P is doubled
   (b) the concentration of Q is tripled
   (c) the concentrations of P and Q are both tripled?

3. The reaction between ozone, $O_3(g)$, and ethene, $C_2H_4(g)$, has the following rate equation:

   rate $= k[O_3(g)][C_2H_4(g)]$

   During a reaction, $5.0 \times 10^{-8}\,mol\,dm^{-3}$ $O_3(g)$ reacted with $2.4 \times 10^{-10}\,mol\,dm^{-3}$ $C_2H_4(g)$. The initial rate of this reaction was $1.0 \times 10^{-12}\,mol\,dm^{-3}\,s^{-1}$.

   Calculate the rate constant, k, for this reaction and state its units.

(2) Concentration–time graphs

By the end of this topic, you should be able to demonstrate and apply your knowledge and understanding of:

* from a concentration–time graph:
  (i) deduction of the order (0 or 1) with respect to a reactant from the shape of the graph
  (ii) calculation of reaction rates from the measurement of gradients

* from a concentration–time graph of a first order reaction, measurement of constant half-life, $t_{1/2}$

* for a first order reaction, determination of the rate constant, k, from the constant half-life, $t_{1/2}$, using the relationship: $k = \dfrac{\ln 2}{t_{1/2}}$

* the techniques and procedures used to investigate reaction rates by the initial rates method and by continuous monitoring, including use of colorimetry

---

KEY DEFINITION

The **half-life** of a reactant is the time taken for the concentration of the reactant to reduce by half.

Chemists can use concentration–time graphs (see Book 1, topic 3.2.7) to find out about the rate and order of reactions. The shapes of concentration–time graphs depend on the order of the reaction.

Finding rates from concentration–time graphs

Chemists can use many experimental techniques to measure concentrations of reactants or products in a reaction.

For reactions involving acids or bases, we can measure:

* pH changes by carrying out titrations
* pH changes by using a pH meter.

For reactions that produce gases, we can measure:

* the change in volume or pressure
* the loss in mass of reactants.

For reactions that produce visual changes, we can observe:

* the formation of a precipitate
* a colour change.

Visual changes are usually monitored using a colorimeter, as the intensity of colour is directly related to the concentration of a coloured substance.

Consider the example of sulfur dichloride dioxide, SO_2Cl_2. This decomposes to produce sulfur dioxide and chlorine:

$$SO_2Cl_2(g) \rightarrow SO_2(g) + Cl_2(g)$$

In an experiment, the concentration of $SO_2Cl_2(g)$ was monitored every 500 seconds. The results were then plotted as a graph (see Figure 1).

The rate of reaction can be found by taking the gradient of the plotted purple line. As the data have produced a curve, tangent lines must be drawn against the curve.

The graph shows tangent lines at $t = 0$ (the initial rate) and at $t = 3000$ s.

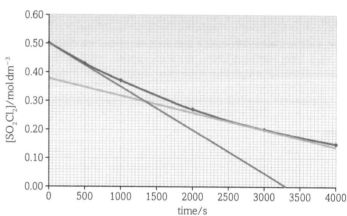

Figure 1 A concentration–time graph for SO_2Cl_2.

The rates are then calculated as follows:

* After $t = 0$ s (shown by the red, steeper tangent line):

$$\text{initial rate} = \frac{\text{change in concentration of } SO_2Cl_2}{\text{time for the change to take place}}$$
$$= \frac{(0.50 - 0.00)}{(3300 - 0)}$$
$$= 1.5 \times 10^{-4} \, \text{mol dm}^{-3} \, \text{s}^{-1}$$

* After $t = 3000$ s (shown by the green, shallower tangent line):

$$\text{rate} = \frac{\text{change in concentration of } SO_2Cl_2}{\text{time for the change to take place}}$$
$$= \frac{(0.38 - 0.14)}{(4000 - 0)}$$
$$= 6.0 \times 10^{-5} \, \text{mol dm}^{-3} \, \text{s}^{-1}$$

Half-life

The time taken for the concentration of a reactant to decrease by half is known as the **half-life** of the reactant. Half-life is given the symbol $t_{1/2}$.

You can find the order of a reaction by looking at the half-life on a concentration–time graph.

Zero order reactions

A concentration–time graph for a zero order reaction will have the following characteristic shape.

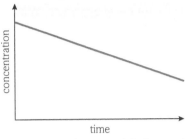

Figure 2 A zero order reaction produces a straight line graph, with the concentration decreasing at a constant rate. The half-life decreases with time.

First order reactions

A concentration–time graph for a first order reaction will have the following characteristic shape.

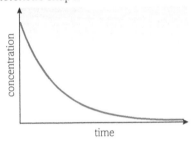

Figure 3 A first order reaction produces a curved graph with constant half-life and halving of the reactant concentration occurring at equal time intervals.

WORKED EXAMPLE 1

Nitrous oxide decomposes when strongly heated, as follows:

$$2N_2O(g) \rightarrow 2N_2(g) + O_2(g)$$

Determine the order of reaction for the decomposition of nitrous oxide using the following graph.

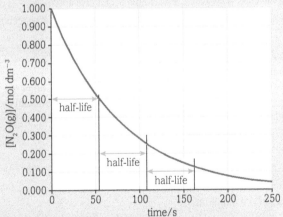

The time is determined for every halving of the concentration. In this example, the initial concentration is 1.000 mol dm⁻³.

- Half of this concentration is 0.500 mol dm⁻³. This concentration occurs at 54 s. The half-life is 54 s.
- When the concentration halves again to 0.250 mol dm⁻³ the time is 108 s. The half-life is again 54 s.
- When the concentration halves again to 0.125 mol dm⁻³ the time is 162 s. The half-life is again 54 s.

Because the half-life is constant, this is an example of a first order reaction.

Using half-lives to find the rate constant

Because first order reactions have constant half-lives, the value of the half-life can be used to determine the value of the rate constant. The half-life and the rate constant are related by the following equation:

$$k = \frac{\ln 2}{t_{1/2}}$$

This only applies to first order reactions. Reactions with other orders do not have constant half-lives.

WORKED EXAMPLE 2

Work out the rate constant for the decomposition of an airborne pollutant that has a half-life of 3 minutes.

$t_{1/2} = 3$ minutes $= 180$ seconds

$$k = \frac{\ln 2}{t_{1/2}}$$
$$= \frac{0.693}{180}$$
$$= 3.85 \times 10^{-3}\,s^{-1}$$

Questions

1 Substance X decomposes spontaneously.
   The concentration of substance X was monitored over time.
   (a) Plot a graph of the following results.

Time/s	0	15	30	60	100	180
[X]/mol dm⁻³	0.40	0.28	0.19	0.07	0.03	0.01

   (b) Suggest whether this reaction is zero or first order, given the shape of your graph.
   (c) Determine the half-life of the reaction and use this to explain your answer to part (b).
   (d) If the initial concentration of substance X had been 1.60 mol dm⁻³, what would its concentration be after 81 seconds?

2 A new medicine is found to have a biological half-life of 380 s.
   (a) How long will it take for the level of the drug in the body to fall to one-sixteenth of its original value?
   (b) Calculate the rate constant, k, for the breakdown of this drug.
   (c) Use your answer to part (a) to suggest how this new drug might need to be given to a patient.

By the end of this topic, you should be able to demonstrate and apply your knowledge and understanding of:

* from a rate–concentration graph:
  (i) deduction of the order (0, 1 or 2) with respect to a reactant from the shape of the graph
  (ii) determination of rate constant for a first order reaction from the gradient

Initial rates

The rate at the very start of a reaction, given the shorthand $t = 0$ (which stands for time = 0), is known as the initial rate. If a tangent line is drawn at $t = 0$ on a concentration–time graph, the gradient of this line is equal to the initial rate.

To determine the order with respect to each reactant, you carry out the reaction several times. Each time, you vary the concentration of one of the reactants. You can then obtain the initial rate for each of these different concentrations and plot the results on a graph to produce a rate–concentration graph for each reactant.

Clock reactions

Some reactions will produce visible changes, such as the formation of a precipitate or a dramatic colour change. Measuring the time taken for such changes to occur can be used to find out how concentration affects the initial rate. Only the initial rate is considered, as the visible changes that occur are assumed to occur as the reaction first happens, when the rate is generally fastest.

In these instances, the time taken for the visible event to occur is inversely proportional to the initial rate: the shorter the time taken for the change, the faster the reaction must be.

This can be expressed mathematically as rate $\propto \frac{1}{t}$.

This means that when a graph is plotted, $\frac{1}{t}$ is taken as a good approximation of the initial rate.

An example of this is the reaction between sodium thiosulfate, $Na_2S_2O_3$, and hydrochloric acid, HCl:

$$Na_2S_2O_3(aq) + 2HCl(aq) \rightarrow 2NaCl(aq) + S(s) + SO_2(aq) + H_2O(l)$$

A cross is drawn on a piece of paper and placed under a beaker. The reactants are then added together in the beaker. The products form a cloudy solution as a precipitate is formed, so a stopwatch is used to time how long it takes for the cross under the beaker to be obscured by the solution.

The experiment is then carried out using various concentrations of both sodium thiosulfate and hydrochloric acid to determine how their concentrations affect the rate. A graph of $\frac{1}{t}$ against concentration is plotted.

Figure 1 The reaction between sodium thiosulfate and hydrochloric acid is an example of a clock reaction. The time taken for the precipitate formed to obscure the cross is measured at varying reactant concentrations. $\frac{1}{t}$ is then used as a good approximation of the initial rate for each set of concentrations.

The iodine clock reaction

If a solution of hydrogen peroxide, H_2O_2, is added to a solution of potassium iodide, KI, in the presence of starch, acid and sodium thiosulfate, $Na_2S_2O_3$, the colourless solution will suddenly turn blue. The reaction occurs in two steps:

$$H_2O_2(aq) + 2I^-(aq) + 2H^+ \rightarrow I_2(aq) + 2H_2O(l)$$
$$2S_2O_3^{2-}(aq) + I_2(aq) \rightarrow S_4O_6^{2-}(aq) + 2I^-(aq)$$

The amount of time that passes before this colour change occurs is a measure of the initial rate.

Plan an investigation to determine the effect that the concentration of each of the reactants has on the initial rate.

Orders and rate–concentration graphs

The order with respect to reactants can be determined by plotting rate–concentration graphs. The resulting graphs will have unique shapes for each order.

Zero order

If the order is 0 with respect to a given reactant, A, the rate–concentration graph will appear as in Figure 2:

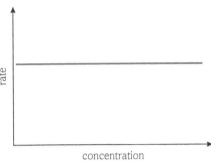

Figure 2 Zero order rate–concentration graph.

- This graph shows that rate $\propto [A]^0$.
- Changes in the concentration of this reactant have no effect on the rate.

First order

If the order is 1 with respect to a given reactant, B, the rate–concentration graph will appear as in Figure 3:

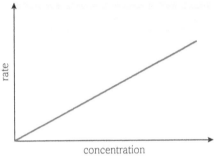

Figure 3 First order rate–concentration graph.

- This graph shows that rate $\propto [B]^1$.
- If the concentration of reactant B is doubled, the rate will double.
- If the concentration of reactant B is tripled, the rate will triple.
- If the concentration of reactant B is increased 100 fold, the rate will increase 100 fold.

Determining the rate constant from first order rate–concentration graphs

If a rate–concentration graph is first order with respect to a given reactant, it is possible to use the graph to determine the rate constant, k.

For first order reactions:
- rate $\propto$ [reactant]
- this means that rate $= k$ [reactant]

If we rearrange this to find k:

$$k = \frac{\text{rate}}{\text{concentration}}$$

Rate is plotted on the y-axis on a rate–concentration graph and concentration is plotted on the x-axis, so calculating the rate is the same as calculating the gradient, as gradient = change in y divided by change in x.

The units for k will depend on the rate equation for the given reaction.

LEARNING TIP

Did you notice that a first order rate–concentration graph is an example of $y = mx + c$ (the equation of a straight line)?

rate $= y$, reactant concentration $= x$, $c = 0$ as a rate–concentration graph always passes through zero, gradient $= k$.

Second order

If the order is 2 with respect to a given reactant, C, the rate–concentration graph will appear as in Figure 4:

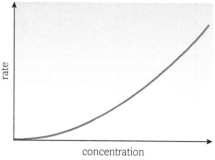

Figure 4 Second order rate–concentration graph.

- This graph shows that rate $\propto [C]^2$.
- If the concentration of reactant C is doubled, the rate will increase by 2^2.
- If the concentration of reactant B is tripled, the rate will increase by 3^2.
- If the concentration of reactant C is increased 100 fold, the rate will increase by 100^2.

Questions

1 A reaction between two reactants, P and Q, gave the rate–concentration graphs shown in Figure 5.

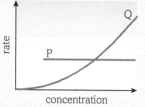

Figure 5

(a) Determine the order with respect to P and Q.

(b) Hence write the rate equation for the reaction.

2 The rate equation for a reaction between R and S is rate = $k [R]^2[S]$.

(a) Determine the order with respect to R and S.

(b) Sketch the rate–concentration graphs for reactants R and S.

3 Figure 6 shows a rate–concentration graph for reactant X in a reaction between X and Y. The order with respect to Y was found to be zero.

(a) State the order with respect to X and write a rate equation.

(b) Use the graph to determine the rate constant for the reaction.

(c) Give the units for the rate constant.

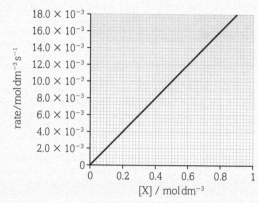

Figure 6

Rate-determining step

By the end of this topic, you should be able to demonstrate and apply your knowledge and understanding of:

* explanation and use of the term: *rate-determining step*

* for a multi-step reaction, prediction of:
  (i) a rate equation that is consistent with the rate-determining step
  (ii) possible steps in a reaction mechanism from the rate equation and the balanced equation for the overall reaction

---

KEY DEFINITIONS

A **reaction mechanism** is a series of steps that, together, make up the overall reaction.

The **rate-determining** step is the slowest step in the reaction mechanism of a multi-step reaction.

An **intermediate** is a species formed in one step of a multi-step reaction that is used up in a subsequent step, and is not seen as either a reactant or a product of the overall equation.

The rate-determining step

Reactions can occur in one step or in many steps. Experimental results can be used to predict how many steps will occur during a reaction. The series of steps that occurs during a reaction is called a **reaction mechanism**.

Some of the steps involved in the reaction mechanism can be slower than others.

* The slowest step in the reaction will dictate how quickly the reaction can proceed. This slowest step is called the **rate-determining step**.

* When you measure the rate of any reaction that has a multi-step reaction mechanism, you are effectively measuring the rate of this rate-determining step.

Predicting reaction mechanisms from rate equations

As you have learnt in previous topics, some reactants have no influence on the rate of reaction. These reactants have an order of zero and do not appear in the rate equation. If they have no influence on the rate, they cannot be involved in the rate-determining step.

Consider the hydrolysis of bromomethane:

$$CH_3Br + OH^- \rightarrow CH_3OH + Br^-$$

The rate equation is:

$$\text{rate} = k\,[CH_3Br]$$

* This shows that only the concentration of CH_3Br influences the rate. Therefore, only this molecule is involved in the rate-determining step.

* This suggests that the rate-determining step must involve the C–Br bond breaking, before the OH^- can take its place:

  Step 1: $CH_3Br \rightarrow CH_3^+ + Br^-$

  Step 2: $CH_3^+ + OH^- \rightarrow CH_3OH$

* In this example, CH_3^+ must be involved in the second step, in order to be 'used up', as it does not appear in the overall equation.

* The species involved in the two steps cancel one another to leave the overall equation.

DID YOU KNOW?

A reaction mechanism is a detailed description of the way in which a reaction occurs. It is a sequence of steps which leads from reactants to products.

Chemists propose mechanisms using rate equations derived from experimental data and balanced equations for the overall reaction.

Mechanisms can be proven wrong, though! They are, at best, 'educated guesses' and tentative explanations of chemistry that fit the experimental evidence.

Orders and the rate-determining step

For any reactant shown in the rate equation, the order attached to it tells us how many molecules of it are involved in the rate-determining step.

If we again consider our example of the hydrolysis of bromomethane, the rate equation is:

$$\text{rate} = k\,[CH_3Br]$$

* $[CH_3Br]$ is raised to the power of 1 (remember a 1 is not shown in the rate equation).

* This means that the rate-determining step involves just one molecule of CH_3Br.

* If the order is 2 with respect to a reactant, e.g. rate $= k\,[X]^2$, there will be two molecules of this reactant involved in the rate-determining step. Doubling the concentration of each of the two molecules will double the rate so overall the rate is quadrupled.

Determining the reaction mechanism

The overall balanced equation does not tell you anything about the reaction mechanism. The reaction mechanism must be worked out by looking at the rate equation and ensuring that:

- a rate-determining step is given that involves the number of molecules shown by the order for that reactant

- subsequent steps are shown that eventually generate the products shown in the balanced equation

- any **intermediate** generated is not present in the overall equation (i.e. it is used up within subsequent steps).

LEARNING TIP

Writing reaction mechanisms can feel a bit uncomfortable as you may need to use species that are not present in the overall equation and steps that you are not familiar with. Use the reactants you have identified that are in the rate-determining step in your first step and then introduce other molecules from the overall equation in the later steps. In each step, try to generate at least one product which appears in the overall equation.

It is unusual for more than two molecules to collide during a reaction so try to limit steps to:

- one molecule (e.g. decomposing)

- two molecules (e.g. colliding).

WORKED EXAMPLE

Nitrogen dioxide, NO_2, reacts with carbon monoxide, CO, to form nitrogen monoxide, NO, and carbon dioxide, CO_2:

$$NO_2(g) + CO(g) \rightarrow NO(g) + CO_2(g)$$

The overall balanced equation tells you that 1 mol $NO_2(g)$ reacts with 1 mol CO(g) to produce 1 mol NO(g) and 1 mol $CO_2(g)$.

The results of rate experiments carried out on this reaction show that the reaction is:

- second order with respect to NO_2

- zero order with respect to CO.

This gives the rate equation, rate = $k [NO_2]^2$.

This rate equation tells us that two molecules of NO_2 will be involved in the rate-determining step.

The two molecules of NO_2 go on the left-hand side of the equation for the rate-determining step. Use this as the first step in your reaction mechanism.

$NO_2 + NO_2 \rightarrow ?$ = the slow, rate-determining step

- The rate-determining step must be followed by further fast steps.

- Together, the sum of all the steps must add up to give the overall equation.

We will propose a two-step mechanism for this reaction. We first summarise what we know so far.

- *1st step* $NO_2 + NO_2 \rightarrow ?$ slow, rate-determining step
- *2nd step* $?$ $\rightarrow ?$ fast

- *Overall equation* $NO_2 + CO \rightarrow NO + CO_2$

CO and CO_2 must be involved in the second step because they are in the overall equation:

- The CO must be used up in order to produce CO_2.

- Any intermediate we make from the two molecules of NO_2 must also be used up.

NO_2 must be a product of the second step because one molecule of NO_2 is in the overall equation.

First, add in CO:

- *1st step* $NO_2 + NO_2 \rightarrow ?$ slow, rate-determining step
- *2nd step* $? + CO \rightarrow NO_2 + CO_2$ fast

- *Overall equation* $NO_2 + CO \rightarrow NO + CO_2$

Now work out what was formed in the first (rate-determining) step.

- NO_3 must be present in the second step to react with CO to form NO_2 and CO_2.

- So, to generate NO_3 for the second step, the first step must be $NO_2 + NO_2 \rightarrow NO_3 + NO$.

- This gives us the products of the first step, NO and NO_3.

The completed mechanism is:

- *1st step* $NO_2 + NO_2 \rightarrow NO + NO_3$ slow, rate-determining step
- *2nd step* $NO_3 + CO \rightarrow NO_2 + CO_2$ fast

- *Overall equation* $NO_2 + CO \rightarrow NO + CO_2$

Notice that NO_3 was generated as an intermediate. Intermediates are short-lived species. The NO_3 intermediate in this example is used up and therefore does not appear in the overall equation.

LEARNING TIP

The rate-determining step is not necessarily the first step in a reaction mechanism.

Questions

1 The rate-determining step of a reaction between X and Y is:

$2X \rightarrow Z$

Predict the rate equation for this reaction.

2 Nitrogen monoxide and oxygen react together as in the overall equation below:

$$2NO(g) + O_2(g) \rightarrow 2NO_2(g)$$

The rate equation for this reaction is rate = $k [NO]^2$.

(a) Explain what is meant by rate-determining step.

(b) What does the rate equation tell you about the rate-determining step?

(c) Suggest a possible two-step mechanism for this reaction.

5 The effect of temperature on rate constants

By the end of this topic, you should be able to demonstrate and apply your knowledge and understanding of:

* a qualitative explanation of the effect of temperature change on the rate of a reaction and hence the rate constant

* the Arrhenius equation:
  (i) the exponential relationship between the rate constant, k, and temperature, T, given by the Arrhenius equation, $k = Ae^{-E_a/RT}$
  (ii) determination of E_a and A graphically using: $\ln k = -E_a/RT + \ln A$ derived from the Arrhenius equation

---

KEY DEFINITION

An **Arrhenius plot** is a graph of $\ln k = \ln A - \dfrac{E_a}{R} \times \dfrac{1}{T}$, where $\ln k$ is plotted against $\dfrac{1}{T}$.

The rate constant, k

Reaction rate depends on both the rate constant and the concentrations of the reactants present in the rate equation:

$$\text{rate} = k[A]^x[B]^y$$

The larger the value of k, the faster the reaction.

The effect of temperature on the rate constant, k

An increase in temperature gives more energy to the molecules. This means that collisions are more frequent, and more of the collisions exceed the activation energy of the reaction. This was explained in terms of the Boltzmann distribution (see Book 1, topic 3.2.9).

The key factor affecting the reaction rate is the number of collisions that exceed the activation energy. This means that rate increases with temperature by much more than can be explained solely from any increased frequency of collisions.

Look at the rate equation above – if the rate increases with increasing temperature when the concentrations are the same, then the rate constant must increase with temperature.

* Raising the temperature speeds up the rate of most reactions by increasing the rate constant, k.

* For many reactions, the rate approximately doubles for each 10 °C increase in temperature. This reflects the greater number of reacting particles that exceed the activation energy.

* Typically, doubling the rate will double the value of the rate constant, k.

Let us look at an example reaction. Sodium thiosulfate and hydrochloric acid react as follows:

$$Na_2S_2O_3(aq) + 2HCl(aq) \rightarrow 2NaCl(aq) + S(s) + SO_2(aq) + H_2O(l)$$

The change in k with temperature for this reaction is shown in Figure 1.

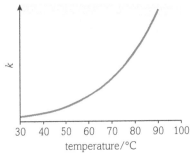

Figure 1 The rate constant for the reaction between sodium thiosulfate and hydrochloric acid increases with increasing temperature.

The Arrhenius equation

The Arrhenius equation is used to describe, mathematically, the exponential relationship between the rate constant and the temperature. The Arrhenius equation is:

$$k = Ae^{-E_a/RT}$$

where:

* k = rate constant
* E_a = activation energy
* T = temperature (in kelvin)
* e = mathematical constant with the value 2.71828. This is known as the inverse natural log.
* A = pre-exponential factor
* R = gas constant

LEARNING TIP

The Arrhenius equation and the gas constant will be provided for you on data sheets if you need to use them.

You will not need to explain what the pre-exponential factor, A, is but you may need to calculate it from information on a graph.

You can convert temperatures in °C into K by adding 273.

What can the Arrhenius equation tell us?

To be able to react, molecules have to collide with enough energy to overcome the activation energy. It has been found that at any given temperature, T, the rate constant, k, is proportional to $e^{-E_a/RT}$. This is expressed mathematically as $k \propto e^{-E_a/RT}$.

This can be turned into an equation by adding a constant, the 'pre-exponential factor', A, giving $k = Ae^{-E_a/RT}$. This is the Arrhenius equation.

The Arrhenius equation tells us that:

- temperature, T, and the rate constant, k, are related exponentially
- as temperature increases, the rate constant increases.

Adding a catalyst provides an alternative reaction path with a lower activation energy, E_a. If you follow this through mathematically, a lower value of E_a increases the rate constant, k.

Arrhenius plots

Taking logarithms of the Arrhenius equation:

$$\ln k = \ln A - \frac{E_a}{R} \times \frac{1}{T}$$

If a given reaction is carried out at varying temperatures and the value of the rate constant is calculated for each temperature, a graph of $\ln k$ can be plotted against $\frac{1}{T}$. Such graphs are referred to as **Arrhenius plots**.

This logarithmic version of the Arrhenius equation follows the general pattern $y = mx + c$:

$$\ln k = \ln A - \frac{E_a}{R} \times \frac{1}{T}$$
$$y = \quad c \quad\quad m \quad\ x$$

An Arrhenius plot can be used to identify:

- A: the intercept on the graph is equal to $\ln A$
- the activation energy, E_a: the gradient is equal to $\frac{-E_a}{R}$, where R is the known gas constant, so this can be rearranged easily to find E_a.

An example is shown in Figure 2.

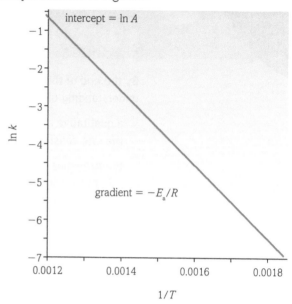

Figure 2 A typical graph of $\ln k$ against $1/T$. Using the expression $y = mx + c$, y is equal to $\ln k$, x is equal to $1/T$, m is equal to the gradient, $-E_a/R$, and c is equal to the intercept, $\ln A$.

WORKED EXAMPLE

Use the data in Figure 2 to calculate:
(a) the activation energy for the reaction
(b) the pre-exponential factor, A, for the reaction.

Solution

(a) Using two points on the plot, the gradient has been calculated as -9.1×10^3 K. (The slope has the units K as $\frac{1}{T}$ is plotted on the x-axis.)

Because gradient $= \frac{-E_a}{R}$, $E_a = \frac{-R}{\text{gradient}}$.

Therefore, $E_a = -8.3145$ J K^{-1} mol^{-1} $\times -9.1 \times 10^3$ K $= \frac{75\,662.0\,\text{J mol}^{-1}}{1000}$

$= 75.7$ kJ mol^{-1}

(b) Intercept on the plot $= -0.7$. Therefore $\ln A = -0.7$ and $A = e^{-0.7} = 0.5$

Questions

1. A reaction between X and Y was carried out at various temperatures and the rate constant determined for each temperature.

   Use the results from the experiment to draw an Arrhenius plot.

Temperature/°C	Rate constant, k/ dm^3 mol^{-1} s^{-1}
10	2.49×10^{-4}
20	7.10×10^{-4}
30	1.84×10^{-3}
40	4.75×10^{-3}
45	7.45×10^{-3}

2. Use your Arrhenius plot to determine the activation energy for the reaction between X and Y. Show all your workings.

3. Calculate the pre-exponential factor, A, if the intercept on the Arrhenius plot y-axis occurs at 22.4.

By the end of this topic, you should be able to demonstrate and apply your knowledge and understanding of:

* the calculation of quantities present at equilibrium, given appropriate data
* the techniques and procedures used to determine quantities present at equilibrium
* expressions for K_c and K_p for homogeneous and heterogeneous equilibria
* calculations of K_c and K_p, or related quantities, including determination of units

> **KEY DEFINITIONS**
>
> The **equilibrium law** states that for the equilibrium $aA + bB \rightleftharpoons cC + dD$
>
> $$K_c = \frac{[C]^c[D]^d}{[A]^a[B]^b}$$
>
> A **homogeneous equilibrium** is an equilibrium in which all the species making up the reactants and products are in the same physical state.
>
> A **heterogeneous equilibrium** is an equilibrium in which species making up the reactants and products are in different physical states.

Dynamic equilibrium and K_c

Not all reactions go to completion. An equilibrium may be established. This is when the rates of the forward and reverse reactions are equal.

Equilibrium does not occur straight away. The reverse reaction cannot take place until the forward reaction has produced a high enough concentration of the product.

Consider the decomposition of dinitrogen tetraoxide, N_2O_4 (a colourless gas) to nitrogen oxide, NO_2 (a brown gas):

$$N_2O_4(g) \rightleftharpoons 2NO_2(g)$$

* At first the concentration of N_2O_4 is high and the concentration of NO_2 is zero.
* As the reaction proceeds, the concentration of N_2O_4 falls and the concentration of NO_2 rises. The reaction mixture appears colourless initially, then a pale brown colour will appear.
* Once equilibrium is reached the concentration of each substance remains unchanged – the reaction has a brown appearance.

This is shown in Figure 1.

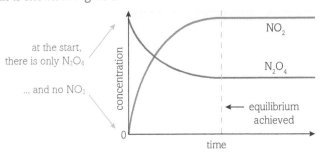

Figure 1 Graph to show how the concentrations of N_2O_4 and NO_2 change until equilibrium is achieved.

* If conditions change, the equilibrium will need to be re-established. A colour change will be observed until the equilibrium is restored.

Remember, according to the **equilibrium law**, for the reaction:

$$aA + bB \rightleftharpoons cC + dD$$

the equilibrium constant, K_c, can be calculated as follows:

$$K_c = \frac{[C]^c[D]^d}{[A]^a[B]^b}$$

* Once an equilibrium is established, K_c remains constant unless the temperature of the reaction changes.

> **LEARNING TIP**
>
> When you are discussing the equilibrium constant, make sure you use a capital K, followed by a subscript letter to indicate which equilibrium constant you are referring to. Be careful not to muddle this K with the symbol for potassium.
>
> A lower-case k is the rate constant, so do not get the three different $k/K/K$ symbols muddled up!

Units of K_c

The units for K_c expressions have to be worked out afresh for each different reaction. Each concentration term in the K_c expression is replaced by its units.

* For the equilibrium described above, the expression for K_c is:

$$K_c = \frac{[NO_2(g)]^2}{[N_2O_4(g)]}$$

Units: $\dfrac{(mol\ dm^{-3})^2}{(mol\ dm^{-3})} = mol\ dm^{-3}$

Note that one of the concentration terms, $mol\ dm^{-3}$, cancels.

* For the reaction: $2SO_2(g) + O_2(g) \rightleftharpoons 2SO_3(g)$

$$K_c = \frac{[SO_3(g)]^2}{[SO_2(g)]^2\ [O_2(g)]}$$

Units: $\dfrac{(mol\ dm^{-3})^2}{(mol\ dm^{-3})^2\ (mol\ dm^{-3})} = dm^3\ mol^{-1}$

Note that positive indices are placed before negative indices, so the units are $dm^3\ mol^{-1}$ rather than $mol^{-1}\ dm^3$.

* For the reaction: $H_2(g) + I_2(g) \rightleftharpoons 2HI(g)$

$$K_c = \frac{[HI(g)]^2}{[H_2(g)]\ [I_2(g)]}$$

Units: $\dfrac{(mol\ dm^{-3})^2}{(mol\ dm^{-3})(mol\ dm^{-3})} = no\ units$

Note that all the units cancel.

In Book 1 you met Le Chatelier's principle. Remember this can be used to predict how changes in conditions will affect equilibrium.

Determining concentrations at equilibrium

In order to determine a value for K_c, the concentrations of the reactants and products at equilibrium need to be known. Chemists can determine the concentrations of reactants and products in a number of ways. The most common ways are:

- titration

- using a colorimeter.

Titrating one of the reactants or products against a substance with a known concentration will enable us to find out how much of the reactant or product is present. There are drawbacks with this method however. Reactant or product has to be removed from the reaction mixture, and this will alter the position of the equilibrium. Also, other substances may be present that can affect the results of titrations. For example, if an alkali is used to titrate against the equilibrium substance in question and an acid catalyst is also used, both equilibrium substance and catalyst will react with the alkali. The amount of any such substance must be carefully determined when the titration results are used to determine equilibrium concentrations.

A colorimeter can be used to monitor the concentration of a reactant or product throughout. This works because coloured substances alter the amount of light that can pass through a substance; more intense colours mean the substance has a higher concentration and absorbs more light.

Often the equilibrium concentration of one substance only is determined and compared with the initial concentration. Other concentrations can then be calculated using the balanced equation.

Figure 2 A colorimeter monitors the concentration of a reactant or product by measuring the amount of light passing through a substance.

INVESTIGATION

Equilibrium investigation

Ethanol and ethanoic acid react, in the presence of an acid catalyst, to form the ester ethyl ethanoate through the following equilibrium reaction:

$$CH_3COOH + CH_3CH_2OH \rightleftharpoons CH_3COOCH_2CH_3 + H_2O$$

Choose suitable starting amounts and concentrations and set up the above reaction. Leave it in a sealed reaction vessel (to prevent evaporation) for at least a week to allow the equilibrium to establish. Using a standard solution of sodium hydroxide and a phenolphthalein indicator, carry out a titration to find out how much ethyl ethanoate has formed and hence the equilibrium concentrations of all the substances involved.

Calculate a value for K_c for the reaction.

Calculating K_c from equilibrium concentrations

WORKED EXAMPLE 1

$H_2(g)$ and $I_2(g)$ were mixed in a sealed container of volume $2\,dm^3$ and allowed to reach equilibrium.

$$H_2(g) + I_2(g) \rightleftharpoons 2HI(g)$$

The number of moles at equilibrium was:

- $H_2(g)$ 0.280 mol
- $I_2(g)$ 0.080 mol
- $HI(g)$ 0.640 mol

Calculate K_c and work out the units for this reaction.

Firstly, calculate the concentrations at equilibrium:

- $H_2(g)$ 0.280 mol/2 dm³ = 0.140 mol dm⁻³
- $I_2(g)$ 0.080 mol/2 dm³ = 0.040 mol dm⁻³
- $HI(g)$ 0.640 mol/2 dm³ = 0.320 mol dm⁻³

Write the expression for K_c and work out the units. Then calculate K_c under these conditions:

- $K_c = \dfrac{[HI(g)]^2}{[H_2(g)]\,[I_2(g)]}$

- Units: $\dfrac{(mol\,dm^{-3})^2}{(mol\,dm^{-3})\,(mol\,dm^{-3})}$ = no units (all cancel)

- Use the equilibrium concentrations to calculate a value for K_c:

$K_c = \dfrac{(0.320)^2}{0.140 \times 0.040} = 18.3$

Calculating unknown equilibrium concentrations

In Book 1 you carried out calculations involving stoichiometry and reacting quantities.

From the balanced equation you can find:

- the reacting quantities needed to prepare a required quantity of a product
- the quantities of products formed by reacting together known quantities of reactants.

For reactions that go to completion, you can assume that all the reactants are converted into products. This is not the case in equilibrium reactions. We still use the balanced equation, but we need to do a little more work to find the amounts of reactants and products present at equilibrium.

The equilibrium constant, K_c, can be determined from experimental results. Worked example 2 shows how to determine the equilibrium concentrations of the components in an equilibrium mixture.

WORKED EXAMPLE 2

$H_2(g)$, $I_2(g)$ and $HI(g)$ exist in equilibrium: $H_2(g) + I_2(g) \rightleftharpoons 2HI(g)$

0.60 mol of $H_2(g)$ was mixed with 0.40 mol of $I_2(g)$ in a sealed container with a volume of 1 dm³. The mixture was allowed to reach equilibrium and it was found that 0.28 mol of $H_2(g)$ remained.
Find the concentrations of all three components in the equilibrium mixture.

- It is useful to summarise the information in a table. At the start, only $H_2(g)$ and $I_2(g)$ are present and these values have been added to the table below in blue.
- You are also told how much $H_2(g)$ is present at equilibrium and this value has been added to the second row of the table in green.

Component	$H_2(g)$	$I_2(g)$	$HI(g)$
Initial amount/mol	0.60	0.40	0
Equilibrium amount/mol	0.28		

We now need to find the equilibrium amounts of I_2 and HI.

- You can use the information in the table to find out how much $H_2(g)$ has reacted.
- Amount of H_2 reacted = initial amount − equilibrium amount
  = 0.60 − 0.28 = 0.32 mol

You now use the balanced equation to find out how much I_2 has reacted and how much HI has formed.

- From the equation: 1 mol H_2 + 1 mol I_2 → 2 mol HI
- Therefore: 0.32 mol H_2 + 0.32 mol I_2 → 0.64 mol HI

You can now work out the equilibrium amount of I_2 using the same method as for H_2 above.

- Amount of I_2 remaining = initial amount − amount of I_2 reacted
  = 0.40 − 0.32 = 0.08 mol
- From the equation, the equilibrium amount of HI = 0.64 mol

Finally, convert the amounts to concentrations. Here the volume of the container is 1.00 dm³, so we already have the amount in 1 dm³. The concentration numbers are the same as the equilibrium amounts.

- Therefore, at equilibrium, $[H_2(g)] = 0.28$ mol dm⁻³;
  $[I_2(g)] = 0.08$ mol dm⁻³; $[HI(g)] = 0.64$ mol dm⁻³

Homogeneous and heterogeneous reactions

All the equilibrium reactions we have considered above have involved reactants and products in the same state (solid, liquid, gas). They are examples of **homogeneous equilibrium** reactions.

If more than one state is present then the reaction is called a **heterogeneous equilibrium**, for example $H_2O(l) \rightleftharpoons H_2O(g)$

When a heterogeneous equilibrium is present, molar concentrations for solids and pure liquids do not change because their volume remains constant.

This means the expression for K_c changes:

- the concentrations of solid substances are not included in the expression
- the concentrations of pure liquids are not included in the expression.

For example, when phosphorus reacts with oxygen, the following heterogeneous equilibrium is established:

$$P_4(s) + 5O_2(g) \rightleftharpoons P_4O_{10}(s)$$

Hence $K_c = [O_2]^5$

Questions

1. 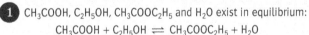 CH_3COOH, C_2H_5OH, $CH_3COOC_2H_5$ and H_2O exist in equilibrium:
   $$CH_3COOH + C_2H_5OH \rightleftharpoons CH_3COOC_2H_5 + H_2O$$
   2.00 moles of CH_3COOH were mixed with 3.00 moles of C_2H_5OH and the mixture allowed to reach equilibrium. At equilibrium 0.43 moles of CH_3COOH remained.
   (a) How many moles of CH_3COOH have reacted?
   (b) Determine the equilibrium concentrations of CH_3COOH, C_2H_5OH, $CH_3COOC_2H_5$ and H_2O. (You will need to use V to represent the volume, but this will cancel out in your K_c calculation.)
   (c) Hence calculate K_c.

2. 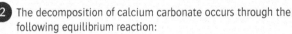 The decomposition of calcium carbonate occurs through the following equilibrium reaction:
   $$CaCO_3(s) \rightleftharpoons CaO(s) + CO_2(g)$$
   (a) Explain whether this reaction is heterogeneous or homogeneous.
   (b) Write an expression for K_c for the reaction.

7 Equilibrium and K_p

By the end of this topic, you should be able to demonstrate and apply your knowledge and understanding of:

* use of the terms *mole fraction* and *partial pressure*

* expressions for K_c and K_p for homogeneous and heterogeneous equilibria

* calculations of K_c and K_p, or related quantities, including determination of units

Mole fractions, partial pressures and K_p

Mole fraction

The amount of a given component within a reaction mixture is called the **mole fraction** and is given the symbol X.

The mole fraction of a substance, A, is calculated as follows:

$$\text{mole fraction, } X_A = \frac{\text{number of moles of substance A}}{\text{total number of moles of all substances}}$$

Partial pressures

The pressure of any reaction mixture involving gases (within a sealed vessel) is effectively the sum of the pressures being exerted by each of the gaseous substances involved.

The amount of pressure being exerted by an individual species within a reaction vessel is called the **partial pressure** and is given the symbol P, followed by a subscript denoting the species to which it refers. (Be careful not to confuse this with the chemical symbol for phosphorus or the generic term for pressure of that substance.)

If we know the total pressure of a reaction along with the mole fraction of a given substance, we can calculate the partial pressure of that substance.

* For substance A: partial pressure, P_A = mole fraction × total pressure

The concentration of a substance is proportional to its partial pressure.

LEARNING TIP

There are many units of pressure that are used within chemistry. You are most likely to come across:

* atmospheres, atm
* pascals, Pa (this is the SI unit for pressure)
* newtons per square metre, $N\,m^{-2}$

$1\,atm = 1.01 \times 10^5\,Pa$, or $101\,kPa$
$1\,Pa = 1\,N\,m^{-2}$

LEARNING TIP

Make sure you do not confuse the type of brackets you use in equilibrium expressions:
* round brackets are used in K_p expressions to denote partial pressures
* square brackets are used in K_c expressions to denote concentrations.

K_p

For reactions involving gases, equilibrium expressions can be written using partial pressures instead of concentrations. This gives a new equilibrium constant, K_p.

For the reaction $a\text{A} + b\text{B} \rightleftharpoons c\text{C} + d\text{D}$, the equilibrium constant, K_p, can be calculated as follows:

$$K_p = \frac{(P_C)^c\,(P_D)^d}{(P_A)^a\,(P_B)^b}$$

For the reaction $H_2(g) + I_2(g) \rightleftharpoons 2HI(g)$

$$K_p = \frac{(P_{HI})^2}{(P_{H_2})\,(P_{I_2})}$$

WORKED EXAMPLE

The following equilibrium was established in a 1 dm³ vessel. The total pressure was 4.1×10^7 Pa.

* $H_2(g) + I_2(g) \rightleftharpoons 2HI(g)$

The concentrations of the individual substances were found to be:

* $[H_2(g)] = 0.28 \text{ mol dm}^{-3}$
* $[I_2(g)] = 0.08 \text{ mol dm}^{-3}$
* $[HI(g)] = 0.64 \text{ mol dm}^{-3}$

As the volume was 1 dm³, the number of moles of each substance is:

* $H_2 = 0.28$ mol
* $I_2 = 0.08$ mol
* $HI = 0.64$ mol

The mole fractions are as follows:

* $H_2 = \dfrac{\text{amount in mol of } H_2}{\text{total number of mol}} = \dfrac{0.28}{1.0} = 0.28$
* $I_2 = \dfrac{0.08}{1} = 0.08$
* $HI = \dfrac{0.64}{1} = 0.64$

The partial pressures are:

* $P_{H_2} = \text{mole fraction} \times \text{total pressure} = 0.28 \times 4.1 \times 10^7 \text{ Pa} = 1.15 \times 10^7$ Pa
* $P_{I_2} = 0.08 \times 400 = 3.3 \times 10^6$ Pa
* $P_{HI} = 0.64 \times 400 = 2.6 \times 10^7$ Pa

$$K_p = \frac{(P_{HI})^2}{(P_{H_2})(P_{I_2})}$$

$$= \frac{(2.6 \times 10^7 \text{ Pa})^2}{(1.15 \times 10^7 \times 3.3 \times 10^6)} = 18.09$$

$$\text{units} = \frac{\text{Pa} \times \text{Pa}}{\text{Pa} \times \text{Pa}} = \text{no units}$$

LEARNING TIP

You will notice that this K_p value is the same as the value for K_c worked out for the same reaction in an earlier worked example. This is not always the case.

K_p expressions for heterogeneous reactions

The expression for K_p when a reaction is heterogeneous will alter as follows:

* solids will not be included in the expression

* pure liquids will not be included in the expression.

Questions

1. Write an equilibrium expression for K_p for the following reaction: $Ni(s) + 4CO(g) \rightleftharpoons Ni(CO)_4(g)$

2. The following equilibrium was established: $PCl_5(g) \rightleftharpoons PCl_3(g) + Cl_2(g)$

   At a pressure of 60 795 Pa, the equilibrium amounts were found to be:

   $PCl_5 = 1.00 \times 10^{-2}$ mol

   $Cl_2 = 2.03$ mol

   $PCl_3 = 0.47$ mol

   (a) Determine the mole fraction for each substance.

   (b) Determine the partial pressure for each substance.

   (c) Write an expression for K_p.

   (d) Calculate K_p.

(8) Equilibrium constants and their significance

By the end of this topic, you should be able to demonstrate and apply your knowledge and understanding of:

* the qualitative effect on equilibrium constants of changing temperature for exothermic and endothermic reactions

* the constancy of equilibrium constants with changes in concentration, pressure or in the presence of a catalyst

* explanation of how an equilibrium constant controls the position of equilibrium on changing concentration, pressure and temperature

* application of the principles for K_c, K_p to other equilibrium constants, where appropriate

The significance of equilibrium constants

The magnitude of the equilibrium constants, K_c and K_p, indicates the extent of a chemical reaction.

An equilibrium constant, K, with a value of 1 would indicate that the position of equilibrium is halfway between reactants and products.

When K is greater than 1:

* the reaction is product-favoured
* the products on the right-hand side predominate at equilibrium.

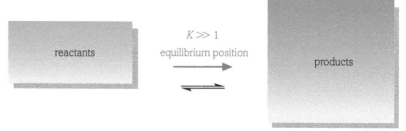

Figure 1 When K is greater than 1, the equilibrium lies to the right.

When K is less than 1:

* the reaction is reactant-favoured
* the reactants on the left-hand side predominate at equilibrium.

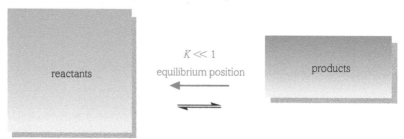

Figure 2 When K is less than 1, the equilibrium lies to the left.

How do changes in temperature affect K?

Previously you will have used Le Chatelier's principle (Book 1, topic 3.2.10) to predict how changes in temperature affect the position of equilibrium. You used the sign of ΔH, which means change in enthalpy, to make predictions.

- An increase in temperature shifts the position of equilibrium in the endothermic direction (ΔH +ve).

- A decrease in temperature shifts the position of equilibrium in the exothermic direction (ΔH −ve).

- The ΔH values for the forward and reverse reactions in an equilibrium have the same magnitude, but have opposite signs.

The control of the equilibrium constant

Shifts in the position of equilibrium are actually controlled by the rate constant, which changes its value only with changes in temperature. When any other changes occur to conditions, K will remain constant. The way that K changes is linked to ΔH.

Forward reaction endothermic: ΔH +ve

If the forward reaction is endothermic, then K increases as temperature increases:

- the equilibrium yield of the products, on the right-hand side, increases

- the equilibrium yield of the reactants, on the left-hand side, decreases.

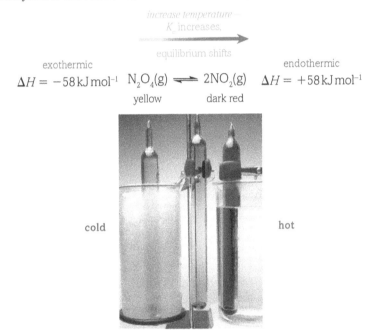

increase temperature —
K_c increases.

equilibrium shifts

exothermic   endothermic

$\Delta H = -58\,\text{kJ}\,\text{mol}^{-1}$ $N_2O_4(g) \rightleftharpoons 2NO_2(g)$ $\Delta H = +58\,\text{kJ}\,\text{mol}^{-1}$

yellow   dark red

cold   hot

Figure 3 Nitrogen dioxide equilibrium: a change in temperature in tubes of N_2O_4 and NO_2 gases shifts the equilibrium between the two species; when more NO_2 is produced, the gas mixture inside the tube becomes darker.

Forward reaction exothermic: ΔH −ve

If the forward reaction is exothermic, then K decreases as temperature increases:

- the equilibrium yield of the products, on the right-hand side, decreases

- the equilibrium yield of the reactants, on the left-hand side, increases.

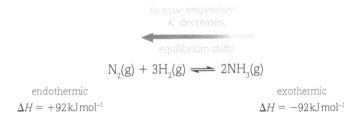

increase temperature —
K_c decreases.

equilibrium shifts

$N_2(g) + 3H_2(g) \rightleftharpoons 2NH_3(g)$

endothermic   exothermic

$\Delta H = +92\,\text{kJ}\,\text{mol}^{-1}$ $\Delta H = -92\,\text{kJ}\,\text{mol}^{-1}$

Figure 4 Ammonia equilibrium: the forward reaction is exothermic, so an increase in temperature decreases the value of K.

DID YOU KNOW?

Changes in temperature are the only changes that affect the equilibrium constants, K_c and K_p.

- K increases as the temperature rises if the forward reaction is endothermic (ΔH +ve).

- K decreases as the temperature rises if the forward reaction is exothermic (ΔH −ve).

How do changes in concentration and pressure affect K?

Although the equilibrium constant, K, is altered by changes in temperature, the value of K is unaffected by changes in concentration and pressure.

Changes in concentration

For a reaction mixture of the equilibrium $N_2O_4(g) \rightleftharpoons 2NO_2(g)$, the equilibrium concentrations are:

$[NO_2(g)] = 1.60 \, \text{mol dm}^{-3}$

$[N_2O_4(g)] = 0.200 \, \text{mol dm}^{-3}$

Therefore $K_c = \dfrac{[NO_2(g)]^2}{[N_2O_4(g)]} = \dfrac{(1.60)^2}{0.200} = 12.8 \, \text{mol dm}^{-3}$

If $[N_2O_4(g)]$ is doubled from $0.200 \, \text{mol dm}^{-3}$ to $0.400 \, \text{mol dm}^{-3}$:

- the concentrations in the K_p expression now give the ratio

$\dfrac{(1.60)^2}{0.400} = 6.4 \, \text{mol dm}^{-3}$

- the system is no longer in equilibrium.

The equilibrium position must shift to restore this ratio to the K_c value of $12.8 \, \text{mol dm}^{-3}$. The system must:

- increase $[NO_2(g)]$ (on the top of the expression for K_c)

- decrease $[N_2O_4(g)]$ (on the bottom of the expression for K_c).

This causes a shift in the equilibrium position from left to right.

The same will be true for the rate constant K_p as this will also remain constant when concentrations, or the equivalent partial pressures, are changed.

> **LEARNING TIP**
>
> Le Chatelier's principle tells us that a reaction will change to minimise any changes made to a reaction at equilibrium. This is true because the reaction shifts to keep the value of K constant.

Changes in pressure

If the pressure is doubled, the concentrations of both $[NO_2(g)]$ and $[N_2O_4(g)]$ will also effectively double (for example, there will be the same number of particles in half the space).

> **LEARNING TIP**
>
> Pressure and volume are inversely proportional to one another, so if pressure is increasing and there are no changes in temperature or concentration then the volume must be decreasing.
>
> Remember that concentration and volume can be linked by $n = cV$
>
> If the number of moles stays the same and volume decreases, concentration must increase.
>
> The molar volume of any gas is proportional to its partial pressure, so increased concentration = increased partial pressure.

The concentrations in the K_c expression now give the ratio:

$\dfrac{(2 \times 1.60)^2}{2 \times 0.200} = \dfrac{(3.20)^2}{0.400} = 25.6 \, \text{mol dm}^{-3}$

The system is no longer in equilibrium so the equilibrium position must shift to restore this ratio to the K_c value of $12.8 \, \text{mol dm}^{-3}$. The system must:

- decrease $[NO_2(g)]$ on the top

- increase $[N_2O_4(g)]$ on the bottom.

This causes a shift in the equilibrium position from right to left. The same will be true for the rate constant K_p.

How does the presence of a catalyst affect K?

The short answer to this is 'It doesn't', for both K_c and K_p.

Catalysts affect the *rate* of a chemical reaction, but not the position of equilibrium.

Catalysts speed up both the forward and reverse reactions in the equilibrium by the same factor. Equilibrium is reached more quickly, but the equilibrium position, and hence the value of the equilibrium constant, is unchanged by the action of a catalyst.

Questions

1. Predict whether the two reactions, A and B, are exothermic or endothermic. Explain your answer.

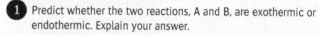

Temperature/K	Numerical value of K_c	
	Reaction A	Reaction B
500	5000	2.8×10^{-3}
800	0.3	4.5×10^{-2}

2. In a closed system, $NO(g)$, $O_2(g)$ and $NO_2(g)$ exist in equilibrium:

$2NO(g) + O_2(g) \rightleftharpoons 2NO_2(g)$

(a) Write the expressions for K_c and K_p.

(b) The concentration of $NO(g)$ is increased. Explain in terms of K_c and K_p what happens to the equilibrium composition.

(c) The pressure on the system is decreased. Explain in terms of K_c and K_p what happens to the equilibrium composition.

3. The equilibrium position does not change with changes of pressure when there are the same number of gaseous moles on both sides. Explain this statement, in terms of K_c and K_p.

(9) Brønsted–Lowry acids and bases

By the end of this topic, you should be able to demonstrate and apply your knowledge and understanding of:

* a Brønsted–Lowry acid as a species that donates a proton and a Brønsted–Lowry base as a species that accepts a proton

* use of the term *conjugate acid–base pairs*

* monobasic, dibasic and tribasic acids

KEY DEFINITIONS

A **Brønsted–Lowry acid** is a proton, H^+, donor.
A **Brønsted–Lowry base** is a proton, H^+, acceptor.

Conjugate acid–base pairs are two species related to each other by the loss or gain of a proton.

Brønsted–Lowry acids and bases

The Brønsted–Lowry model of acids and bases states that:

* reactions between acids and bases involve the transfer of H^+ ions (known as protons)

* a **Brønsted–Lowry acid** is any substance that can donate a proton

* a **Brønsted–Lowry base** is any substance that can accept a proton.

LEARNING TIP

Acids or bases are referred to as Brønsted–Lowry acids or bases because they behave in the way described by the Brønsted–Lowry model. Acids and bases you will come across are examples of this – do not be put off by their full descriptive title!

Models for acids and bases over time

Chemists have studied the behaviour of acids and bases over many years. Our understanding, or model, of how they behave has changed as new research has been conducted.

* In the 1770s many chemists were investigating air and the gases within it. Joseph Priestley and Karl Scheele both reported findings that suggested the existence of the gas we now know as oxygen. Another chemist, Antoine-Laurent de Lavoisier, found, among other things, that this gas was important in the rusting of metal and in 1778 he proposed that it was the source of acidity. Lavoisier was wrong, but it was an early step towards the understanding of acids.

* In 1815 Humphrey Davy showed that some acidic substances, such as HCl, did not actually contain oxygen.

* In 1832 Justus Liebig defined an acid as a substance containing hydrogen that could be replaced by a metal – very close to today's definition.

* In the late 1880s Svante Arrhenius proposed that acids dissociated in water to form hydrogen ions, H^+, and that bases dissociated in water to form hydroxide ions, OH^-. This is true for many acids and bases, and they can be referred to as Arrhenius acids and bases. However, the model breaks down when the acids and bases are not solutions in water or when bases are not soluble hydroxides.

* In 1929 the Brønsted–Lowry definition discussed above was proposed by the Danish chemist Johannes Brønsted and the British chemist Thomas Lowry.

Even the Brønsted–Lowry model can break down, e.g. for some solvents. The American chemist Gilbert Lewis proposed a more general theory that describes an acid (known as a Lewis acid) as an electron-pair acceptor and a base (a Lewis base) as an electron-pair donor. This model is beyond the scope of your A level studies.

Examples of Brønsted–Lowry acids and bases in action

When an acid is added to water, the acid dissociates (splits up), releasing H^+ ions (protons) into solution.

* For hydrogen chloride, HCl: $HCl(g) \rightarrow H^+(aq) + Cl^-(aq)$
  The HCl has donated an H^+ ion – it is a Brønsted–Lowry acid.

* For sulfuric acid, H_2SO_4: $H_2SO_4(l) \rightarrow H^+(aq) + HSO_4^-(aq)$
  The H_2SO_4 has donated an H^+ ion – it is a Brønsted–Lowry acid. HSO_4^+ can actually go on to donate another H^+ ion (sulfuric acid is dibasic) – this will be covered later.

Bases are able to accept protons.

* For ammonia, NH_3: $NH_3(aq) + H^+(aq) \rightarrow NH_4^+(aq)$
  The ammonia has accepted a proton (H^+) – it is a Brønsted–Lowry base.

Ionic equations

Because acid–base reactions involve the transfer of H^+ ions, they are usually represented by ionic equations.

Ionic equations show:

* the H^+ ions being donated
* the ion that is accepting the H^+ ion
* the products formed
* all charges, correctly balanced (i.e. they cancel out)
* state symbols for all substances, including ions.

For the reaction between hydrochloric acid, HCl, and sodium hydroxide (NaOH):

$$HCl(aq) + NaOH(aq) \rightarrow NaCl(aq) + H_2O(l)$$

the ionic equation would be written as follows:

$$H^+(aq) + OH^-(aq) \rightarrow H_2O(l)$$

> **LEARNING TIP**
>
> The Na^+ and Cl^- ions are not involved in the actual acid–base reaction. They are referred to as spectator ions and do not need to be included in the ionic equation.

Mono, di and tribasic acids

Depending on their formulae and bonding, different acids can release different numbers of protons.

HCl is a monobasic acid because each molecule can release one proton:

$$HCl(aq) \rightarrow H^+(aq) + Cl^-(aq)$$

H_2SO_4 is a dibasic acid because each molecule can release two protons. This is done in two stages:

$$H_2SO_4(aq) \rightarrow H^+(aq) + HSO_4^-(aq)$$
$$HSO_4^-(aq) \rightarrow H^+(aq) + SO_4^{2-}(aq)$$

H_3PO_4 is a tribasic acid because each molecule can release three protons. This is done in three stages:

$$H_3PO_4(aq) \rightarrow H^+(aq) + H_2PO_4^-(aq)$$
$$H_2PO_4^-(aq) \rightarrow H^+(aq) + HPO_4^{2-}(aq)$$
$$HPO_4^{2-}(aq) \rightarrow H^+(aq) + PO_4^{3-}(aq)$$

Conjugate acid–base pairs

An acid–base pair is a set of two species that transform into each other by gain or loss of a proton. Figure 1 gives an example of a **conjugate acid–base pair**, for the dissociation of nitrous acid, HNO_2. It also shows how the acid and base in an acid–base pair are linked by H^+.

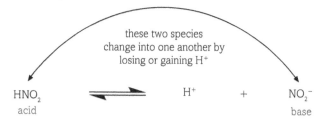

Figure 1 An example of a conjugate acid–base pair.

Acid–base equilibria involve two acid–base pairs. The equilibrium below shows the dissociation of nitrous acid, HNO_2, in water:

$$HNO_2(aq) + H_2O(l) \rightleftharpoons H_3O^+(aq) + NO_2^-(aq)$$

In the forward direction:

- the acid HNO_2 releases a proton to form its conjugate base NO_2^-
- the base H_2O accepts the proton to form its conjugate acid H_3O^+.

In the reverse direction:

- the acid H_3O^+ releases a proton to form its conjugate base H_2O
- the base NO_2^- accepts the proton to form its conjugate acid HNO_2.

HNO_2 and NO_2^- differ by H^+ and make up one acid–base pair: acid 1 and base 1.

H_3O^+ and H_2O differ by H^+ and make up a second acid–base pair: acid 2 and base 2.

$$\underset{\text{acid 1}}{HNO_2(aq)} + \underset{\text{base 2}}{H_2O(l)} \rightleftharpoons \underset{\text{acid 2}}{H_3O^+(aq)} + \underset{\text{base 1}}{NO_2^-(aq)}$$

Questions

1 Write equations for dissociation of the following acids in water:
(a) nitric acid, HNO_3
(b) chromic acid, H_2CrO_4.

2 Identify and explain whether the acids in question 1 are monobasic, dibasic or tribasic.

3 Ammonia is an example of a base. It reacts with boron trifluoride as follows:

$$NH_3 + BF_3 \rightarrow NH_3BF_3$$

Is ammonia acting as a Brønsted–Lowry base in this reaction? Explain your answer.

4 Identify the acid–base pairs in the acid–base equilibria below:
(a) $HIO_3 + H_2O \rightleftharpoons H_3O^+ + IO_3^-$
(b) $CH_3COOH + H_2O \rightleftharpoons CH_3COO^- + H_3O^+$
(c) $NH_3 + H_2O \rightleftharpoons NH_4^+ + OH^-$

(10) Acid–base reactions and K_a

By the end of this topic, you should be able to demonstrate and apply your knowledge and understanding of:

* the role of H⁺ in the reactions of acids with metals and bases (including carbonates, metal oxides and alkalis), using ionic equations
* the acid dissociation constant, K_a, for the extent of acid dissociation
* the relationship between K_a and pK_a
* the application of the principles for K_c, K_p to other equilibrium constants, where appropriate

Typical acid–base reactions

Aqueous acids take part in typical acid–base reactions with carbonates, bases and **alkalis**.

In all of these reactions, a **neutralisation** reaction occurs and water is formed as one of the products.

LEARNING TIP

When acids release protons in water, they are usually accepted by the water to form a hydronium ion, H_3O^+ (sometimes called the oxonium ion).

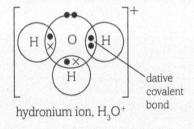

hydronium ion, H_3O^+

dative covalent bond

Figure 1 The hydronium ion.

For example, when hydrochloric acid is added to water the following reaction occurs:

$$HCl(aq) + H_2O(l) \rightarrow H_3O^+(aq) + Cl^-(aq)$$

H^+ and H_3O^+ can be used interchangeably when describing aqueous acid solutions, and H^+ is usually used in ionic equations for acid–base reactions. That is why the above equation would be simplified to:

$$HCl(aq) \rightarrow H^+(aq) + Cl^-(aq)$$

Reactions with carbonates

In the reactions of acids that follow, ionic equations are used to show the important role of the H^+ ion.

* The ionic equations show only $H^+(aq)$ from the acid. This H^+ could be provided by hydrochloric acid, HCl, nitric acid, HNO_3, or sulfuric acid, H_2SO_4, for example.

* These acids have the same ionic equation for each type of reaction because they all release H^+ ions when they dissolve in water.

Aqueous acids react with solid carbonates, forming a salt, carbon dioxide and water:

* full equation:
  $$2HCl(aq) + CaCO_3(s) \rightarrow CaCl_2(aq) + CO_2(g) + H_2O(l)$$

* all ions:
  $$2H^+(aq) + \cancel{2Cl^-(aq)} + CaCO_3(s)$$
  $$\rightarrow Ca^{2+}(aq) + \cancel{2Cl^-(aq)} + CO_2(g) + H_2O(l)$$

* ionic equation:
  $$2H^+(aq) + CaCO_3(s) \rightarrow Ca^{2+}(aq) + CO_2(g) + H_2O(l)$$

LEARNING TIP

The sum of the charges on either side of an ionic equation need not be zero, but must be the same on the two sides. In this example, two H^+ ions are balanced by one Ca^{2+} ion – there is a charge of 2+ on each side of the equation.

If the carbonate is in solution, the final ionic equation simplifies because the carbonate is dissociated:

* full equation:
  $$2HCl(aq) + Na_2CO_3(aq) \rightarrow 2NaCl(aq) + CO_2(g) + H_2O(l)$$

* all ions:
  $$2H^+(aq) + \cancel{2Cl^-(aq)} + \cancel{2Na^+(aq)} + CO_3^{2-}(aq)$$
  $$\rightarrow \cancel{2Na^+(aq)} + \cancel{2Cl^-(aq)} + CO_2(g) + H_2O(l)$$

* ionic equation:
  $$2H^+(aq) + CO_3^{2-}(aq) \rightarrow CO_2(g) + H_2O(l)$$

Reactions with bases

Aqueous acids react with bases, forming a salt and water:

- full equation:

$$2HNO_3(aq) + MgO(s) \rightarrow Mg(NO_3)_2(aq) + H_2O(l)$$

- all ions:

$$2H^+(aq) + \cancel{2NO_3^-(aq)} + MgO(s)$$
$$\rightarrow Mg^{2+}(aq) + \cancel{2NO_3^-(aq)} + H_2O(l)$$

- ionic equation:

$$2H^+(aq) + MgO(s) \rightarrow Mg^{2+}(aq) + H_2O(l)$$

Reactions with alkalis

Aqueous acids react with alkalis, forming a salt and water:

- full equation:

$$H_2SO_4(aq) + 2KOH(aq) \rightarrow K_2SO_4(aq) + 2H_2O(l)$$

- all ions:

$$2H^+(aq) + \cancel{SO_4^{2-}(aq)} + \cancel{2K^+(aq)} + 2OH^-(aq)$$
$$\rightarrow \cancel{2K^+(aq)} + \cancel{SO_4^{2-}(aq)} + 2H_2O(l)$$

- ionic equation:

$$2H^+(aq) + 2OH^-(aq) \rightarrow 2H_2O(l)$$

- which cancels to:

$$H^+(aq) + OH^-(aq) \rightarrow H_2O(l)$$

Redox reactions of acids with metals

The reaction of an acid with a metal is a redox reaction, so does not fit in with the acid–base model. However, they are a common acid reaction. The general equation is:

$$\text{acid} + \text{metal} \rightarrow \text{salt} + \text{hydrogen}$$

Figure 2 Zinc reacting with hydrochloric acid, an example of a metal–acid reaction. The bubbles are hydrogen gas and the salt is zinc chloride, $ZnCl_2$.

Aqueous acids react with metals, forming a salt and hydrogen:

- full equation:

$$2HCl(aq) + Mg(s) \rightarrow MgCl_2(aq) + H_2(g)$$

- all ions:

$$2H^+(aq) + \cancel{2Cl^-(aq)} + Mg(s) \rightarrow Mg^{2+}(aq) + \cancel{2Cl^-(aq)} + H_2(g)$$

- ionic equation:

$$2H^+(aq) + Mg(s) \rightarrow Mg^{2+}(aq) + H_2(g)$$

Although most acids react with metals in this way, you have to be careful. Some acids, such as sulfuric and nitric acids, are powerful oxidising agents and other reactions may also take place, especially when the acids are concentrated. It is best to steer clear of H_2SO_4 and HNO_3 when giving examples of acid reactions with metals.

Strong acids and weak acids

In aqueous solution, acids dissociate and an equilibrium is set up. The equilibrium below shows the dissociation of an acid, HA, in water:

$$HA(aq) + H_2O(l) \rightarrow H_3O^+(aq) + A^-(aq)$$

or more simply:

$$HA(aq) \rightarrow H^+(aq) + A^-(aq)$$

The strength of an acid HA is the extent of its dissociation into H^+ and A^- ions.

Strong acids

Some acids are strong – they are 100% dissociated in aqueous solution. There are comparatively few **strong acids**; the most common are listed in Table 1. Almost every other acid is weak.

HCl	hydrochloric acid
HNO_3	nitric acid
H_2SO_4	sulfuric acid
HBr	hydrobromic acid
HI	hydriodic acid
$HClO_4$	chloric(VII) (perchloric) acid

Table 1 Strong acids.

Weak acids

A **weak acid** only partially dissociates in aqueous solution. Many naturally occurring acids are weak.

The equilibrium that is set up when ethanoic acid, CH_3COOH, dissociates in water is shown below:

$$CH_3COOH(aq) \rightleftharpoons H^+(aq) + CH_3COO^-(aq)$$

- The equilibrium position lies well over to the left.
- There are only small concentrations of dissociated ions, $H^+(aq)$ and $CH_3COO^-(aq)$, compared with the concentration of undissociated $CH_3COOH(aq)$.

We can actually say that CH_3COO^- is a very good base – it is very good at accepting the dissociated H^+ back.

The acid dissociation constant, K_a

In topic 5.1.8 you learned about equilibrium constants and what they can tell you about the extent of a reaction.

The actual extent of acid dissociation is measured by an equilibrium constant called the **acid dissociation constant**, K_a.

A weak acid HA has the following equilibrium in aqueous solution:

$$HA(aq) \rightleftharpoons H^+(aq) + A^-(aq)$$

The expression for the acid dissociation constant is:

$$K_a = \frac{[H^+(aq)][A^-(aq)]}{[HA(aq)]}$$

The units of K_a are always $mol\,dm^{-3}$. You can show this by cancelling the units in the K_a expression:

$$K_a = \frac{(mol\,dm^{-3})\,(mol\,dm^{-3})}{(mol\,dm^{-3})} = mol\,dm^{-3}$$

The terms 'strong' and 'weak' describe the extent of dissociation of an acid given by the K_a value:

- a large K_a value indicates a large extent of dissociation – the acid is strong
- a small K_a value indicates a small extent of dissociation – the acid is weak.

K_a and pK_a

The range of values for K_a is vast. For this reason a logarithmic scale is used to describe them. You will learn more about this in topic 5.1.11 on the pH scale.

pK_a is a more manageable number than K_a. The two quantities are related as follows:

- $\mathbf{pK_a = -\log_{10} K_a}$
- $\mathbf{K_a = 10^{-pK_a}}$
- A low value of K_a matches a high value of pK_a.
- A high value of K_a matches a low value of pK_a.
- The smaller the pK_a value, the stronger the acid.

Table 2 compares K_a and pK_a values for some weak acids.

Acid	Relative strength	K_a/mol dm^{-3}	pK_a
Phosphoric acid H_3PO_4	Stronger acid	7.9×10^{-3}	$-\log(7.9 \times 10^{-3}) = 2.10$
Sulfurous acid H_2SO_3		1.5×10^{-3}	$-\log(1.5 \times 10^{-2}) = 2.82$
Methanoic acid HCOOH		1.6×10^{-4}	$-\log(1.6 \times 10^{-4}) = 3.80$
Ethanoic acid CH_3COOH	Weaker acid	1.7×10^{-5}	$-\log(1.7 \times 10^{-5}) = 4.77$

Table 2 K_a and pK_a values of some weak acids.

Questions

1. Write full and ionic equations for the following acid–base reactions:
   (a) sulfuric acid and solid magnesium carbonate
   (b) sulfuric acid and aqueous potassium carbonate
   (c) hydrochloric acid and solid calcium oxide
   (d) nitric acid and aqueous sodium hydroxide
   (e) hydrochloric acid and aluminium.

2. For each of the following acid–base equilibria, write down the expression for K_a:
   (a) $HCOOH(aq) \rightleftharpoons H^+(aq) + HCOO^-(aq)$
   (b) $CH_3CH_2COOH(aq) \rightleftharpoons H^+(aq) + CH_3CH_2COO^-(aq)$

3. Calculate pK_a from the following K_a values. Give your answers to two decimal places.
   (a) $K_a = 2.3 \times 10^{-1}\,mol\,dm^{-3}$
   (b) $K_a = 2.5 \times 10^{-3}\,mol\,dm^{-3}$
   (c) $K_a = 4.8 \times 10^{-11}\,mol\,dm^{-3}$

4. Calculate K_a from the following pK_a values. Give your answers to three significant figures.
   (a) $pK_a = 2.90$
   (b) $pK_a = 7.20$
   (c) $pK_a = 10.60$

5.1 ⑪ Calculating pH of strong and weak acids

By the end of this topic, you should be able to demonstrate and apply your knowledge and understanding of:

- the use of the expression for pH as: $pH = -\log[H^+]$
  $[H^+] = 10^{-pH}$
- calculations of pH, or related quantities, for strong monobasic acids
- calculations of pH, K_a or related quantities, for a weak monobasic acid using approximations
- limitations of using approximations to K_a related calculations for 'stronger' weak acids

KEY DEFINITIONS

pH $= -\log[H^+(aq)]$
$[H^+(aq)]$ $= 10^{-pH}$

The pH scale

Søren Sørensen, a Danish chemist, carried out research that involved measuring hydrogen ion concentrations. These numbers could be cumbersome to measure and record.

Aqueous solutions have concentrations of $H^+(aq)$ ions in the range:

- from about 10 mol dm^{-3} (10^1 mol dm^{-3})
- to about 0.000 000 000 000 001 mol dm^{-3} (10^{-15} mol dm^{-3}).

Sørensen came up with an easier way to measure and denote these concentrations – a logarithmic scale called the pH scale.

The pH is the negative log (to base 10) of the concentration of $H^+(aq)$ ions. This is expressed as:

- **pH** $= -\log[H^+(aq)]$

The related expression for the concentration of $H^+(aq)$ ions is:

- **$[H^+(aq)]$** $= 10^{-pH}$

LEARNING TIP

Remember that the shorthand for 'the concentration of' is to use square brackets [ ].

Table 1 shows the $[H^+]$ values and corresponding pH values that make up the pH scale.

pH	−1	0	1	2	3	4	5	6
$[H^+(aq)]$/mol dm^{-3}	10	1	10^{-1}	10^{-2}	10^{-3}	10^{-4}	10^{-5}	10^{-6}

pH	7	8	9	10	11	12	13	14	15
$[H^+(aq)]$/mol dm^{-3}	10^{-7}	10^{-8}	10^{-9}	10^{-10}	10^{-11}	10^{-12}	10^{-13}	10^{-14}	10^{-15}

Table 1 pH values and corresponding H^+ values.

Converting between pH and $[H^+(aq)]$

Sometimes it will be necessary to work with $[H^+(aq)]$ concentrations that are not whole numbers. For example, readings from a pH probe may not be whole numbers.

- If pH = 4.52, $[H^+(aq)] = 10^{-4.52}$
- If pH = 10.63, $[H^+(aq)] = 10^{-10.63}$

We need to express the concentration of H^+ ions in standard form. This will involve the use of a scientific calculator, as shown in Figure 1.

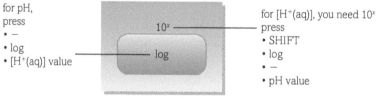

for pH,
press
• –
• log
• [H⁺(aq)] value
10ˣ
log
for [H⁺(aq)], you need 10ˣ
press
• SHIFT
• log
• –
• pH value

Figure 1 The calculator key for pH calculations may look like this; refer to your calculator manual to make sure you know how to carry out this type of calculation.

LEARNING TIP

The 10^x button is usually the 'shift function' on the log button. Check your own calculator to make sure you know where each button is.

WORKED EXAMPLE 1

Calculate [H$^+$(aq)] if the pH of a solution is 13.41.

- Press shift log (i.e. the 10^x function).
- Type in: −13.41
- Press =

The answer shown is 3.89×10^{-14}

[H$^+$(aq)] = 3.89×10^{-14} mol dm^{-3}

WORKED EXAMPLE 2

Calculate the pH if [H$^+$(aq)] = 1.62×10^{-6}

- Press −
- Press log
- Type in 1.62×10^{-6}
- Press =

The answer shown is 5.79

pH = 5.79

LEARNING TIP

The order in which you need to complete the operations may vary between calculators. If these Worked examples do not work on your calculator, refer to the instruction manual to make sure you can carry out this type of calculation.

What does a pH value mean?

The relationship between pH and [H$^+$(aq)] is sometimes called a 'see-saw' relationship – 'when one is up, the other is down'.

- A low pH value means a large [H$^+$(aq)].
- A high pH value means a small [H$^+$(aq)].
- A pH change of 1 changes [H$^+$(aq)] by 10 times.
- An acid with a pH of 2 contains 1000 times the [H$^+$(aq)] of an acid with a pH of 5.

Calculating the pH of strong acids

A strong monobasic acid, HA, has virtually complete dissociation in water:

$$HA(aq) \rightarrow H^+(aq) + A^-(aq)$$

This means that [H$^+$(aq)] of a strong acid is approximately equal to the concentration of the acid, [HA(aq)]:

$$[H^+(aq)] = [HA(aq)]$$

The pH can then be calculated using pH = −log[H$^+$(aq)]

WORKED EXAMPLE 3

A sample of hydrochloric acid, HCl, has a concentration of 1.22×10^{-3} mol dm^{-3}. What is its pH?

HCl(aq) is a strong acid and completely dissociates:

$$HCl(aq) \rightarrow H^+(aq) + Cl^-(aq)$$

Therefore [H$^+$(aq)] = [HCl(aq)] = 1.22×10^{-3} mol dm^{-3}

pH = −log[H$^+$(aq)]

= −log(1.22×10^{-3})

= 2.91

Calculating the pH of weak acids

In aqueous solution, a weak monobasic acid, HA, partially dissociates, setting up the equilibrium:

$$HA(aq) \rightleftharpoons H^+(aq) + A^-(aq)$$

Because we can no longer assume [H$^+$(aq)] is equal to the concentration of the acid, we need to work out the value of K_a for the reaction – to find the extent of dissociation that has occurred, as this will dictate the value of [H$^+$(aq)].

$$K_a = \frac{[H^+(aq)]\,[A^-(aq)]}{[HA(aq)]}$$

- For weak acids, [H$^+$(aq)] is much less than [HA(aq)] because the extent of dissociation is so small.
- When HA molecules dissociate, H$^+$(aq) and A$^-$(aq) ions are formed in equal quantities and it is assumed that any water present will have dissociated to such a negligible amount that it will not affect the concentration of H$^+$.
- Therefore [H$^+$(aq)] and [A$^-$(aq)] can be considered equal.
- This means the expression '[H$^+$(aq)] [A$^-$(aq)]' in the K_a calculation becomes [H$^+$(aq)]^2.

Because some HA molecules have dissociated, [HA(aq)] will have reduced slightly.

- The equilibrium concentration of HA(aq) will be [HA(aq)]$_{undissociated}$ − [H$^+$(aq)].

The K_a expression now becomes:

$$K_a = \frac{[H^+(aq)]^2}{[HA(aq)]_{undissociated} - [H^+(aq)]}$$

Finally, we make an approximation to simplify the pH calculation. We assume that such a small amount of acid has dissociated that:

$$[HA(aq)]_{undissociated} - [H^+(aq)] = [HA(aq)]_{undissociated}$$

Using this approximation, the K_a expression now becomes:

$$K_a = \frac{[H^+(aq)]^2}{[HA(aq)]}$$

Remember, we are trying to find the pH so we need to rearrange this equation to find [H$^+$(aq)]. The equation is rearranged to give:

$$[H^+(aq)] = \sqrt{K_a \times [HA(aq)]}$$

The pH can then be calculated using pH = −log[H$^+$(aq)].

WORKED EXAMPLE 4

The concentration of a sample of nitrous acid, HNO$_2$, is 0.055 mol dm^{-3}. $K_a = 4.70 \times 10^{-4}$ mol dm^{-3} at 25 °C. Calculate the pH.

First we find [H$^+$(aq)]. HNO$_2$ is a weak acid and partially dissociates:

$$HNO_2(aq) \rightleftharpoons H^+(aq) + NO_2^-(aq)$$

- $K_a = \dfrac{[H^+(aq)]\,[NO_2^-(aq)]}{[HNO_2(aq)]}$

If we assume that the H$^+$(aq) and NO$_2^-$(aq) ions only come from the dissociation of HNO$_2$(aq), then this approximates to: $\dfrac{[H^+(aq)]^2}{[HNO_2(aq)]}$.

Therefore:

- [H$^+$(aq)]$^2 = K_a \times$ [HNO$_2$(aq)]
- [H$^+$(aq)] = $\sqrt{K_a \times [HNO_2(aq)]} = \sqrt{(4.70 \times 10^{-4} \times 0.055)}$

  = 5.08×10^{-3} mol dm^{-3}

Now we calculate pH:

- pH = −log[H$^+$(aq)] = −log (5.08×10^{-3}) = 2.29

For strong acids, HA, $[H^+(aq)] = [HA(aq)]$

For weak acids, HA, $[H^+(aq)] = \sqrt{K_a \times [HA(aq)]}$

The limitations of scientific approximations

When calculating the pH for a weak acid it is assumed that so little of the original acid has dissociated that the concentration of the acid at equilibrium is effectively the same as the concentration of the original amount of acid.

This can be expressed as: $[HA]_{equilibrium} \approx [HA]_{undissociated}$

This is an example of an approximation.

It is actually the case that there will be some acid which does dissociate, even if only a small amount. If this amount is less than 5% of the total for $[HA]_{undissociated}$ then it is deemed safe to use the above approximation. This will be the case for very weak acids, with very low K_a values.

If, however, more than 5% of $[HA]_{undissociated}$ dissociates, then the above approximation cannot be used. This would occur with stronger examples of weak acids, with higher K_a values.

In this instance, the expression for K_a can only be solved using a quadratic equation. This is not required for your A level course but you should be aware of the limitations of the approximation.

Calculating K_a for weak acids

To determine K_a for a weak acid, we need to measure the pH of a solution of the weak acid using a pH meter (see Figure 2). We also need the concentration of the weak acid.

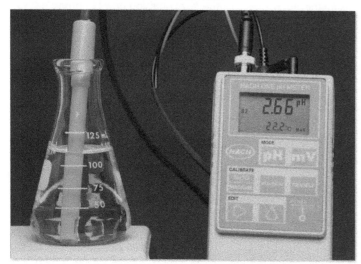

Figure 2 The pH of a solution of the weak acid methanoic acid is being measured using a pH meter. The concentration of the methanoic acid is 0.030 mol dm⁻³ and the pH reading can be seen to be 2.66.

A sample of 0.030 mol dm⁻³ methanoic acid, HCOOH, has a pH of 2.66. Calculate K_a.

- First we find $[H^+(aq)]$ from the pH: $[H^+(aq)] = 10^{-pH} = 10^{-2.66}$
  $= 2.19 \times 10^{-3}$ mol dm⁻³

- Now we can calculate K_a: $K_a = \dfrac{[H^+(aq)][HCOO^-(aq)]}{[HCOOH(aq)]}$

- Which approximates to: $K_a = \dfrac{[H^+(aq)]^2}{[HCOOH(aq)]}$

- Therefore $K_a = \dfrac{(2.19 \times 10^{-3})^2}{0.030} = 1.6 \times 10^{-4}$ mol dm⁻³

1. Calculate the pH of solutions with the following $[H^+(aq)]$ values. Give your answers to two decimal places.
   (a) 3.33×10^{-3} mol dm⁻³
   (b) 4.73×10^{-4} mol dm⁻³
   (c) 2.39×10^{-12} mol dm⁻³
   (d) 2.30 mol dm⁻³

2. Calculate $[H^+(aq)]$ of solutions with the following pH values. Give your answers to three significant figures.
   (a) pH 6.53
   (b) pH 2.87
   (c) pH 9.58
   (d) pH −0.12

3. Find the pH of the following solutions of weak acids.
   (a) 0.65 mol dm⁻³ CH₃COOH ($K_a = 1.7 \times 10^{-5}$ mol dm⁻³)
   (b) 4.4×10^{-2} mol dm⁻³ HClO ($K_a = 3.7 \times 10^{-8}$ mol dm⁻³)

4. Find the values of K_a and pK_a for the following weak acids.
   (a) 0.13 mol dm⁻³ solution with a pH of 3.52
   (b) 7.8×10^{-2} mol dm⁻³ solution with a pH of 5.19

(12) The ionisation of water and K_w

By the end of this topic, you should be able to demonstrate and apply your knowledge and understanding of:

* use of the expression for the ionic product of water, K_w

* calculations of pH, or related quantities, for strong monobasic acids and strong bases using K_w

* application of the principles for K_c, K_p to other equilibrium constants, where appropriate

---

KEY DEFINITION

The **ionic product of water**, K_w, is defined as $K_w = [H^+(aq)] [OH^-(aq)]$
At 25 °C, $K_w = 1.00 \times 10^{-14}$ mol^2 dm^{-6}

Water – acid or base?

Water can act as an acid by donating a proton:

$$H_2O(l) \rightarrow H^+(aq) + OH^-(aq)$$

Water can act as a base by accepting a proton:

$$H_2O(l) + H^+(aq) \rightarrow H_3O^+(aq)$$

This back-and-forth transfer of protons happens continuously in water. In fact, water exists at equilibrium:

$$H_2O(l) \rightleftharpoons H^+(aq) + OH^-(aq)$$

This is known as the ionisation of water.

The equilibrium lies well to the left and only an extremely tiny amount of water is dissociated at any given time.

If we write an expression for K_c for this equilibrium we see that:

$$K_c = \frac{[H^+(aq)] [OH^-(aq)]}{[H_2O(l)]}$$

Because the amount of dissociation, or ionisation, of water is so small, $[H_2O(l)]$ is considered constant, just like K_c. We rearrange the equation as follows to combine the two constants.

$$K_c \times [H_2O(l)] = [H^+(aq)] [OH^-(aq)]$$

This can be further simplified by referring to the expression $K_c \times [H_2O(l)]$ as a new constant, K_w:

$$K_w = K_c \times [H_2O(l)] = [H^+(aq)] [OH^-(aq)]$$

This in turn is simplified to:

$$K_w = [H^+(aq)] [OH^-(aq)]$$

This expression for K_w is called the **ionic product of water**.

The units for K_w are always mol^2 dm^{-6} because:

$$[H^+(aq)] \, \text{mol dm}^{-3} \times [OH^-(aq)] \, \text{mol dm}^{-3} = \text{mol}^2 \, \text{dm}^{-6}$$

At 25 °C, the measured pH of water is 7 and $[H^+(aq)] = 10^{-7}$ mol dm^{-3}.

When pure water ionises, the same number of OH$^-$ and H$^+$ ions are released:

$$[H^+(aq)] = [OH^-(aq)]$$

This means that the concentration of OH$^-$(aq) is also equal to 10^{-7} mol dm^{-3}.

So the calculation for K_w becomes:

$$K_w = [H^+(aq)] [OH^-(aq)] = 1.0 \times 10^{-7} \times 1.0 \times 10^{-7}$$
$$= 10^{-14} \, \text{mol}^2 \, \text{dm}^{-6}$$

Therefore, at 25 °C, $K_w = 1.00 \times 10^{-14}$ mol^2 dm^{-6}

LEARNING TIP

Remember: the values of $[H^+(aq)]$ and $[OH^-(aq)]$ are only equal if the water is pure and neutral.

The significance of K_w

The balance between $[H^+(aq)]$ and $[OH^-(aq)]$ can be affected by the addition of extra H$^+$ or OH$^-$ ions that cause the equilibrium to shift. Remember that, as with all equilibrium reactions and following Le Chatelier's principle, the equilibrium will shift to restore the value of K_w. K_w changes only if the temperature is altered.

K_w controls the balance between $[H^+(aq)]$ and $[OH^-(aq)]$ in all aqueous solutions.

At 25 °C, the pH value of 7 is the neutral point at which H$^+$(aq) and OH$^-$(aq) concentrations are the same, equal to 10^{-7} mol dm^{-3}. This applies to water and also to neutral solutions.

All aqueous solutions contain H$^+$(aq) and OH$^-$(aq) ions:

* in water and neutral solutions, $[H^+(aq)] = [OH^-(aq)]$

* in acidic solutions, $[H^+(aq)] > [OH^-(aq)]$

* in alkaline solutions, $[H^+(aq)] < [OH^-(aq)]$

The relative concentrations of H$^+$(aq) and OH$^-$(aq) ions in solution are determined by K_w.

At 25 °C, $K_w = [H^+(aq)] \times [OH^-(aq)]$ must always be equal to 1.00×10^{-14} mol^2 dm^{-6}. (This value is provided for you on a data sheet.)

K_w in action

At 25 °C rainwater is actually acidic. Carbon dioxide dissolves in rainwater. This forms a weak acid called carbonic acid, H_2CO_3, which is able to dissociate and release H$^+$ ions into the water.

Because [H$^+$(aq)] has increased, [OH$^-$(aq)] must fall until:

$$[H^+(aq)][OH^-(aq)] = 1.00 \times 10^{-14} \, mol^2 \, dm^{-6} \, (at \ 25°C).$$

The overall effect is that the pH decreases. Typical rainwater has a pH value of about 5.6.

Mineral water contains dissolved ions such as carbonate, CO_3^{2-}. These ions typically react with acids and thus remove H$^+$ ions and lower [H$^+$(aq)].

Because [H$^+$(aq)] has decreased, [OH$^-$(aq)] must increase until:

$$[H^+(aq)][OH^-(aq)] = 1.00 \times 10^{-14} \, mol^{-14} \, mol^2 \, dm^{-6} \, (at \ 25°C).$$

The overall effect is that the pH increases. The pH of mineral water is typically in the range 7.0–8.0, meaning that mineral water is alkaline.

The link between [H$^+$(aq)] and [OH$^-$(aq)]

We can easily find the concentrations of H$^+$(aq) and OH$^-$(aq) in solutions with different pH values.

- We need to use $[H^+(aq)][OH^-(aq)] = 1.00 \times 10^{-14} \, mol^2 \, dm^{-6}$.

- Table 1 shows how [H$^+$(aq)] and [OH$^-$(aq)] are related at different pH values at 25°C.

- Note that the indices of [H$^+$(aq)] and [OH$^-$(aq)] always add up to −14.

pH	−1	0	1	2	3	4	5	6	7
[H$^+$(aq)]/mol dm^{-3}	10^1	$10^0 = 1$	10^{-1}	10^{-2}	10^{-3}	10^{-4}	10^{-5}	10^{-6}	10^{-7}
[OH$^-$(aq)]/mol dm^{-3}	10^{-15}	10^{-14}	10^{-13}	10^{-12}	10^{-11}	10^{-10}	10^{-9}	10^{-8}	10^{-7}

pH	8	9	10	11	12	13	14	15
[H$^+$(aq)]/mol dm^{-3}	10^{-8}	10^{-9}	10^{-10}	10^{-11}	10^{-12}	10^{-13}	10^{-14}	10^{-15}
[OH$^-$(aq)]/mol dm^{-3}	10^{-6}	10^{-5}	10^{-4}	10^{-3}	10^{-2}	10^{-1}	$10^0 = 1$	10^1

Table 1 The relationship between pH, [H$^+$(aq)] and [OH$^-$(aq)].

At 25°C, [H$^+$(aq)] and [OH$^-$(aq)] are linked by K_w:

$$K_w = [H^+(aq)][OH^-(aq)] = 1.00 \times 10^{-14} \, mol^2 \, dm^{-6}$$

Calculating pH for strong bases

Just as we measure the strength of acids by how well they dissociate to form H$^+$, we measure the strength of bases by their ability to dissociate in solution to release OH$^-$ ions.

Sodium hydroxide, NaOH, is an example of a strong base. In aqueous solution, NaOH dissociates completely:

$$NaOH(aq) \rightarrow Na^+(aq) + OH^-(aq)$$

Bases that dissociate in water to release hydroxide ions are called alkalis. Strong bases are alkalis – they are 100% dissociated in aqueous solution. Strong bases tend to be hydroxides of the metals in groups 1 and 2 in the Periodic Table.

- The strong bases most commonly met are NaOH, KOH and Ca(OH)$_2$.

Ammonia, NH$_3$, is a weak base. In aqueous solution, an equilibrium is set up. The equilibrium position lies well to the left-hand side:

$$NH_3(aq) + H_2O(l) \rightleftharpoons NH_4^+(aq) + OH^-(aq)$$

Using K_w to calculate the pH of strong bases

To work out the pH of a strong base we need to know [H$^+$(aq)] and this depends on:

- the concentration of the base

- the ionic product of water, $K_w = 1.00 \times 10^{-14} \, mol^2 \, dm^{-6}$ (because this links [H$^+$] with [OH$^-$]).

A strong monobasic alkali, e.g. NaOH, is completely dissociated in aqueous solution. This means that $[OH^-(aq)]$ of a strong base is equal to the concentration of the base.

$$NaOH(aq) \rightarrow Na^+(aq) + OH^-(aq)$$

So $[OH^-(aq)] = [NaOH(aq)]$

We can find $[H^+(aq)]$ from K_w and $[OH^-(aq)]$:

$$K_w = [H^+(aq)] [OH^-(aq)]$$

$$[H^+(aq)] = \frac{K_w}{[OH^-(aq)]}$$

The pH can then be calculated using $pH = -\log_{10} [H^+(aq)]$.

WORKED EXAMPLE

A solution of KOH has a concentration of $0.050 \, mol \, dm^{-3}$. What is the pH?
KOH(aq) is a strong base, so there will be complete dissociation. Therefore:

- $[OH^-(aq)] = [KOH(aq)] = 0.050 \, mol \, dm^{-3}$

Using K_w:
First we find $[H^+(aq)]$ from $[OH^-(aq)]$:
$K_w = [H^+(aq)] [OH^-(aq)]$
$\quad = 1.00 \times 10^{-14} \, mol^2 \, dm^{-6}$

$$[H^+(aq)] = \frac{K_w}{[OH^-(aq)]} = \frac{1.00 \times 10^{-14}}{0.050} = 2.0 \times 10^{-13} \, mol \, dm^{-3}$$

Now we calculate pH:
$pH = -\log_{10} [H^+(aq)] = -\log_{10} (2.0 \times 10^{-13})$
$\quad = 12.70$

LEARNING TIP

When calculating the pH of a strong alkali, e.g. NaOH, $[H^+(aq)]$ depends on both $[OH^-(aq)]$ and K_w.
$[OH^-(aq)] = [strong \, base]$

$$[H^+(aq)] = \frac{K_w}{[OH^-(aq)]}$$

We can use K_w to find either the concentration of $H^+(aq)$ or $OH^-(aq)$, as long as we know the value for K_w and the other concentration.

Questions

1 Some aqueous solutions have the following concentrations of $H^+(aq)$. In each solution, what is the concentration of $OH^-(aq)$?

(a) $[H^+(aq)] = 10^{-6} \, mol \, dm^{-3}$

(b) $[H^+(aq)] = 10^{-2} \, mol \, dm^{-3}$

(c) $[H^+(aq)] = 10^{-11} \, mol \, dm^{-3}$

2 Some aqueous solutions have the following concentrations of $OH^-(aq)$. In each solution, what is the concentration of $H^+(aq)$?

(a) $[OH^-(aq)] = 10^{-13} \, mol \, dm^{-3}$

(b) $[OH^-(aq)] = 10^{-4} \, mol \, dm^{-3}$

(c) $[OH^-(aq)] = 10^{-10} \, mol \, dm^{-3}$

3 Find the pH of the following solutions at $25 \, °C$.

(a) $0.0050 \, mol \, dm^{-3}$ KOH(aq)

(b) $3.56 \times 10^{-2} \, mol \, dm^{-3}$ NaOH(aq)

4 Find the concentration, in $mol \, dm^{-3}$, of $OH^-(aq)$ at $25 \, °C$.

(a) pH = 12.43

(b) pH = 13.82

(13) Buffers

By the end of this topic, you should be able to demonstrate and apply your knowledge and understanding of:

※ a buffer solution as a system that minimises pH changes on addition of small amounts of an acid or a base

※ formation of a buffer solution from:
  (i) a weak acid and a salt of the weak acid, e.g. CH_3COOH/CH_3COONa
  (ii) excess of a weak acid and a strong alkali, e.g. excess $CH_3COOH/NaOH$

※ explanation of the role of the conjugate acid–base pair in an acid buffer solution, e.g. CH_3COOH/CH_3COO^-, in the control of pH

※ calculation of the pH of a buffer solution, from the K_a value of a weak acid and the equilibrium concentrations of the conjugate acid–base pair; calculations of related quantities

※ explanation of the control of blood pH by the carbonic acid–hydrogencarbonate buffer system

---

KEY DEFINITION

A **buffer solution** is a mixture that **minimises** pH changes on addition of small amounts of acid or base. The word 'minimises' is essential to this definition.

Buffer solutions

A **buffer solution** 'resists' changes in pH during the addition of small amounts of acid or alkali. A buffer solution cannot prevent the pH from changing slightly, but pH changes are minimised, at least for as long as some of the buffer solution remains.

Buffer solutions are very important in the control of pH in living systems.

A buffer solution is a mixture of:

- a weak acid, HA
- its conjugate base, A^-.

It can be made from a weak acid and a salt of the weak acid, for example, ethanoic acid (CH_3COOH) and sodium ethanoate (CH_3COONa).

In the CH_3COOH/CH_3COONa buffer system:

- the weak acid, CH_3COOH, dissociates partially:
$$CH_3COOH(aq) \rightleftharpoons H^+(aq) + CH_3COO^-(aq)$$

- the salt dissociates completely, generating the conjugate base, CH_3COO^-:
$$CH_3COO^-Na^+(aq) \rightarrow CH_3COO^-(aq) + Na^+(aq)$$

The equilibrium mixture formed contains a high concentration of the undissociated weak acid, CH_3COOH, and its conjugate base, CH_3COO^-. The high concentration of the conjugate base pushes the equilibrium to the left, so the concentration of $H^+(aq)$ ions is very small.

The resulting buffer solution contains large 'reservoirs' of the weak acid and its conjugate base. You can see this in Figure 1.

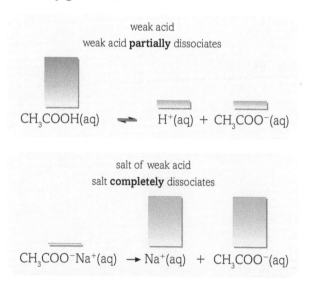

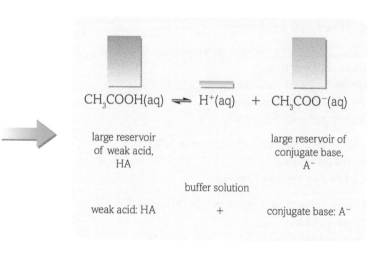

Figure 1 Making a buffer solution from a weak acid and a salt of the weak acid.

Alternatively, buffers can be made from a weak acid and a strong alkali. In this situation, a solution containing a mixture of the salt and the excess of weak acid is formed. For example, a weak acid such as methanoic acid, HCOOH, can be partially neutralised by an aqueous alkali, such as NaOH(aq).

How does a buffer act to control pH?

In a buffer solution, the weak acid, HA, and the conjugate base, A^-, are both responsible for controlling pH. The buffer solution minimises pH changes by using the equilibrium:

$$HA(aq) \rightleftharpoons H^+(aq) + A^-(aq)$$

The overall principle is simple:

- the weak acid, HA, removes added alkali
- the conjugate base, A^-, removes added acid.

On addition of an acid, $H^+(aq)$, to a buffer solution:

- $[H^+(aq)]$ is increased
- the conjugate base, $A^-(aq)$, reacts with $H^+(aq)$ ions
- the equilibrium shifts to the left, removing most of the added $H^+(aq)$ ions.

$$\xleftarrow{\text{Equilibrium shifts to the left}}$$
$$HA(aq) \rightleftharpoons H^+(aq) + A^-(aq)$$

On addition of an alkali, $OH^-(aq)$, to a buffer mixture:

- $[OH^-(aq)]$ is increased
- the small concentration of $H^+(aq)$ ions reacts with the $OH^-(aq)$ ions

$$H^+(aq) + OH^-(aq) \rightarrow H_2O(l)$$

- HA dissociates, shifting the equilibrium to the right to restore most of the $H^+(aq)$ ions that have reacted:

$$\xrightarrow{\text{Equilibrium shifts to the right}}$$
$$HA(aq) \rightleftharpoons H^+(aq) + A^-(aq)$$

These actions of a buffer are summarised in Figure 2.

$$HA(aq) \underset{\text{added alkali}}{\overset{\text{added acid}}{\rightleftharpoons}} H^+(aq) + A^-(aq)$$

Figure 2 Shifting the buffer equilibrium.

LEARNING TIP

Remember, a buffer cannot remove *all* of any acid or alkali that is added – it just minimises the pH changes.

Calculating the pH of buffer solutions

The pH of a buffer solution depends on:

- the acid dissociation constant, K_a, of the buffer system
- the concentration ratio of the weak acid and its conjugate base.

For a buffer consisting of a weak acid, HA, and its conjugate base, A^-:

$$K_a = \frac{[H^+(aq)]\,[A^-(aq)]}{[HA(aq)]}$$

$$\text{Therefore } [H^+(aq)] = K_a \times \frac{[HA(aq)]}{[A^-(aq)]}$$

We need to check on the concentrations of HA(aq) and $A^-(aq)$. Only a very small proportion of HA dissociates so, as discussed in topic 5.1.11 on calculating the pH of weak acids, we can assume that $[HA(aq)]_{equilibrium} = [HA(aq)]_{undissociated}$

The salt of the weak acid is ionic and dissociates completely in aqueous solution, for example:

$$CH_3COO^-Na^+(aq) \rightarrow CH_3COO^-(aq) + Na^+(aq)$$

So, effectively $[CH_3COO^-(aq)] = [CH_3COO^-Na^+(aq)]$

Once $[H^+(aq)]$ is known, the pH can be calculated using $pH = -\log_{10}[H^+(aq)]$.

LEARNING TIP

You can calculate the pH of a buffer using: $[H^+(aq)] = K_a \times \frac{[HA(aq)]}{[A^-(aq)]}$

Then $pH = -\log_{10}[H^+(aq)]$

Remember, this is not the same as the calculation for a weak acid because in a buffer $[H^+]$ and $[OH^-]$ are not the same.

WORKED EXAMPLE

This example shows a method for working out the pH of buffer solutions. Calculate the pH at 25 °C of a buffer solution containing 0.050 mol dm^{-3} $CH_3COOH(aq)$ and 0.10 mol dm^{-3} $CH_3COO^-Na^+(aq)$.

For CH_3COOH, $K_a = 1.7 \times 10^{-5}$ mol dm^{-3} at 25 °C.

$$[H^+(aq)] = K_a \times \frac{[CH_3COOH(aq)]}{[CH_3COO^-(aq)]}$$
$$= 1.7 \times 10^{-5} \times \frac{0.050}{0.10}$$
$$= 8.5 \times 10^{-6}\ \text{mol dm}^{-3}$$

Then:

$pH = -\log_{10}[H^+(aq)]$

$= -\log_{10}(8.5 \times 10^{-6})$

$= 5.07$

The carbonic acid–hydrogencarbonate buffer system

Healthy human blood plasma needs to have a pH between 7.35 and 7.45. If the pH falls below 7.35, a condition called acidosis occurs. If the pH rises above 7.45, then the condition is called alkalosis.

The pH of blood is controlled by a mixture of buffers. The carbonic acid–hydrogencarbonate ion buffer, which is present in blood plasma, is the most important buffer system in the blood:

- carbonic acid, H_2CO_3, acts as the weak acid
- the hydrogencarbonate ion, HCO_3^-, acts as the conjugate base.

$$\underset{\text{weak acid}}{H_2CO_3(aq)} \rightleftharpoons H^+(aq) + \underset{\text{conjugate base}}{HCO_3^-(aq)}$$

Any increase in [H⁺(aq)] ions in the blood is removed by the conjugate base, HCO_3^-(aq). The equilibrium shifts to the left, removing most of the H⁺(aq) ions:

Equilibrium shifts to the left

$$H_2CO_3(aq) \rightleftharpoons H^+(aq) + HCO_3^-(aq)$$

Any increase in OH⁻(aq) ions in the blood is removed by the weak acid, H_2CO_3(aq). The small concentration of H⁺(aq) ions reacts with the OH⁻(aq) ions:

$$H^+(aq) + OH^-(aq) \rightarrow H_2O(l)$$

- H_2CO_3 dissociates, shifting the equilibrium to the right to restore most of the H⁺(aq) ions:

Equilibrium shifts to the right

$$H_2CO_3(aq) \rightleftharpoons H^+(aq) + HCO_3^-(aq)$$

These changes are summarised in Figure 3 below.

added acid

$$H_2CO_3(aq) \rightleftharpoons H^+(aq) + HCO_3^-(aq)$$

added alkali

Figure 3 Acid–hydrogencarbonate ion equilibrium.

The acid dissociation constant, K_a, for this equilibrium is 4.3×10^{-7} mol dm⁻³. Most materials released into the blood are acidic and the hydrogencarbonate ions effectively remove these materials by being converted into H_2CO_3. The carbonic acid is converted into aqueous carbon dioxide through the action of an enzyme. In the lungs, the dissolved carbon dioxide is converted into carbon dioxide gas, which is then exhaled.

The amount of CO_2(aq) in the blood can be controlled by simply changing the rate of breathing – heavy breathing removes more CO_2(g); breathing more slowly removes less CO_2(g).

The acid–base balance of your blood is vital to your survival. If the pH of your blood drops below 7.2 or rises above 7.6, then you are in deep trouble!

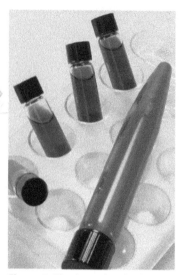

Figure 4 Blood sample – healthy blood has a pH of between 7.35 and 7.45.

Questions

1. A buffer solution is made from propanoic acid and sodium propanoate. Explain, using equations, the role of the conjugate acid–base pair in this buffer solution. You should aim to give at least 3 distinct points in your answer.

2. Calculate the pH of a buffer solution containing 0.15 mol dm⁻³ methanoic acid, HCOOH(aq), and 0.065 mol dm⁻³ sodium methanoate, HCOONa(aq), at 25 °C.
   For HCOOH, $K_a = 1.6 \times 10^{-4}$ mol dm⁻³ at 25 °C.

3. If extra acid is produced in the blood, how do buffers prevent the pH from falling? You should aim to include at least 3 distinct points in your answer.

(14) Neutralisation – titration curves

By the end of this topic, you should be able to demonstrate and apply your knowledge and understanding of:

* pH titration curves for combinations of strong and weak acids with strong and weak bases, including:
  (i) sketch and interpretation of their shapes
  (ii) explanation of the choice of suitable indicators, given the pH range of the indicator
  (iii) explanation of indicator colour changes in terms of equilibrium shift between the HA and A⁻ forms of the indicator

* the techniques and procedures used when measuring pH with a pH meter

---

KEY DEFINITIONS

The **equivalence point** is the point in a titration at which the volume of one solution has reacted exactly with the volume of the second solution. This matches the stoichiometry of the reaction taking place.

The **end point** is the point in a titration at which there are equal concentrations of the weak acid and conjugate base forms of the indicator. The colour at the end point is midway between the colours of the acid and conjugate base forms.

Titrations

When you carry out a titration, you are determining the volume of one solution that reacts exactly with a known volume of another solution. This is called the **equivalence point** of the titration.

At this point, the solution in the conical flask has exactly reacted with the solution in the burette.

This topic considers pH changes from acidic to basic, i.e. a base being added to an acid. Titrations can be carried out from acid to base or from base to acid – the same principles apply to both routes.

Titration curves

pH meters or data loggers can be used to measure the pH of the reaction mixture as solution from the burette is added over time. The varying pH values can then be plotted as a graph – known as an acid–base pH titration curve. The curve may be plotted by hand or by a computer attached to a pH data logger.

A typical acid–base titration pH curve is shown in Figure 1 for the reaction between a strong acid and a strong base. You will notice that this typical curve shows three distinct areas:

* 1 – a slight increase in pH
* 2 – a sharp increase in pH
* 3 – a slight increase in pH.

Indicators

An acid–base indicator is a weak acid, often represented as HIn. An indicator has one colour in its acid form (HIn) and a different colour in its conjugate base form (In⁻).

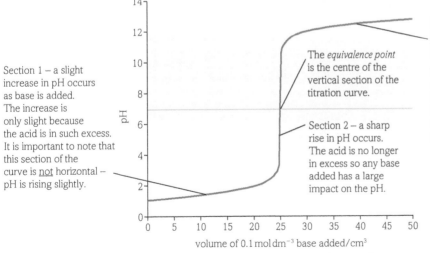

Section 1 – a slight increase in pH occurs as base is added. The increase is only slight because the acid is in such excess. It is important to note that this section of the curve is not horizontal – pH is rising slightly.

The *equivalence point* is the centre of the vertical section of the titration curve.

Section 2 – a sharp rise in pH occurs. The acid is no longer in excess so any base added has a large impact on the pH.

Section 3 – a slight increase in pH occurs as further base is added. The increase is only slight because the base is in excess now and extra base has little impact on the pH. It is important to note however that this section is not horizontal – pH is rising slightly.

Figure 1 Titration curve for a strong acid–strong base titration, showing the vertical section.

Using a pH meter

An indicator such as universal indicator can only give a general indication of the pH of a solution. A pH meter can give an accurate reading for pH, usually to two decimal places.

pH meters contain a protected electrode (the pH probe), which is placed into the solution in question, and a small computer display, which gives the pH reading.

When using a pH probe, it is important to calibrate it first so that the results are accurate.

Usually this is done through a two- or three-step calibration.

- The probe is removed from its storage solution and rinsed with deionised water.

- The probe is blotted dry and then placed into a solution of a known pH (such solutions are usually manufactured professionally just for this purpose), often starting with a solution of pH 4. The pH reading is allowed to settle before checking that a pH of 4 has been registered.

- This process is repeated with other solutions of known pH, often using prepared solutions of pH 7 and pH 10.

This confirms that the pH probe is accurately measuring pH across a range of acidic, neutral and alkaline values.

Some pH meters will have settings or buttons that must be used during the calibration process and you should always check the instructions for the specific pH meter you are using.

For example, for the indicator methyl orange, HIn is red and In⁻ is yellow:

$$HIn \rightleftharpoons H^+ + In^-$$

When there are equal amounts of the weak acid and the conjugate base present:

- $[HIn] = [In^-]$

- the indicator is at its **end point**.

End points are accompanied by a visual colour change that allows a titration to be monitored easily. For example, the colour of the end point for methyl orange is orange, at which point [HIn] (red) and [In⁻] (yellow) are equal.

Most indicators change colour over a range of about two pH units. The pH ranges for the indicators methyl orange, bromothymol blue and phenolphthalein are shown in Figure 2. The end point is typically at the middle of this range.

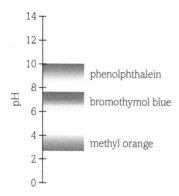

Figure 2 pH colour ranges for some common indicators.

Choosing suitable indicators

For a titration, an indicator is chosen so that the pH value of the end point is as close as possible to the pH value of the titration's equivalence point. In practice, a suitable indicator changes colour within the pH range in the vertical section of the titration curve. This often coincides with just a single drop of the base.

You may notice in the following curves that the ranges of the indicators are not actually on the equivalence point. So how can they show when the reaction has reached this point? Because the equivalence point occurs halfway through an incredibly sharp rise (or fall) in pH, the moment at which the indicator passes through its end point is effectively the same as the moment that the equivalence point takes place.

Strong acid–strong base titrations

An example of a titration curve for a strong acid reacting with a strong base was shown in Figure 1. Let us now look at indicators and their possible suitability. Comparisons between methyl orange and phenolphthalein are shown in Figure 3.

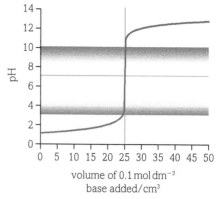

volume of 0.1 mol dm⁻³
base added/cm³

Figure 3 pH titration curve for a titration between a strong acid and a strong base with phenolphthalein and methyl orange indicator ranges.

You should notice the following points from this curve.

- The vertical section of the graph covers a large change in pH, starting around pH 3 and ending around pH 11, with an equivalence point at pH 7.

- Both the indicators, methyl orange and phenolphthalein, have end points that fall within this pH range.

- Either indicator would be suitable to use in a titration between a strong acid and a strong base.

Strong acid–weak base titrations

The curve for a titration between a strong acid and a weak base and the end points of the two indicators are shown in Figure 4.

You should notice the following points.

- The vertical section of the graph covers a smaller change in pH, starting around pH 3 and ending around pH 7.5, with the equivalence point occurring at a more acidic value, i.e. at a pH lower than 7.

- Methyl orange has an end point that falls within this pH range.

- Phenolphthalein does not have an end point that falls within this pH range.

- Only methyl orange would be a suitable indicator for a titration between a strong acid and a weak base.

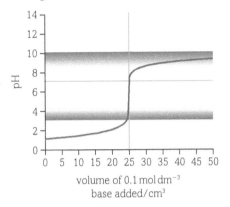

Figure 4 pH titration curve for a titration between a strong acid and a weak base with phenolphthalein and methyl orange indicator ranges.

Weak acid–strong base titrations

The curve for a titration between a weak acid and a strong base and the end points of the two indicators are shown in Figure 5.

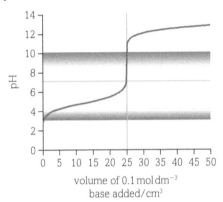

Figure 5 pH titration curve for a titration between a weak acid and a strong base with phenolphthalein and methyl orange indicator ranges.

You should notice the following points.

- The vertical section of the graph covers a smaller change in pH, and occurs further up towards the higher pH values, starting around pH 6.5 and ending around pH 11.5.

- The equivalence point occurs at a more basic pH, i.e. a pH value above 7.

- Methyl orange has an end point that falls outside this pH range.

- Phenolphthalein has an end point that falls within this pH range.

- Only phenolphthalein would be a suitable indicator for a titration between a strong acid and a weak base.

Weak acid–weak base titrations

The titration curve for a weak acid reacting with a weak base is shown in Figure 6.

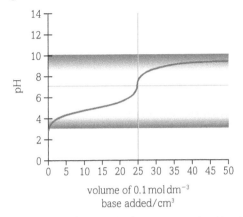

Figure 6 pH titration curve for a titration between a weak acid and a weak base with phenolphthalein and methyl orange indicator ranges.

You will notice the following points.

- There is no real vertical section.

- Neither indicator has an end point near the equivalence point.

- Neither indicator is suitable.

- An indicator would change colour gradually over a few cm³ of base added.

- No indicators are really suitable for a weak acid–weak base titration.

INVESTIGATION

Set up titrations between acids and alkalis of different strengths. You should select some weak acids and bases and some strong acids and bases with guidance from your technician or teacher. Carry out titrations using various combinations of weak and strong acids and bases. Use a pH meter to record the pH throughout the titrations. Use your results to plot acid–base titration curves. Identify the equivalence points for each titration.

Questions

1. Explain why the end point for a titration is not the same as the equivalence point.

2. The pH ranges for two indicators are shown below:
   bromocresol green pH 3.8–5.4
   thymolphthalein pH 9.3–10.5
   Use the titration curves in Figures 3–6 to identify whether these indicators are suitable for the titration of each combination of acid and base.

OCEAN ACIDIFICATION

Ocean acidification impairs mussels' ability to attach to surfaces – alarming commercial growers farming the waters around Puget Sound on the northwest American coastline.

MUSSELS LOSE FOOTING IN MORE ACIDIC OCEAN

Figure 1 Carbon from greenhouse gas emissions has steadily turned seawater more acidic, disrupting organisms accustomed to the slightly alkaline waters of the past 20 million years.

PENN COVE, Wash. – Cookie tray in hand and lifejacket around chest, Laura Newcomb looks more like a confused baker than a marine biologist. But the University of Washington researcher is dressed for work. Her job: testing how mussels in this idyllic bay, home to the nation's largest harvester of mussels, are affected by changing ocean conditions, especially warmer and more acidic waters. It's a question critical to the future of mussel farmers in the region. More important, it's key to understanding whether climate change threatens mussels around the world, as well as to the food chains mussels support and protect in the wild.

'Along the West Coast, mussels are well-known ecosystem engineers,' said Bruce Menge, an Oregon State University researcher who studies how climate impacts coastal ecosystems. 'They provide habitat for dozens of species, they provide food for many predators and occupy a large amount of space, so are truly a "dominant" species.'

20 million years

Carbon from greenhouse gas emissions has steadily turned seawater more acidic, disrupting organisms accustomed to the slightly alkaline waters of the past 20 million years. In the case of mussels, an earlier University of Washington lab study found that increased carbon dioxide weakens the sticky fibers, called byssus, that mussels use to survive by clinging to objects like shorelines or the ropes used by commercial harvesters.

'If byssal thread weakening does eventually become important,' Menge added, 'the consequences would be major if not catastrophic.'

Newcomb's goal now is to apply in the real world what was learned in the lab. 'Instead of spending a lot of time tightly controlling the temperature and pH conditions mussels grow in, I use the natural seasonal variation to try to answer the same questions,' Newcomb said.

Newcomb's field office is the rear deck of a harvesting boat – right between the toilet and the microwave. The quarters are cramped, but the view is grand: The blue waters of Penn Cove on Washington State's Whidbey Island are set against rolling bluffs and snow-capped mountains.

30 percent increase in acidity

The University of Washington marine biologist is there courtesy of Penn Cove Shellfish, which is also the oldest and best-known mussel operation in the United States. If you're a mussel fan, you've probably had a few – they're sold at Costco as well as upscale restaurants across the country.

Placing just-harvested mussels on her tray, Newcomb samples for size, thickness and strength. The mussels are grown on 21-foot-long ropes hanging from several dozen rafts in the bay, and Newcomb takes samples from two depths: 3 feet and 21 feet. She also samples water temperature and pH levels at those depths.

Prior to the Industrial Revolution and the explosion of manmade CO_2, ocean pH averaged 8.2. Today it's 8.1, a 30 percent increase in acidity on the logarithmic scale. Computer models peg ocean acidity at 7.8 to 7.7 by the end of the century at the current rate of greenhouse gas emissions.

Washington State is a bit ahead of that curve because ancient carbon stores in the deep ocean are periodically churned up by local currents. The surprising lab discovery was that mussel byssus weakened by 40 percent when exposed to a pH of 7.5. At Penn Cove, low pH levels are not uncommon – Newcomb has even seen 7.4 in the year that she's been sampling.

'We're worried they're going to see it more frequently,' said Emily Carrington, Newcomb's graduate adviser and leader of the University of Washington team that published the earlier lab results.

Source
- http://www.dailyclimate.org/tdc-newsroom/2014/09/acidification-mussels. Sep 9, 2014. By Miguel Llanos and *The Daily Climate*.

Where else will I encounter these themes?

Let us start by considering the nature of the writing in the article.

1. Who is the intended audience for the article? How does the writing style reflect this?

Now we will look at the chemistry in, or connected to, this article. Do not worry if you are not ready to give answers to these questions yet. You may like to return to the questions once you have covered other topics later in the book. Use the timeline at the bottom of the page to help you put this work in context with what you have already learned and what is ahead in your course.

April 2014 was the first month in which the average atmospheric carbon dioxide levels reached the 400 ppm milestone. But carbon dioxide is not alone in causing climate change. Once oxides of nitrogen and methane and minute trace amounts of other gases such as SF_6 are factored in, the value is equivalent to about 480 ppm carbon dioxide.

2. Write an equation for the reaction of carbon dioxide with water to form hydrogencarbonate ions and $H^+(aq)$ ions.

3. a. Using the relationship $pH = -\log_{10}[H^+]$, show that a change in pH from 8.2 to 8.1 is approximately equivalent to a 30% increase in acidity.

   b. Calculate the pH of the ocean if the acidity increases by 100% from a starting pH of 8.2.

4. Many sea organisms form shells from calcium carbonate. Use the equation below to explain why increasing levels of carbon dioxide in the air are making this process increasingly difficult.

$$CO_2(aq) + CO_3^{2-}(aq) + H_2O(l) \rightleftharpoons 2HCO_3^-(aq)$$

5. Many organisms use intracellular HCO_3^- ions to buffer changes in pH. Show, with the aid of equations, how HCO_3^- ions can buffer small changes in $H^+(aq)$ and $OH^-(aq)$.

6. The equation for the dissociation of carbonic acid in water can be represented as

$$H_2CO_3(aq) \rightleftharpoons H^+(aq) + HCO_3^-(aq)$$

   a. Write an expression for K_a for carbonic acid.

   b. Given that the value of this $pK_a = 6.3$ and that the pH of blood is 7.4, calculate the ratio of hydrogencarbonate ions to non-dissociated carbonic acid molecules in blood serum.

   c. What assumptions have you made in 6.b.?

Remember that because pH is a minus $\log_{10}$ scale, a ×10 increase in $[H^+]$ will result in a decrease of 1 pH unit.

Activity

The molecule Histidine is one of the 20 amino acids that make up all proteins. It is argued that Histidine residues in haemoglobin molecules are also involved in buffering the blood.

Research and create a presentation addressing the questions below.

1. What does Histidine look like?

2. Identify the part of the molecule that acts as the proton donor/acceptor.

3. Given that the pK_a value of the protonated Histidine side chain has a value of 6.0, what fraction of Histidine residues in haemoglobin would you expect to be in the protonated form at the pH of blood (7.4)?

4. Why is this value unlikely to agree with the true value *in vivo*?

5. Histidine can frequently be found at the catalytic centres of enzymes. Can you suggest why?

6.1   6.2   6.3

1. An investigation into the reaction $A(g) + B(g) \rightarrow C(g)$ concluded that the rate equation for the reaction is:

    rate $= k[A]^2$.

    Which of the following statements about the reaction is true? [1]

    A. Reactant B is not involved in the reaction.

    B. Doubling the concentration of both A and B will quadruple the reaction rate.

    C. The units for k are mol dm^{-3} s^{-1}.

    D. A catalyst would not affect this reaction.

2. The radioactive decay of ^{14}C shows first-order kinetics with a half-life of 5730 years. What can we deduce from this? [1]

    A. The decay constant k has a value of $6.21 \times 10^{-3}\,y^{-1}$.

    B. k will increase as temperature increases.

    C. It will take approximately 9079 years for the initial amount of ^{14}C to drop to $\frac{1}{3}$ of its original value.

    D. It will take approximately 9079 years for the initial amount of ^{14}C to drop to $\frac{1}{4}$ of its original value.

3. Consider the following gaseous equilibrium:

    $$2CO_2(g) \rightleftharpoons 2CO(g) + O_2(g)$$

    Which of the following statements is true? [1]

    A. The units of K_p for the reverse reaction could be Pa^{-1}.

    B. Pressure will favour the forward reaction.

    C. A platinum catalyst will increase the rate of the forward reaction only.

    D. The units of K_p for the forward reaction are Pa^{-1}.

4. Consider the following equilibrium for ethanoic acid in water. $(K_a = 1.8 \times 10^{-6})$

    $$CH_3COOH(aq) + H_2O(l) \rightleftharpoons CH_3COO^-(aq) + H_3O^+(aq)$$

    Which of the following is true? [1]

    A. $CH_3COO^-(aq)$ is a conjugate acid.

    B. The pH of a 0.02 M solution of ethanoic acid is 1.98.

    C. Diluting the acid will increase the value of K_a.

    D. Increasing the temperature will change the value of K_a.

    [Total: 4]

5. The chemicals that we call 'acids' have been known for thousands of years. However, modern theories of acids have been developed comparatively recently. It wasn't until the early 1900s that the concept of dissociation became accepted by the scientific community and the concept of pH was introduced.

    A student carried out a series of experiments with acids and alkalis.

    (a) Propanoic acid, CH_3CH_2COOH, is a naturally occurring weak acid. The equation for the dissociation of propanoic acid is shown below.

    $$CH_3CH_2COOH(aq) \rightleftharpoons H^+(aq) + CH_3CH_2COO^-(aq)$$

    The student wanted to prove that propanoic acid is a weak acid. The student had access to a pH meter and 0.100 mol dm^{-3} propanoic acid.

    - Explain how the student could prove that propanoic acid is a weak acid by taking a single pH measurement.

    - Show how the student could then calculate the acid dissociation constant, K_a, for propanoic acid. [4]

    (b) The student measured the pH of a solution of sodium hydroxide at 25 °C. The measured pH was 13.46.

    Calculate the concentration of the aqueous sodium hydroxide. [2]

    (c) A student made a buffer solution by adding an excess of propanoic acid to an aqueous solution of sodium hydroxide at 25 °C. This buffer solution contains an equilibrium system that minimises changes in pH when small amounts of acids and alkalis are added.

    - Explain why a buffer solution formed when an excess of propanoic acid was mixed with aqueous sodium hydroxide.

    - Explain how this buffer solution controls pH when an acid or an alkali is added.

    In your answer you should explain how the equilibrium system allows the buffer solution to control the pH. [7]

(d) A student added nitric acid to propanoic acid. A reaction took place to form an equilibrium mixture containing two acid–base pairs.

Complete the equilibrium below and label the two conjugate acid–base pairs.

$$HNO_3 + CH_3CH_2COOH \rightleftharpoons \ldots\ldots + \ldots\ldots \quad [2]$$

(e) Finally, the student reacted an aqueous solution of propanoic acid with a reactive metal and with a carbonate.

(i) Write an equation for the reaction of aqueous propanoic acid with magnesium. [1]

(ii) Write an ionic equation for the reaction of aqueous propanoic acid with aqueous sodium carbonate. [1]

[Total: 17]

[Q3, F325 June 2010]

6. A chemist carries out an investigation on the equilibrium system shown below.

$$2CO(g) + 2NO(g) \rightleftharpoons 2CO_2(g) + N_2(g) \quad \Delta H = -788\,kJ\,mol^{-1}$$

The chemist mixes 0.46 mol of CO with 0.45 mol of NO. The mixture is left to reach equilibrium at constant temperature.

The student analyses the equilibrium mixture and finds that 0.25 mol NO remains. The total volume of the equilibrium mixture is 1.0 dm^3.

(a) (i) Write the K_c expression for this equilibrium. [1]

(ii) What are the units of this equilibrium constant? [1]

(iii) Determine the value of K_c for this equilibrium mixture. Show all your working. [4]

(iv) What does your value of K_c suggest about the position of equilibrium in this experiment? [1]

(b) The chemist increases **both** the temperature and the pressure of the equilibrium mixture. The mixture is left to reach equilibrium again.

(i) What is the effect, if any, on the value of K_c? Explain your answer. [1]

(ii) Explain why it is difficult to predict what would happen to the position of equilibrium after these changes in temperature and pressure. [2]

[Total: 10]

[Q2, F325 June 2012]

7. Butanoic acid, $CH_3(CH_2)_2COOH$, is the 'butter acid', formed when butter turns rancid and tastes sour. A student prepares an aqueous solution of butanoic acid with a concentration of 0.250 mol dm^{-3}.

The K_a of butanoic acid is 1.51×10^{-5} mol dm^{-3}.

(a) (i) Write the expression for the acid dissociation constant of butanoic acid. [1]

(ii) Calculate the pK_a of butanoic acid. [1]

(iii) Calculate the pH of the 0.250 mol dm^{-3} butanoic acid. Give your answer to **two** decimal places. [3]

(b) The student adds aqueous butanoic acid to magnesium.

The student then adds aqueous butanoic acid to aqueous sodium carbonate.

(i) Write the ionic equation for the reaction between aqueous butanoic acid and magnesium. [1]

(ii) Write the ionic equation for the reaction between aqueous butanoic acid and aqueous sodium carbonate. [1]

(c) The student adds 50.0 cm^3 of 0.250 mol dm^{-3} butanoic acid to 50.0 cm^3 of 0.0500 mol dm^{-3} sodium hydroxide. A buffer solution forms.

(i) Explain why a buffer solution forms. [2]

(ii) Calculate the pH of the buffer solution.

The K_a of butanoic acid is 1.51×10^{-5} mol dm^{-3}.

Give your answer to **two** decimal places. [5]

(d) The student adds methanoic acid, HCOOH ($K_a = 1.82 \times 10^{-4}$ mol dm^{-3}), to butanoic acid. A reaction takes place to form an equilibrium mixture containing two acid–base pairs.

Complete the equilibrium below and label the conjugate acid–base pairs.

$$HCOOH + CH_3(CH_2)_2COOH \rightleftharpoons \ldots\ldots + \ldots\ldots$$

$$\ldots\ldots \qquad \ldots\ldots\ldots\ldots \qquad \ldots\ldots \quad \ldots\ldots \quad [2]$$

[Total: 16]

[Q3, F325 June 2012]

Physical chemistry and transition elements

ENERGY

Introduction

Much of our lives rely upon energy. Thanks to chemistry, energy can often be stored and used at our convenience – we use batteries to power gadgets such as our mobile phones, cars are being developed that can be run by fuel cells and even spacecraft use hydrogen fuel cells to provide their electricity.

Chemists need to be able to measure exactly how much energy is required or released by reactions, so that they can utilise this energy or know how much energy will be needed to make something happen. The area of chemistry that studies energy and how it is transferred during chemical processes is known as **thermodynamics**. Much of the work done in thermodynamics is theoretical, for example determining the energy of forming an ionic solid directly from gaseous ions. Thermodynamics can be used to predict which reactions will occur and explain why, or suggest how to make, unlikely reactions take place – all of huge importance when trying to utilise the natural resources and processes around us.

In this chapter you will learn how to construct Born–Haber cycles to find energy changes that cannot be measured directly and you will learn what entropy and free energy are and how they can help us predict why some reactions occur spontaneously and others do not. You will also take a more in-depth look at redox reactions and redox titrations, along with processes that occur within fuel cells and learn how to calculate electrode and standard potentials, and about the uses these may have in the future.

All the maths you need

To unlock the puzzles of this chapter you need the following maths:

- Recognise and use units appropriately (*e.g. the units for enthalpy and entropy*)
- Changing the subject of an equation (*e.g. finding missing enthalpy changes using Born–Haber cycles*)
- Solving algebraic equations (*e.g. finding the free energy for a reaction*)
- Use ratios (*e.g. balancing half-equations using ratios of charges*)

What have I studied before?

- How to calculate oxidation numbers and predict charges on ions
- How to recognise and interpret redox reactions
- How to determine enthalpy changes, including from experimental results using $q = mc\Delta T$
- The definitions of standard enthalpy changes
- How to construct an enthalpy cycle using Hess' law

What will I study later?

- How to use half-equations to represent the reactions of transition elements
- The redox reactions that occur between different oxidation states of transition metals

What will I study in this chapter?

- What lattice enthalpy is
- How to construct and use a Born–Haber cycle
- How ionic size and charge impact on the energy changes associated with dissolving
- What entropy and free energy are and how they are calculated
- How the feasibility of a reaction can be predicted
- How to carry out redox titrations and represent redox reactions using half-equations
- How to calculate standard cell potentials from electrode potentials
- What storage and fuel cells are and their possible uses

1 Lattice enthalpy

By the end of this topic, you should be able to demonstrate and apply your knowledge and understanding of:

* explanation of the term *lattice enthalpy* (formation of 1 mol of ionic lattice from gaseous ions, $\Delta_{LE}H^{\ominus}$) and use as a measure of the strength of ionic bonding in a giant ionic lattice

* use of the lattice enthalpy of a simple ionic solid (i.e. NaCl, $MgCl_2$) and relevant energy terms for the construction of Born–Haber cycles

Why do ionic substances form?

As you learned in Book 1, topic 2.2.4, ionic substances are formed when ions attract one another and bond through electrostatic interactions. However, in order to form ions a huge amount of energy is required. So why do stable elements go on to form ions if so much energy is needed?

When oppositely charged ions attract one another, forming a giant ionic lattice, there is a huge lowering of the energy through very strong attraction. So, although the amount of initial energy required to form ions is large, the lowering of the energy on forming the lattice more than compensates for this.

This is the reason that ionic substances have:

* very strong ionic bonds

* very high melting and boiling points.

We measure the amount of energy involved in the formation of an ionic lattice using **lattice enthalpy**, $\Delta_{LE}H^{\ominus}$. This is the enthalpy change 'when one mole of a solid ionic lattice forms from its gaseous ions under standard conditions'.

> **LEARNING TIP**
>
> You will need to be able to give the exact definition for lattice enthalpy and for first ionisation energy.

For example, when the ionic solid potassium chloride is formed:

$\Delta_{LE}H^{\ominus}$ for potassium chloride is:

$$K^+(g) + Cl^-(g) \rightarrow KCl(s) \qquad \Delta_{LE}H^{\ominus} = -711 \, kJ \, mol^{-1}$$

Notice the following points:

* The ions are both gaseous.

* One mole of the substance is formed.

* The enthalpy change is negative – energy is released to the surroundings.

* Ionic lattice formation is exothermic.

Lattice enthalpy and bond strength

Lattice enthalpy values indicate the relative strength of ionic bonds.

* More exothermic lattice enthalpy values mean stronger ionic bonds (stronger electrostatic interactions).

* More exothermic lattice enthalpy values mean higher melting and boiling points as more energy is required to overcome the interactions present.

* The most exothermic lattice enthalpies arise when ions are small and have large charges – as the charges cause large electrostatic forces and smaller ions can get closer together.

Determining lattice enthalpy

Born–Haber cycles

It is not possible to measure lattice enthalpy directly. This is because it is impossible to form one mole of an ionic solid from its gaseous ions.

We need to construct a special Hess' cycle called a Born–Haber cycle, which shows a series of intermediate steps between the elements that form the ionic substance and the ionic substance itself. The key features of a Born–Haber cycle are as follows:

- A continuous cycle is formed that can start at the elements and end at the elements.
- It includes one step that shows the formation of one mole of the solid ionic lattice from the gaseous ions – this corresponds to the lattice enthalpy.
- The remaining steps show intermediate changes that correspond to key enthalpy changes that can be measured.
- The lattice enthalpy can be calculated by applying Hess' law. If a reaction can take place via more than one route and the initial and final conditions are the same, the total enthalpy change for each route is the same.

Key enthalpy changes

Elements have to go through a series of steps before they are ready to form ionic lattices. These changes have enthalpy changes associated with them. The key enthalpy changes are given below.

The standard enthalpy change of formation, $\Delta_f H^\ominus$

For the standard enthalpy change of formation, $\Delta_f H^\ominus$, one mole of a compound is formed from its constituent elements in their standard states.

This is usually an exothermic process for an ionic compound because bonds are formed.

$\Delta_f H^\ominus$ for potassium chloride is:

$$K(s) + \tfrac{1}{2}Cl_2(g) \rightarrow KCl(s) \qquad \Delta_f H^\ominus = -437\,kJ\,mol^{-1}$$

The standard enthalpy change of atomisation, $\Delta_a H^\ominus$

For the standard enthalpy change of atomisation, $\Delta_a H^\ominus$, one mole of gaseous atoms is formed from its element in its standard state.

$\Delta_a H^\ominus$ is always an endothermic process because bonds have to be broken.

For potassium, metallic bonds are broken:

$$K(s) \rightarrow K(g) \qquad \Delta_a H^\ominus = +89\,kJ\,mol^{-1}$$

For chlorine, covalent bonds are broken:

$$\tfrac{1}{2}Cl_2(g) \rightarrow Cl(g) \qquad \Delta_a H^\ominus = +121\,kJ\,mol^{-1}$$

The first ionisation energy, $\Delta_{I1} H^\ominus$

For the first ionisation energy, $\Delta_{I1} H^\ominus$, one mole of gaseous 1+ ions is formed from gaseous atoms.

$\Delta_{I1} H^\ominus$ is an endothermic process because the electron being lost has to overcome attraction from the nucleus in order to leave the atom.

$\Delta_{I1} H^\ominus$ for potassium is:

$$K(g) \rightarrow K^+(g) + e^- \qquad \Delta_{I1} H^\ominus = +419\,kJ\,mol^{-1}$$

The second ionisation energy, $\Delta_{I2} H^\ominus$

For the second ionisation energy, $\Delta_{I2} H^\ominus$, one mole of gaseous 2+ ions are formed from one mole of gaseous 1+ ions.

As with the first ionisation energy, this is an endothermic process because the electron being lost has to overcome attraction from the nucleus.

$\Delta_{I2} H^\ominus$, for calcium is:

$$Ca^+(g) \rightarrow Ca^{2+}(g) + e^- \qquad \Delta_{I2} H^\ominus = +1145\,kJ\,mol^{-1}$$

The first electron affinity, $\Delta_{EA1} H^\ominus$

Electron affinity is effectively the opposite of ionisation energy – the addition of electrons rather than the removal of electrons. For the first electron affinity, $\Delta_{EA1} H^\ominus$, one mole of gaseous 1− ions is formed from gaseous atoms. This is an exothermic process because the electron is attracted into the outer shell of an atom by the nucleus.

$\Delta_{EA1} H^\ominus$ for chlorine is:

$$Cl(g) + e^- \rightarrow Cl^-(g) \qquad \Delta_{EA1} H^\ominus = -346\,kJ\,mol^{-1}$$

The second electron affinity, $\Delta_{EA2} H^\ominus$

For the second electron affinity, $\Delta_{EA2} H^\ominus$, one mole of gaseous 2− ions is formed from gaseous 1− ions.

This is an endothermic process because the electron is repelled by the 1− ion. This repulsion has to be overcome.

$\Delta_{EA2} H^\ominus$, for oxygen is:

$$O^-(g) + e^- \rightarrow O^{2-}(g) \qquad \Delta_{EA2} H^\ominus, = +790\,kJ\,mol^{-1}$$

> **LEARNING TIP**
>
> Make sure you know these definitions, in particular how many moles are used/formed and the states involved – without knowing this, you will not be able to construct and use Born–Haber cycles.

Constructing Born–Haber cycles

Step 1

Draw a zero line (called a datum line). This is the line that starts your Born–Haber cycle. Onto this line, write the elements *in their standard states* at zero energy.

Under this line, draw another line. Place the ionic solid on this line. Join the first line with this new line, using a downward arrow to represent an exothermic change.

This change represents the enthalpy of formation, one mole of the ionic solid being formed from its constituent elements in their standard states. An example is shown in Figure 1 below.

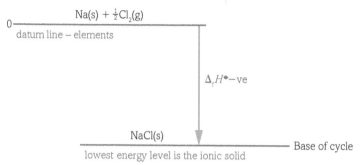

Figure 1 Step 1 in constructing the Born–Haber cycle for sodium chloride.

Step 2

Draw a line near the top of your cycle, slightly to the right-hand side. Onto this line place each constituent element needed to form the ionic solid in the gaseous state. Join this line with the line showing the ionic solid, using a downward pointing arrow to represent an exothermic change.

This change represents the lattice enthalpy. This second step is shown for sodium chloride in Figure 2 below.

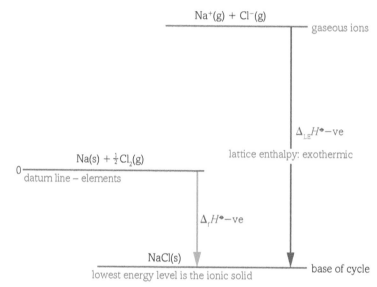

Figure 2 Adding the second step to the Born–Haber cycle for sodium chloride.

Step 3

This third step will involve a number of small steps. The overall aim will be to:

- make sure each original element is gaseous and on its own as separated atoms (i.e. enthalpy changes of atomisation for each element)

- ionise relevant elements to give the appropriate positive charge needed (i.e. first and second ionisation energies)

- ionise relevant elements to give the appropriate negative charge needed (i.e. first and second electron affinity).

The specific steps you take will depend on the solid ionic lattice being formed. For sodium chloride, the steps required are:

- atomisation of sodium: $Na(s) \rightarrow Na(g)$ endothermic

- atomisation of chlorine: $\frac{1}{2}Cl_2(g) \rightarrow Cl(g)$ endothermic

- ionisation of sodium: $Na(g) \rightarrow Na^+(g) + e^-$ endothermic

- electron affinity of chlorine: $Cl(g) + e^- \rightarrow Cl^-(g)$ exothermic

The completed Born–Haber cycle for sodium chloride is shown in Figure 3.

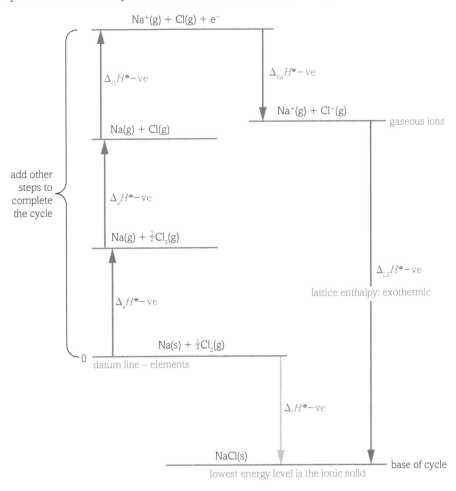

Figure 3 Completed Born–Haber cycle for sodium chloride.

Born–Haber cycles, such as the one shown above, can be used to calculate an unknown enthalpy change. We can apply Hess' law (see Book 1, topic 3.2.6) to the cycle such that:

$\Delta_f H^\ominus$ = the sum of all the other enthalpy changes

or

sum of anticlockwise enthalpy changes = sum of clockwise enthalpy changes
        (*green arrows*)   (*purple arrows*)

Questions

1 Write equations to represent the following enthalpy changes:
   (a) the first ionisation energy of lithium
   (b) the enthalpy change of formation of calcium chloride
   (c) the lattice enthalpy of calcium chloride.

2 (a) Write the equations for making the ionic solid $MgBr_2$:
   (i) from its constituent elements under standard conditions
   (ii) from its constituent gaseous ions.
   (b) Name the enthalpy changes represented by your two equations.

3 Explain why the atomisation of chlorine is a positive enthalpy change.

4 Explain why you have to use a Born–Haber cycle to determine the lattice enthalpy of an ionic solid.

2 Born–Haber cycle calculations

By the end of this topic, you should be able to demonstrate and apply your knowledge and understanding of:

* use of the lattice enthalpy of a simple ionic solid (i.e. NaCl, MgCl₂) and relevant energy terms for:
  * (i) the construction of Born–Haber cycles
  * (ii) related calculations

WORKED EXAMPLE 1

Construct a Born–Haber cycle from the following data and calculate the lattice enthalpy for caesium chloride.

* Enthalpy change of formation of CsCl = −433 kJ mol⁻¹ $Cs(s) + \frac{1}{2}Cl_2(g) \rightarrow CsCl(s)$
* Enthalpy change of atomisation of Cs = +79 kJ mol⁻¹ $Cs(s) \rightarrow Cs(g)$
* First ionisation energy of Cs = +376 kJ mol⁻¹ $Cs(g) \rightarrow Cs^+(g) + e^-$
* Enthalpy change of atomisation of Cl = +121 kJ mol⁻¹ $\frac{1}{2}Cl_2(g) \rightarrow Cl(g)$
* First electron affinity of Cl = −346 kJ mol⁻¹ $Cl(g) + e^- \rightarrow Cl^-(g)$

Before starting to draw your cycle, you should construct the equations alongside the data as shown above in blue.

The question asks you to calculate the lattice enthalpy for $Cs^+(g) + Cl^-(g) \rightarrow CsCl(s)$

The cycle is drawn in steps:

* place the elements at zero energy and add the enthalpy change of formation
* add the lattice enthalpy
* complete the cycle and insert the enthalpy values.

The complete Born–Haber cycle for caesium chloride is shown in Figure 1.

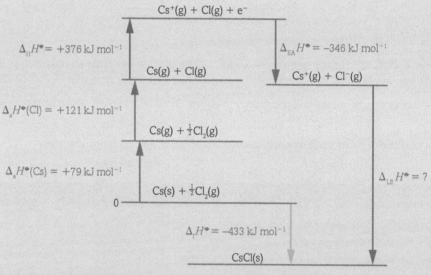

Figure 1 Born–Haber cycle for caesium chloride.

Calculating the lattice enthalpy is the final stage.

Using Hess' law:

> sum of anticlockwise enthalpies = sum of clockwise enthalpies
>
> $\Delta_f H^\ominus(CsCl) = \Delta_a H^\ominus(Cs) + \Delta_a H^\ominus(Cl) + \Delta_{I1} H^\ominus(Cs) + \Delta_{EA} H^\ominus(Cl) + \Delta_{LE} H^\ominus(CsCl)$
>
> $-433 = +79 + 121 + 376 + (-346) + \Delta_{LE}H^\ominus(CsCl)$

Rearranging this to find $\Delta_{LE}H^\ominus(CsCl)$:

> $\Delta_{LE}H^\ominus(CsCl) = -433 - (79 + 121 + 376 - 346)$

Therefore lattice enthalpy of CsCl = −663 kJ mol⁻¹

WORKED EXAMPLE 2

Construct a Born–Haber cycle from the following data, and calculate the lattice enthalpy for sodium oxide.

- Enthalpy change of formation of sodium oxide = −414 kJ mol⁻¹ $2Na(s) + \frac{1}{2}O_2(g) \rightarrow Na_2O(s)$
- Enthalpy change of atomisation of sodium = +108 kJ mol⁻¹ $Na(s) \rightarrow Na(g)$
- Enthalpy change of atomisation of oxygen = +249 kJ mol⁻¹ $\frac{1}{2}O_2(g) \rightarrow O(g)$
- First ionisation energy of sodium = +496 kJ mol⁻¹ $Na(g) \rightarrow Na^+(g) + e^-$
- First electron affinity of oxygen = −141 kJ mol⁻¹ $O(g) + e^- \rightarrow O^-(g)$
- Second electron affinity of oxygen = +790 kJ mol⁻¹ $O^-(g) + e^- \rightarrow O^{2-}(g)$

The question requires the calculation of the lattice enthalpy for: $2Na^+(g) + O^{2-}(g) \rightarrow Na_2O(s)$

LEARNING TIP

Note that in the Born–Haber cycle shown in Worked example 2, the atomisation of sodium has been doubled. This is because 2 moles of sodium are atomised, $2Na(s) \rightarrow 2Na(g)$.
The ionisation energy of sodium has been doubled because 2 moles of sodium are ionised, $2Na(g) \rightarrow 2Na^+(g) + 2e^-$.
The second electron affinity of oxygen is a positive energy change, so there is an additional upwards step in the cycle as shown.

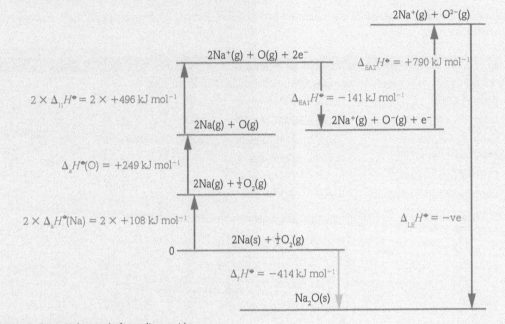

Figure 2 Born–Haber cycle for sodium oxide.

Calculating the lattice enthalpy is the final stage.

Using Hess' law:

sum of clockwise enthalpies = sum of anticlockwise enthalpies

$\Delta_f H^\ominus(Na_2O) = 2 \times \Delta_a H^\ominus(Na) + \Delta_a H^\ominus(O) + 2 \times \Delta_{I1} H^\ominus(Na) + \Delta_{EA1} H^\ominus(O) + \Delta_{EA2} H^\ominus(O) + \Delta_{LE} H^\ominus(Na_2O)$

$-414 = (+108 \times 2) + 249 + (+496 \times 2) + (-141) + 790 + \Delta_{LE} H^\ominus(Na_2O)$

Rearranging the equation to find $\Delta_{LE} H^\ominus(Na_2O)$:

$\Delta_{LE} H^\ominus(Na_2O) = -414 - ((+108 \times 2) + 249 + (+496 \times 2) - 141 + 790)$

Therefore, lattice enthalpy of Na₂O = −2520 kJ mol⁻¹

Question

1 Calculate the lattice enthalpy for rubidium chloride given the following data.

ΔH(formation) of RbCl = −435 kJ mol⁻¹
ΔH(atomisation) of Rb = +81 kJ mol⁻¹
First ionisation energy of Rb = +403 kJ mol⁻¹
ΔH(atomisation) of Cl = +121 kJ mol⁻¹
First electron affinity of Cl = −346 kJ mol⁻¹

By the end of this topic, you should be able to demonstrate and apply your knowledge and understanding of:

* the use of the lattice enthalpy of a simple ionic solid (i.e. NaCl, $MgCl_2$) and relevant energy terms for:
  (i) the construction of Born–Haber cycles
  (ii) related calculations

This topic will cover some more complicated examples of Born–Haber cycles and their associated calculations.

LEARNING TIP

Notice that in Worked example 1, the atomisation enthalpy of chlorine has been doubled. Atomisation refers to the formation of 1 mole of gaseous atoms. In this example, 2 moles of gaseous chlorine atoms are obtained, so the enthalpy change is:

* $Cl_2(g) \rightarrow 2Cl(g)$
  $\Delta H = +242 \text{ kJ mol}^{-1}$

This is double the value of $\Delta_a H^{\ominus}(Cl)$:

* $\frac{1}{2}Cl_2(g) \rightarrow Cl(g)$
  $\Delta H = +121 \text{ kJ mol}^{-1}$

As with atomisation, a factor of 2 is included for the electron affinity of chlorine. Two moles of gaseous chloride ions are obtained from gaseous chlorine atoms for the enthalpy change:

$2Cl(g) + 2e^- \rightarrow 2Cl^-(g)$
$\Delta H = 2 \times \Delta_{EA} H^{\ominus}$

WORKED EXAMPLE 1

The following data is provided for use in the Born–Haber cycle for calcium chloride:

* $Ca(s) \rightarrow Ca(g)$ $\Delta H = +190 \text{ kJ mol}^{-1}$
* $Ca(g) \rightarrow Ca^+(g) + e^-$ $\Delta H = +590 \text{ kJ mol}^{-1}$
* $Ca^+(g) \rightarrow Ca^{2+}(g) + e^-$ $\Delta H = +1145 \text{ kJ mol}^{-1}$
* $\frac{1}{2}Cl_2(g) \rightarrow Cl(g)$ $\Delta H = +121 \text{ kJ mol}^{-1}$
* $Ca(s) + Cl_2(g) \rightarrow CaCl_2(s)$ $\Delta H = -795 \text{ kJ mol}^{-1}$
* $Ca^{2+}(g) + 2Cl^-(g) \rightarrow CaCl_2(s)$ $\Delta H = -2237 \text{ kJ mol}^{-1}$

Draw a Born–Haber cycle for calcium chloride, and determine the electron affinity of chlorine.

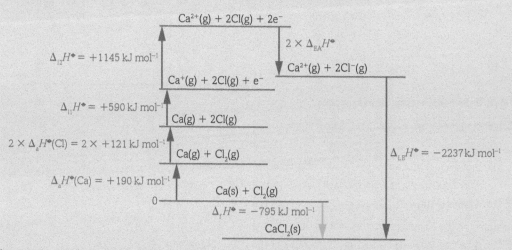

Figure 1 Born–Haber cycle for calcium chloride.

Calculating the electron affinity of chlorine is the final stage. The equation for electron affinity is:

$Cl(g) + e^- \rightarrow Cl^-(g)$

Using Hess' law:

sum of anticlockwise enthalpies = sum of clockwise enthalpies

$\Delta_f H^{\ominus}(CaCl_2) = \Delta_a H^{\ominus}(Ca) + 2 \times \Delta_a H^{\ominus}(Cl) + \Delta_{i1} H^{\ominus}(Ca) + \Delta_{i2} H^{\ominus}(Ca) + 2 \times \Delta_{EA} H^{\ominus}(Cl) + \Delta_{LE} H^{\ominus}(CaCl_2)$

$-795 = + 190 + 242 + 590 + 1145 + 2 \times \Delta_{EA} H^{\ominus}(Cl) + (-2237)$

Rearranging to find $\Delta_{EA} H^{\ominus}(Cl)$:

$-795 - (190 + 242 + 590 + 1145 - 2237) = 2 \times \Delta_{EA} H^{\ominus}(Cl)$

Therefore:

$2 \times \Delta_{EA} H^{\ominus}(Cl) = -725 \text{ kJ mol}^{-1}$

Therefore:

$\Delta_{EA} H^{\ominus}(Cl) = \dfrac{-725}{2}$

$= -363 \text{ kJ mol}^{-1}$

WORKED EXAMPLE 2

Using the enthalpy data below, draw a Born–Haber cycle and calculate a value for the lattice enthalpy of copper(II) oxide.

Process	$\Delta H^{\ominus}$/kJ mol^{-1}
Atomisation of copper	+339
First ionisation of copper	+745
Second ionisation of copper	+1960
Atomisation of oxygen	+249
First electron affinity of oxygen	−141
Second electron affinity of oxygen	+790
Formation of copper(II) oxide	−155

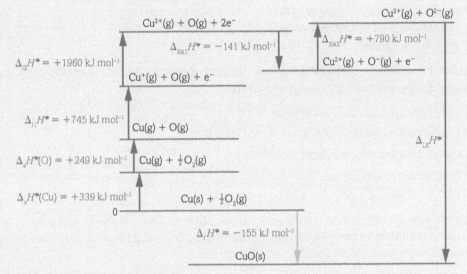

Figure 2 Born–Haber cycle for copper(II) oxide.

Calculating the lattice enthalpy is the final stage.

Using Hess' law:

sum of anticlockwise enthalpies = sum of clockwise enthalpies

$$\Delta_f H^{\ominus}(\text{CuO}) = \Delta_a H^{\ominus}(\text{Cu}) + \Delta_a H^{\ominus}(\text{O}) + \Delta_{i1} H^{\ominus}(\text{Cu}) + \Delta_{i2} H^{\ominus}(\text{Cu}) + \Delta_{EA1} H^{\ominus}(\text{O}) + \Delta_{EA2} H^{\ominus}(\text{O}) + \Delta_{LE} H^{\ominus}(\text{CuO})$$

$$-155 = +339 + 249 + 745 + 1960 - 141 + 790 + \Delta_{LE} H^{\ominus}(\text{CuO})$$

Rearranging to find $\Delta_{LE} H^{\ominus}(\text{CuO})$:

$$\Delta_{LE} H^{\ominus}(\text{CuO}) = -155 - (339 + 249 + 745 + 1960 - 141 + 790)$$

Therefore:

lattice enthalpy of copper(II) oxide = −4097 kJ mol^{-1}

Question

 (a) Draw a labelled Born–Haber cycle for the formation of solid $MgCl_2$ showing the species present at each stage.

(b) Use the data below to find the value of the electron affinity of chlorine.

Enthalpy change of formation of $MgCl_2$	−642 kJ mol^{-1}
Lattice enthalpy of $MgCl_2$	−2493 kJ mol^{-1}
First ionisation energy of Mg	+736 kJ mol^{-1}
Second ionisation energy of Mg	+1450 kJ mol^{-1}
Enthalpy change of atomisation of Cl	+121 kJ mol^{-1}
Enthalpy change of atomisation of Mg	+150 kJ mol^{-1}

4 Enthalpy change of solution and hydration

By the end of this topic, you should be able to demonstrate and apply your knowledge and understanding of:

* explanation and use of the terms:
  (i) *enthalpy change of solution* (dissolving of 1 mol of solute, $\Delta_{sol}H^{\ominus}$)
  (ii) *enthalpy change of hydration* (dissolving of 1 mol of gaseous ions in water, $\Delta_{hyd}H^{\ominus}$)

* use of the enthalpy change of solution of a simple ionic solid (i.e. NaCl, MgCl$_2$) and relevant energy terms (*enthalpy change of hydration* and *lattice enthalpy*) for:
  (i) the construction of enthalpy cycles
  (ii) related calculations

* qualitative explanation of the effect of ionic charge and ionic radius on the exothermic value of a lattice enthalpy and enthalpy change of hydration

KEY DEFINITIONS

The **standard enthalpy change of solution, $\Delta_{sol}H^{\ominus}$**, is the enthalpy change that takes place when one mole of a *solute* is completely dissolved in water under standard conditions.

The **standard enthalpy change of hydration, $\Delta_{hyd}H^{\ominus}$**, is the enthalpy change that takes place when dissolving one mole of gaseous ions in water.

What happens when a solid dissolves?

When a solid dissolves, two processes take place.

* The ionic lattice breaks down.

* The free ions become part of the solution (hydration).

A change in enthalpy occurs when this overall process happens. The enthalpy of this process is known as the **standard enthalpy change of solution, $\Delta_{sol}H^{\ominus}$**, and it corresponds to one mole of solute dissolving in water under standard conditions.

The dissolving of a substance can be endothermic (for example, when ammonium nitrate dissolves in water) or exothermic (for example, when calcium chloride dissolves in water). The substance that is dissolved is known as the solute, and the substance it dissolves in is known as the solvent. In order to understand why this is possible, we need to consider the stages of dissolving in more detail.

DID YOU KNOW?

Standard conditions are considered as 100 kPa and a temperature of 298 K.

The breakdown of the ionic lattice

In topic 5.2.1 we learned that the lattice enthalpy is the enthalpy change that accompanies the formation of one mole of an ionic compound from its gaseous ions. The process that occurs when the lattice breaks down during dissolving is the reverse of this process. We imagine the lattice becoming free gaseous ions.

Therefore the enthalpy change of ionic lattice breakdown is equal to $-\Delta_{LE}H^{\ominus}$.

For example, when the ionic lattice of potassium chloride forms:

$$K^+(g) + Cl^-(g) \rightarrow KCl(s)$$

$$\Delta_{LE}H^{\ominus} = -711\,kJ\,mol^{-1}$$

When the ionic lattice of potassium chloride breaks down during dissolving:

$$KCl(s) \rightarrow K^+(g) + Cl^-(g)$$

$$\Delta H = +711\,kJ\,mol^{-1}$$

Notice that:

* the processes are identical but the reverse of one another

* the enthalpy change has the same value but different signs

* lattice enthalpy has a negative sign and is *exothermic*

* the breakdown of the ionic lattice has a positive sign and is *endothermic*.

What dictates the size of the lattice enthalpy?

The magnitude of any lattice enthalpy is dependent on:

- the size of the ions involved

- the charges on the ions

- ionic bond strength (which is dependent on ionic size and charge).

Smaller ions, i.e. ions with a smaller ionic radius, can get closer together. They will attract one another more strongly and give rise to more exothermic lattice enthalpy values, i.e. more negative values.

This is illustrated in Figure 1 using the sodium halides NaCl, NaBr and NaI as examples.

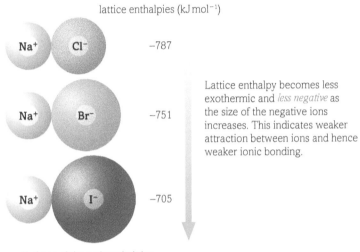

lattice enthalpies (kJ mol^{-1})

−787

−751

−705

Lattice enthalpy becomes less exothermic and *less negative* as the size of the negative ions increases. This indicates weaker attraction between ions and hence weaker ionic bonding.

Figure 1 The lattice enthalpies of the sodium halides.

Ions with higher charges cause greater electrostatic attraction and in turn more exothermic lattice enthalpy values. The most exothermic lattice enthalpy values arise from small, highly charged ions.

This is illustrated in Figure 2, using ions of period 3 elements as examples.

Na^+ Mg^{2+} Al^{3+}

The smallest, most highly charged ions will give rise to the largest lattice enthalpies, as they can pack closer to oppositely charged ions, with higher attraction. Lattice enthalpies will become more exothermic – more negative.

Figure 2 The effect of charge on lattice enthalpies.

WORKED EXAMPLE

State and explain which compound has the most exothermic lattice enthalpy: $MgCl_2$, $MgBr_2$ or MgI_2.
Magnesium chloride has the most exothermic lattice enthalpy because the chloride ion is smaller than both bromide ions and iodide ions. This means the Mg^{2+} and Cl^- ions in the $MgCl_2$ lattice can pack closer together and exert a greater attraction on each other than the ions in $MgBr_2$ or MgI_2.

Hydration of ions

Once the ionic lattice has broken down into its constituent ions, these have to become part of the solution. This is known as hydration. The ions are able to do this if the solvent (water in the case of hydration) can interact with them in similar ways to the bonding in the lattice – remember '*like dissolves like*'. Ionic solids are therefore able to dissolve in polar solvents, such as water.

During hydration:

- the positive ions will be attracted to the slightly negative (δ–) oxygen in the water molecules

- the negative ions will be attracted to the slightly positive (δ+) hydrogens in the water molecules

- the water molecules will completely surround the ions.

The hydration of potassium ions, K^+, and chloride ions, Cl^- is shown in Figure 3.

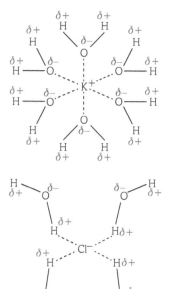

Figure 3 The hydration of K^+ and Cl^- ions occurs when potassium chloride is dissolved.

An enthalpy change occurs when ions become hydrated. Energy is released when new bonds are formed between ions and water molecules. This is the **standard enthalpy change of hydration**, $\Delta_{hyd}H^\ominus$, and is the enthalpy change when one mole of aqueous ions are formed from their gaseous ions, under standard conditions.

For example $\Delta_{hyd}H^\ominus$ for potassium ions is:

$$K^+(g) \xrightarrow{(aq)} K^+(aq)$$

$$\Delta_{hyd}H^\ominus = -322 \, kJ \, mol^{-1}$$

$\Delta_{hyd}H^\ominus$ for chloride ions is:

$$Cl^-(g) \xrightarrow{(aq)} Cl^-(aq)$$

$$\Delta_{hyd}H^\ominus = -363 \, kJ \, mol^{-1}$$

Notice that both have negative values for $\Delta_{hyd}H^\ominus$. This is because

hydration is an exothermic process.

What dictates the magnitude of the enthalpy of hydration?

The magnitude of the enthalpy of hydration is dependent on:

- the size of the ions involved

- the charges on the ions.

Ions with smaller ionic radii can get closer to the water molecules and are able to attract them more strongly. This means that on hydration, more energy is released and $\Delta_{hyd}H^\ominus$ becomes more exothermic. For more information on reduced charge density in larger ions, see the 'Did you know?' box on the previous page.

This trend is illustrated in Figure 4 for the hydration of halide ions.

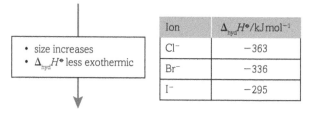

	Ion	$\Delta_{hyd}H^\ominus$/kJ mol^{-1}
• size increases	Cl^-	-363
• $\Delta_{hyd}H^\ominus$ less exothermic	Br^-	-336
	I^-	-295

Figure 4 Enthalpy changes for the hydration of halide ions.

The higher the charge on an ion, the greater the attraction it will have with the water molecule. This will give a more negative and hence more exothermic value for the enthalpy of hydration.

This trend is illustrated for some common period 3 ions in Figure 5. The enthalpies in these examples will also be affected slightly by the size of the ions as there is a decrease in size across the period.

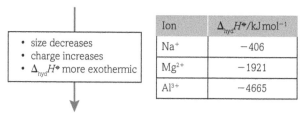

	Ion	$\Delta_{hyd}H^\ominus$/kJ mol^{-1}
• size decreases	Na^+	-406
• charge increases	Mg^{2+}	-1921
• $\Delta_{hyd}H^\ominus$ more exothermic	Al^{3+}	-4665

Figure 5 Enthalpy changes of hydration of some common cations.

Calculating a lattice enthalpy from enthalpy changes of solution and hydration

The energy changes associated with dissolving of ionic substances can be shown using Born–Haber cycles.

The ionic solid and the gaseous ions can be linked by either:

- the lattice enthalpy

- the standard enthalpy of solution *along with* the standard enthalpy of hydration.

This means that if we know the standard enthalpies of solution and hydration, we can calculate the lattice enthalpy for an ionic solid – which, as we learned in topic 5.2.1, cannot be measured directly.

Figure 6 Determining the enthalpy change of solution of potassium chloride, using simple apparatus sometimes known as a coffee cup calorimeter.

An example Born–Haber cycle is shown in Figure 7. This shows how the Born–Haber cycle for the dissolution of potassium chloride, KCl, can be used to determine lattice enthalpy.

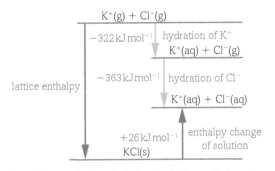

Figure 7 Born–Haber cycle for calculating the lattice enthalpy of KCl(s).

Notice that:

- the ionic solid is at the bottom of the cycle

- the gaseous ions are at the top of the cycle

- the route via lattice enthalpy is shown on the left

- the route via enthalpies of solution and hydration is shown on the right

- the enthalpy change of solution (upwards arrow) does not correspond to the overall enthalpy change (downwards arrow): upwards = endothermic, downwards = exothermic.

The processes that form the Born–Haber cycle for the dissolving of potassium chloride in water are:

$$KCl(s) \rightarrow K^+(aq) + Cl^-(aq) \quad \Delta_{sol}H^\ominus = +26\,\text{kJ mol}^{-1}$$

$$K^+(g) \rightarrow K^+(aq) \quad \Delta_{hyd}H^\ominus = -322\,\text{kJ mol}^{-1}$$

$$Cl^-(g) \rightarrow Cl^-(aq) \quad \Delta_{hyd}H^\ominus = -363\,\text{kJ mol}^{-1}$$

$$K^+(g) + Cl^-(g) \rightarrow KCl(s) \quad \Delta_{LE}H^\ominus = \text{unknown}$$

Using the Born–Haber cycle and Hess' law, we can say that:

sum of clockwise enthalpy changes = sum of anticlockwise enthalpy changes

$$\Delta_{hyd}H^\ominus(K^+) + \Delta_{hyd}H^\ominus(Cl^-) = \Delta_{LE}H^\ominus(KCl) + \Delta_{sol}H^\ominus(KCl)$$

$$-322 + (-363) = \Delta_{LE}H^\ominus(KCl) + 26$$

Rearranging to find $\Delta_{LE}H^\ominus$:

$$-322 - 363 - 26 = \Delta_{LE}H^\ominus(KCl)$$

Therefore:

$$\Delta_{LE}H^\ominus(KCl) = -711\,\text{kJ mol}^{-1}$$

Questions

1 Lattice enthalpy is used to compare the strengths of ionic bonds. Describe and explain the effect of ionic charge and ionic radius on the magnitude of lattice enthalpy.

2 Describe how, and explain why, the enthalpy change of hydration of sodium ions differs from that of rubidium ions.

3 You are provided with the following enthalpy changes.

$$Na^+(g) + F^-(g) \rightarrow NaF(s) \quad \Delta H = -918\,\text{kJ mol}^{-1}$$

$$Na^+(g) \xrightarrow{(aq)} Na^+(aq) \quad \Delta H = -390\,\text{kJ mol}^{-1}$$

$$NaF(s) \xrightarrow{(aq)} Na^+(aq) + F^-(aq) \quad \Delta H = +71\,\text{kJ mol}^{-1}$$

(a) Name these *three* enthalpy changes.

(b) Write an equation, including state symbols, for the change that accompanies the enthalpy change of hydration of F⁻ ions.

(c) Calculate the enthalpy change of hydration of F⁻ ions.

By the end of this topic, you should be able to demonstrate and apply your knowledge and understanding of:

* explanation that entropy is a measure of the dispersal of energy in a system which is greater, the more disordered a system

* explanation of the difference in magnitude of the entropy of a system:
  (i) of solids, liquids and gases
  (ii) for a reaction in which there is a change in the number of gaseous molecules

* calculation of the entropy change of a system, ΔS, and related quantities for a reaction given the entropies of the reactants and products

What is entropy?

Entropy, given the symbol S, is a measure of disorder in a system. *System* is used to describe the actual particles involved in a reaction or process. *Surroundings* describes everything outside the system. The more disordered the particles in the system are, the higher the entropy of the system. If you think of solids compared to gases, we would say that the solid has a lower disorder compared to the gas – the gas is more disordered. Entropy is a key thermodynamic factor that can be used to describe chemical processes, usually alongside enthalpy.

The **standard entropy, $S^{\ominus}$**, of a substance is the entropy content of one mole of the substance under standard conditions. Standard entropies have units of $J\,K^{-1}\,mol^{-1}$.

All substances above 0 K possess a certain degree of disorder because they are in constant motion. This means that:

* entropy, S, is always a positive number above 0

* at 0 K, entropy is zero for perfect crystals.

Most substances are thermodynamically stable at the lowest energy state, which would correspond to a low entropy. However, entropy always tends to increase; liquid water naturally evaporates into gaseous water, increasing entropy; heat energy spreads out from hot objects, increasing entropy; salt particles dissolve in water, increasing entropy. It is always more probable that a more disordered system will be found instead of a more ordered system. This is illustrated in Figure 1.

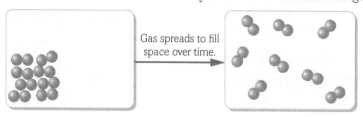

Gas spreads to fill space over time.

Pile of bricks dropped from a lorry.
Which arrangement is the more probable?

Disorder is more probable than order.

Figure 1 Increasing entropy.

Often a useful way to think about entropy is:

- more spreading out of energy = higher entropy
- more random arrangements of particles = higher entropy.

Entropy can change during the course of a chemical process or reaction. These changes would be described in terms of the change in entropy, ΔS.

Although there is always a tendency towards higher entropy (this is the second law of thermodynamics) it is possible for entropy to decrease, i.e. for ΔS to be negative. For example, water freezing would represent a decrease in entropy as a liquid has become a more ordered solid, with lower levels of energy dispersal.

How entropy is affected by temperature

Entropy of pure substances increases with increasing temperature. The values of entropy are given as per kelvin (K^{-1}) because entropy values are dependent on the temperature.

- Particles at higher temperatures have higher energy and move more.
- The arrangement of particles at higher temperatures becomes more random.
- Entropy of solids < entropy of liquids < entropy of gases.

When water boils and becomes gaseous its entropy rises:

$$H_2O(l)\ S = +70\,kJ\,mol^{-1}$$

$$H_2O(g)\ S = +189\,kJ\,mol^{-1}$$

Entropy changes when dissolving ionic solids

If a reaction results in products that allow more disorder, i.e. more ways for energy to be arranged or dispersed, then entropy will increase.

If a solid ionic lattice dissolves, ions can spread out and the positions of the ions are far more disordered than within the lattice. This means entropy increases, for example:

$$CuSO_4 \cdot 5H_2O(s) \xrightarrow{(aq)} Cu^{2+}(aq) + SO_4^{2-}(aq) + 5H_2O(l)$$

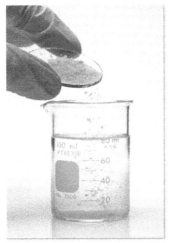

Figure 2 Crystals of $CuSO_4 \cdot 5H_2O(s)$ and a solution of $CuSO_4(aq)$ with some undissolved crystals. The blue crystals are soluble and dissolving them produces a blue solution – copper(II) sulfate ions are spreading out and entropy is increasing.

How the number of gas molecules affects entropy

If the number of gas molecules changes during a reaction, entropy changes.

- An increase in the number of gas molecules causes an increase in entropy.

    E.g. $CaCO_3(s) + 2H^+(aq) \rightarrow Ca^{2+}(aq) + CO_2(g) + H_2O(l)$

- A decrease in the number of gas molecules causes a decrease in entropy.

    E.g. $Mg(s) + O_2(g) \rightarrow MgO(s)$

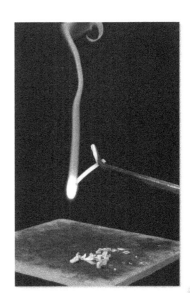

Figure 3 Entropy decreases when magnesium burns in oxygen, as there is a decrease in the number of gas particles present.

Calculating entropy changes

The **standard entropy change for a reaction** is the change in entropy that occurs when a reaction takes place under standard conditions, with molar quantities of each reactant reacting in their standard states. This is given the symbol $\Delta S^\ominus$.

The standard entropy of a reaction is calculated using the entropies of the reactants and products as follows:

$$\Delta S = \Sigma S^\ominus_{products} - \Sigma S^\ominus_{reactants}$$

(where Σ = 'sum of')

Entropy is a measure of randomness. Systems that are more chaotic have a higher entropy value.

- If a change makes a system more random, ΔS is positive.
- If a change makes a system more ordered, ΔS is negative.

WORKED EXAMPLE

This equation shows the formation of ammonia from nitrogen and hydrogen.

$$N_2(g) + 3H_2(g) \rightarrow 2NH_3(g)$$

Use the data in the table to calculate ΔS for this reaction.

	$N_2(g)$	$H_2(g)$	$NH_3(g)$
$S^\ominus/J\,K^{-1}\,mol^{-1}$	+192	+131	+193

$$\Delta S = \Sigma S^\ominus_{products} - \Sigma S^\ominus_{reactants}$$
$$= (2 \times 193) - (+192 + (3 \times 131))$$
$$= -199\,J\,K^{-1}\,mol^{-1}$$

This decrease in entropy is to be expected because a more ordered system with fewer gaseous molecules is produced: 2 moles rather than 4 moles.

Questions

1 What is the sign of ΔS for the following reactions?

(a) $NaCl(s) \rightarrow Na^+(aq) + Cl^-(aq)$

(b) $2SO_2(g) + O_2(g) \rightarrow 2SO_3(g)$

(c) $C_8H_{18}(l) + 12\frac{1}{2}O_2(g) \rightarrow 8CO_2(g) + 9H_2O(l)$

2 Use the data in the table below to calculate ΔS for these reactions.

(a) $2NO(g) + O_2(g) \rightarrow N_2O_4(g)$

(b) $C_6H_6(l) + 7\frac{1}{2}O_2(g) \rightarrow 6CO_2(g) + 3H_2O(l)$

	$NO(g)$	$O_2(g)$	$N_2O_4(g)$	$C_6H_6(l)$	$CO_2(g)$	$H_2O(l)$
$S^\ominus/J\,K^{-1}\,mol^{-1}$	+211	+205	+304	+173	+214	+70

(6) Free energy

By the end of this topic, you should be able to demonstrate and apply your knowledge and understanding of:

* the explanation that the feasibility of a process depends upon the entropy change and temperature in the system, $T\Delta S$, and the enthalpy change of the system, ΔH

* the explanation, and related calculations, of the free energy change, ΔG, as: $\Delta G = \Delta H - T\Delta S$ (the Gibbs' equation) and that a process is feasible when ΔG has a negative value

* the limitations of predictions made by ΔG about feasibility, in terms of kinetics

---

KEY DEFINITION

The **free energy change, ΔG**, is the balance between enthalpy, entropy and temperature for a process:

$$\Delta G = \Delta H - T\Delta S$$

A process can take place spontaneously when $\Delta G < 0$.

Spontaneous changes

For a process to happen spontaneously, i.e. instantaneously, entropy must increase. But there are many examples of reactions that happen spontaneously that would seem to lower entropy, for example a metal rusting in oxygen or a gas condensing.

To understand why these are able to occur, we need to consider the entropy change of both the system and the surroundings.

Total change in entropy

We calculate the total change in entropy as follows.

$$\Delta S^{\ominus}_{total} = \Delta S^{\ominus}_{system} + \Delta S^{\ominus}_{surroundings}$$

For spontaneous changes to occur, the total change in entropy must be positive.

Reactions that have a decrease in entropy can occur spontaneously if the change in entropy of the surroundings is positive enough to make the total change in entropy positive.

Free energy

Changes in entropy are linked with energy changes that occur during a reaction or change; remember, entropy is a measure of the dispersal of energy. A reaction that would have a positive entropy change still may only be spontaneous at certain temperatures.

Reactions and processes have to be considered in terms of enthalpy changes ΔH, total entropy changes ΔS, and temperature T (measured in K).

The Gibbs' equation

The energy that becomes 'free' during a reaction is known as Gibbs' free energy after the theoretical physicist and mathematician, Josiah Willard Gibbs, and is given the symbol G.

It is dependent on the total entropy and enthalpy changes that occur, as well as being dependent on temperature.

The **free energy change** is calculated using the Gibbs' equation:

$$\Delta G = \Delta H - T\Delta S$$

For a reaction to occur spontaneously, ΔG must be negative.

LEARNING TIP

$$\Delta S_{surroundings} = -\frac{\Delta H}{T}$$

This means that $\Delta S_{total} = \Delta S - \frac{\Delta H}{T}$

This can be rearranged to give $-T\Delta S_{total} = \Delta H - T\Delta S$

This can then be converted to the Gibbs' equation as $\Delta G = -T\Delta S$

Using the Gibbs' equation to predict feasibility of reactions

Using the Gibbs' equation, we can make some assumptions about conditions that will favour spontaneous changes.

* Large increases in entropy will cause decreases in ΔG, because the term $-T\Delta S$ will become larger.

* Large negative values for ΔH (i.e. highly exothermic reactions) will result in more negative values for ΔG.

However, there is a balance to be struck. If, for example, a highly exothermic reaction causes a large decrease in entropy, ΔG may be positive, in which case the change would not be spontaneous.

The effect of enthalpy and entropy changes on ΔG are summarised in Table 1.

ΔH	ΔS	ΔG	Feasibility of spontaneous change
negative	positive	always negative	reaction feasible
positive	negative	always positive	reaction never feasible
negative	negative	negative at low temperatures	feasible at low temperatures
positive	positive	negative at high temperatures	feasible at high temperatures

Table 1 The effect of enthalpy and entropy changes on ΔG.

In summary:

- exothermic reactions are generally spontaneous – the negative value of ΔH is usually still able to make ΔG negative, even if the entropy change is positive

- endothermic changes are only spontaneous if the entropy is positive and the temperature is high enough to make $T\Delta S$ large and positive, i.e. greater than ΔH.

WORKED EXAMPLE

The thermal decomposition of zinc carbonate, $ZnCO_3$, is represented by:

$$ZnCO_3(s) \rightarrow ZnO(s) + CO_2(g)$$

$$\Delta H = +71\ kJ\ mol^{-1}$$

Use the data in the table to calculate the minimum temperature, in °C, for thermal decomposition to take place.

	$ZnCO_3(s)$	$ZnO(s)$	$CO_2(g)$
$S^{\ominus}/J\ K^{-1}\ mol^{-1}$	+82	+44	+214

Step 1: Calculate ΔS in units of $kJ\ K^{-1}\ mol^{-1}$.

$$\Delta S = \Sigma S^{\ominus}_{products} - \Sigma S^{\ominus}_{reactants}$$

$$= 44 + 214 - 82$$

$$= 176\ J\ K^{-1}\ mol^{-1}$$

$$= 0.176\ kJ\ K^{-1}\ mol^{-1}$$

Step 2: Calculate T when $\Delta G = 0$

$$\Delta G = \Delta H - T\Delta S$$

When:

$$\Delta G = 0$$

$$\Delta H - T\Delta S = 0$$

Therefore:

$$T = \frac{\Delta H}{\Delta S}$$

$$= \frac{71}{0.176}$$

$$= 403\ K$$

$$= 403 - 273 = 130\ °C$$

LEARNING TIP

In free energy-related calculations, you need to get the units of joules for entropy and enthalpy the same.

- ΔH is usually given in $kJ\ mol^{-1}$
- ΔS is usually given in $J\ K^{-1}\ mol^{-1}$

First get ΔS into $kJ\ K^{-1}\ mol^{-1}$:

- to convert J to kJ, divide by 1000.

Also remember that entropy is worked out using temperature in K:

- to convert °C to K, add 273.
- to convert K to °C, subtract 273.

Limitations of using ΔG to predict feasibility of reactions

Calculating the value of the free energy, ΔG, gives a theoretical answer for whether a reaction or process will react spontaneously, i.e. whether it is thermodynamically possible.

If a reaction is found to have a negative value of ΔG, it does not mean it will actually go on to react. For example, burning petrol in a car involves combusting a fuel; it is a very exothermic reaction and produces gaseous products, hence an increase in entropy. Thermodynamically, it should react spontaneously. Thankfully for all the drivers in the world filling their cars with petrol, it does not – it needs to be ignited first!

Whether or not a reaction proceeds spontaneously also depends on kinetic factors.

- The reaction may have a high activation energy – energy needs to be initially supplied to overcome this, e.g. igniting a fuel.

- The rate of the reaction may be extremely slow.

Equally, reactions that have positive values of ΔG are considered not feasible. They can be made to take place however, usually by changing the temperature of the reaction.

Figure 1 Some reactions are thermodynamically feasible and have a negative value of ΔG but do not occur spontaneously due to kinetic factors such as activation barriers or slow rates of reactions. An example is the burning of a fuel such as petrol.

Questions

1. The combustion of ethanol is represented by:

   $$C_2H_5OH(l) + 3O_2(g) \rightarrow 2CO_2(g) + 3H_2O(l)$$

   $$\Delta H = -786\ kJ\ mol^{-1}$$

   $$\Delta S = +68\ J\ K^{-1}\ mol^{-1}$$

   Calculate ΔG for this reaction at 25 °C.

2. The equation for the thermal decomposition of ammonium chloride, NH_4Cl, is:

   $$NH_4Cl(s) \rightarrow NH_3(g) + HCl(g)$$

   $$\Delta H = +176\ kJ\ mol^{-1}$$

   Calculate the minimum temperature, in °C, for thermal decomposition to take place spontaneously.

	$NH_4Cl(s)$	$NH_3(g)$	$HCl(g)$
$S^{\ominus}/J\ K^{-1}\ mol^{-1}$	+95	+192	+187

Redox

By the end of this topic, you should be able to demonstrate and apply your knowledge and understanding of:

* explanation and use of the terms *oxidising agent* and *reducing agent*
* construction of redox equations using half-equations and oxidation numbers
* interpretation and prediction of reactions involving electron transfer

---

KEY DEFINITIONS

Oxidation is loss of electrons, or an increase in oxidation number.
Reduction is gain of electrons, or a decrease in oxidation number.
An **oxidising agent** is the species that is reduced in a reaction and causes another species to be oxidised.
A **reducing agent** is the species that is oxidised in a reaction and causes another species to be reduced.

Redox

You will recall from Book 1, topic 2.1.19, that redox reactions involve both oxidation and reduction.

* **Oxidation** is the loss of electrons – an increase in oxidation number.
* **Reduction** is the gain of electrons – a decrease in oxidation number.

LEARNING TIP

Refer back to Book 1, topic 2.1.19 to recap how to assign oxidation numbers to species involved in redox reactions.

Oxidising and reducing agents

Electrons must be transferred during a redox reaction – they cannot just disappear, they must be transferred between species.

Oxidising agents

When a species is oxidised, it loses electrons. These electrons are gained by the species being reduced. If a chemical is readily reduced it will gain the electrons in the reaction and is known as the **oxidising agent**.

Reducing agents

When a species is reduced, it gains electrons. These electrons have been removed from the species being oxidised. If a chemical is readily oxidised it will provide the electrons in the reaction and is known as the **reducing agent**.

Redox half-equations

Redox is essentially two reactions occurring simultaneously:

* the reduction of a species
* the oxidation of a species.

It is often helpful to consider each of these reactions separately. We can use half-equations to do this. Half-equations show 'half' the reaction.

If we consider the redox reaction that occurs when a metal, such as magnesium, reacts with an acid:

$$Mg(s) + 2HCl(aq) \rightarrow MgCl_2(aq) + H_2(g)$$

this overall redox equation can be broken down into two half-equations.

$$Mg \rightarrow Mg^{2+} + 2e^-$$ this is the *oxidation* half-equation

$$2H^+ + 2e^- \rightarrow H_2$$ this is the *reduction* half-equation

Notice that the number of electrons *lost* in the oxidation half-equation is the same as the number of electrons *gained* in the reduction half-equation. This must *always* be the case – the electrons must balance.

Notice also that Cl^- is not included in either of the half-equations. This is because chlorine ions are not involved in the redox reaction – its oxidation number remains -1 throughout the reaction. It is known as a spectator ion and does not need to be included in the half-equation.

LEARNING TIP

Oxidation is the loss of electrons and you might be inclined to write a half-equation with the electrons being lost on the left hand side of the arrow, for example:

$$Mg - 2e^- \rightarrow Mg^{2+}$$

As no particles are gained or lost in a chemical reaction it is better to write the half-equation as:

$$Mg \rightarrow Mg^{2+} + 2e^-$$

Using half-equations to construct full redox equations

We can construct an overall equation for a redox reaction if we know the oxidation and reduction half-equations.

To construct a balanced equation, we must follow a series of steps. These are illustrated in the first worked example.

WORKED EXAMPLE 1

Iron reacts with aqueous copper(II) ions to form iron(III) ions and copper metal.

Step 1: Identify the oxidation and reduction half-equations.

- Equations for the separate oxidation and reduction reactions are shown lined up.
- $Fe \rightarrow Fe^{3+} + 3e^+$ oxidation (loss of electrons)
- $Cu^{2+} + 2e^- \rightarrow Cu$ reduction (gain of electrons)

Step 2: Balance the electrons.

The half-equations are scaled up or down (by multiplying the entire half-equation) so that the number of electrons is the same in each. Here, we need to multiply the oxidation half-equation by 2 and the reduction half-equation by 3:

- $2Fe \rightarrow 2Fe^{3+} + 6e^-$ oxidation
- $3Cu^{2+} + 6e^- \rightarrow 3Cu$ reduction

Step 3: Add the two half-equations together and cancel the electrons.

- $3Cu^{2+} + 2Fe + \cancel{6e^-} \rightarrow 2Fe^{3+} + 3Cu + \cancel{6e^-}$

So the balanced equation is:

- $3Cu^{2+} + 2Fe \rightarrow 2Fe^{3+} + 3Cu$

Using oxidation numbers to construct redox equations

Assigning oxidation numbers to each substance involved in a redox reaction allows you to track the movement of electrons.

- An increase in oxidation number shows a substance has been oxidised.
- A decrease in oxidation number shows a substance has been reduced.
- The overall increase in oxidation number will equal the overall decrease in oxidation number.

By knowing how each species involved in a reaction will behave (i.e. its tendency towards losing or gaining electrons, the ions it forms, and so on) it is possible to predict the whole reaction that will occur and write a full redox equation to represent reactions. For example:

- metals generally will be the species that lose their electrons
- non-metals usually gain electrons
- the group number (number of outer electrons) can be used to predict the ion charge for most elements – along with the other rules for assigning oxidation numbers to species involved in redox reactions.

We use this information to identify half-equations that can then be joined together to form a full redox equation.

An example of using oxidation numbers to construct redox equations is shown in the second worked example.

WORKED EXAMPLE 2

Hydrogen iodide, HI, is oxidised to iodine, I_2, by concentrated sulfuric acid, H_2SO_4, which is reduced to hydrogen sulfide, H_2S.

Step 1: Identify the reactants and products from the information given.

- The reactants and products are written down to give an incomplete draft equation:

     $HI + H_2SO_4 \rightarrow H_2S + I_2$

- Do not worry that there is no 'O' on the right-hand side of the equation – we will sort that out later.

Step 2: Identify the oxidation number changes.

- Write oxidation numbers underneath the atoms that change oxidation number.
- Balance any atom that changes oxidation number.
- Identify the total changes in oxidation numbers.

     $2HI + H_2SO_4 \rightarrow H_2S + I_2$

	+6 −2	S has decreased oxidation number by 8, from +6 to −2
−1(×2)	0(×2)	I has increased oxidation number by 2, from −1 to 0, twice

Step 3: Balance the oxidation number changes.

- The equation must be balanced so that:

     total increase in oxidation number
       = total decrease in oxidation number

- Here, we need to multiply 2HI and I_2 by 4 to balance the total oxidation number change for S, so that the number of electrons being lost when HI becomes I_2 is the same as the number being gained when H_2SO_4 becomes H_2S:

     $8HI + H_2SO_4 \rightarrow H_2S + 4I_2$

	+6 −2	*total gain of 8 electrons*
−1(×8)	0(×8)	*total loss of 8 electrons*
+8		

Step 4: Check to see if anything else needs to be balanced.

- Here there are 8H and 4O extra on the left-hand side. We can easily complete the equation by adding $4H_2O$ to the right-hand side.
- This gives the final equation:

     $8HI + H_2SO_4 \rightarrow H_2S + 4I_2 + 4H_2O$

LEARNING TIP

In Worked example 2, extra substances need to be added into the equation to make it balance. This will be the case sometimes. The most common species you may need to add are:

- H^+ usually when reactions are carried out in acid conditions
- OH^- usually when reactions are carried out in alkaline conditions
- H_2O usually when the equation needs extra O and H to be added.

Simply add the number of ions needed to balance the equation.

Questions

1   Construct redox equations from the following half-equations.

   (a) $Al \rightarrow Al^{3+} + 3e^-$ $Cu^{2+} + 2e^- \rightarrow Cu$

   (b) $2Br^- \rightarrow Br_2 + 2e^-$ $MnO_4^- + 8H^+ + 5e^- \rightarrow Mn^{2+} + 4H_2O$

2   Use oxidation states to construct redox equations for the following reactions.

   (a) Hydrogen bromide, HBr, is oxidised to bromine, Br_2, by concentrated sulfuric acid, H_2SO_4, which is reduced to sulfur dioxide, SO_2.

   (b) Hydrogen sulfide, H_2S, is oxidised to sulfur, S, by nitric acid, HNO_3, which is reduced to nitrogen monoxide, NO.

(8) Redox titrations

By the end of this topic, you should be able to demonstrate and apply your knowledge and understanding of:

* the techniques and procedures used when carrying out redox titrations including those involving Fe^{2+}/MnO_4^- and $I_2/S_2O_3^{2-}$

* structured and non-structured titration calculations, based on experimental results of redox titrations involving:
  (i) Fe^{2+}/MnO_4^- and $I_2/S_2O_3^{2-}$
  (ii) non-familiar redox systems

Redox titrations

Titrations are a way of determining amounts of substances, for example the concentration of an unknown acid. They can also be used to determine the amounts of species being oxidised or reduced – this is known as a redox titration.

A known concentration of either a reducing agent or oxidising agent is placed in a burette and titrated against an unknown concentration of the chemical that is being oxidised or reduced, respectively.

Whilst for acid–base reactions we use an indicator to identify the end point, often for redox titrations this is not necessary. Many redox titrations involve species that self-indicate – they change colour between different oxidation states.

We will look at some examples of redox titrations below.

Redox titrations between Fe^{2+} and MnO_4^-

Manganate(VII) MnO_4^- is a common oxidising agent, usually obtained from potassium permanganate(VII), $KMnO_4$. It has a deep purple colour but becomes colourless when it is reduced from +7 to +2 oxidation states:

$$MnO_4^-(aq) \rightarrow Mn^{2+}(aq)$$

(purple)   (colourless)

This usually occurs in the presence of H^+ ions, i.e. in acidic solutions (typically sulfuric acid is used, as hydrochloric acid reacts with MnO_4^-):

$$MnO_4^-(aq) + 8H^+(aq) + 5e^- \rightarrow Mn^{2+}(aq) + 4H_2O(l)$$

MnO_4^- is commonly used to oxidise solutions containing iron(II), but can be used as an oxidising agent in many other chemical reactions.

The reaction between iron(II) and acidified manganate(VII) is shown below:

$$MnO_4^-(aq) + 8H^+(aq) + 5Fe^{2+}(aq)$$
$$\rightarrow Mn^{2+}(aq) + 5Fe^{3+}(aq) + 4H_2O(l)$$

The end point is seen when excess MnO_4^- ions are present – indicated by a faint pink colour appearing. This occurs because all the Fe^{2+} ions have reacted and the MnO_4^- can no longer be reduced to the colourless Mn^{2+}.

We can use this reaction to determine the concentration of iron in an unknown solution, or indeed other ions that can be oxidised by MnO_4^-, or to calculate the percentage composition of a metal in a solid sample of a compound or alloy.

When calculating the mass of iron in a substance, the molar mass of an Fe^{2+} ion is taken to be the same as the molar mass of Fe – therefore the titration results can be used directly to convert the number of moles into the mass present.

Figure 1 Colours of solutions containing MnO_4^-(aq) (left) and Mn^{2+}(aq) ions.

Figure 2 Measuring the concentration of iron(II) ions in a solution of iron(II) sulfate. You can see the purple colour of the MnO_4^- before it reacts with the acidified Fe^{2+}(aq) solution in the conical flask.

WORKED EXAMPLE 1

$25.0\ cm^3$ of a solution of an iron(II) salt required $23.00\ cm^3$ of $0.0200\ mol\ dm^{-3}$ potassium permanganate for complete oxidation in acid solution.

(a) Calculate the amount, in moles, of permanganate, MnO_4^-, used in the titration.

$$\frac{n(MnO_4^-) = c \times V}{1000} = \frac{(0.0200 \times 23.00)}{1000} = 4.60 \times 10^{-4}\ mol$$

(b) Deduce the amount, in moles, of iron(II), Fe^{2+}, in the solution that was titrated.

From the equation: 1 $MnO_4^-(aq)$ reacts with $5Fe^{2+}(aq)$
So, the actual moles: $4.60 \times 10^{-4}\ mol\ MnO_4^-$ $2.30 \times 10^{-3}\ mol\ Fe^{2+}$
$MnO_4^-(aq) + 8H^+(aq) + 5Fe^{2+}(aq) \rightarrow Mn^{2+}(aq) + 5Fe^{3+}(aq) + 4H_2O(l)$

(c) Calculate the concentration, in mol dm^{-3}, of Fe^{2+} in the solution that was titrated.

$$[Fe^{2+}] = \frac{n\ (mol)}{V\ (dm^3)} = \frac{2.30 \times 10^{-3}}{(25.0/1000)} = 0.0920\ mol\ dm^{-3}$$

LEARNING TIP

Notice that to solve the problem given in Worked example 1, we need to use the equation $n = cV$ for parts (a) and (c) and the balanced equation for part (b).

Redox titrations between I_2 and $S_2O_3^{2-}$

Iodine is a useful substance to use in redox titrations as it has a dark blue–black colour in the presence of starch but when it is reduced to iodide ions this colour disappears.

$$I_2(aq) \qquad \rightarrow \qquad 2I^-(aq) + 2e^-$$
(blue–black colour with starch)   (blue–black colour disappears)

This reduction of iodine to iodide ions will occur in the presence of thiosulfate ions, $S_2O_3^{2-}$

$$2S_2O_3^{2-}(aq) + I_2(aq) \rightarrow S_4O_6^{2-}(aq) + 2I^-(aq)$$
thiosulfate   tetrathionate

We can use aqueous iodide ions and aqueous thiosulfate to determine the concentration of unknown reducible species.

Often this involves an initial reaction between the unknown oxidising agent and iodide ions (for example by mixing it with potassium iodide), which has iodine as a product.

For example:

$$Cl_2(aq) + 2I^-(aq) \rightarrow 2Cl^-(aq) + I_2(aq)$$

This liberated iodine then goes on to react with thiosulfate ions, from a solution with a known concentration being added from a burette.

$$2S_2O_3^{2-}(aq) + I_2(aq) \rightarrow S_4O_6^{2-}(aq) + 2I^-(aq)$$

In the presence of starch, a blue–black colour will remain present as long as there is any iodine. Once this has all reacted with the thiosulfate ions, this will disappear, marking the end point of the reaction.

Figure 3 Iodine has a blue–black appearance in the presence of starch. During a titration between iodine and thiosulfate ions, this colour will remain as long as iodine is present.

WORKED EXAMPLE 2

Chlorine was reacted with potassium iodide and the solution then titrated against a $0.27\ mol\ dm^{-3}$ solution of sodium thiosulfate in the presence of starch.

$18.0\ cm^3$ of sodium thiosulfate was required for the blue–black colour to disappear.

Calculate the number of moles of chlorine that reacted.

Step 1: Calculate the number of moles of thiosulfate used.

$$n = cV$$
$$n(S_2O_3^{2-}) = 0.27 \times \left(\frac{18.0}{1000}\right) = 4.86 \times 10^{-3}\ mol$$

Step 2:
Write balanced equations for the reaction between the species being investigated and iodine and the reaction between iodine and thiosulfate (you may need to deduce these).

$$Cl_2 + 2I^- \rightarrow 2Cl^- + I_2$$
$$2S_2O_3^{2-} + I_2 \rightarrow S_4O_6^{2-} + 2I^-$$

Step 3:
Work out the stoichiometry of each reaction and the number of moles present for known substances. Then calculate the number of moles for unknown substances.

- 1 mol Cl_2 produces 1 mol I_2: therefore, $2.43 \times 10^{-3}\ mol\ I_2$ must have been produced from $2.43 \times 10^{-3}\ mol\ Cl_2$
- 1 mol I_2 reacts with 2 mol $S_2O_3^{2-}$: therefore, $4.86 \times 10^{-3}\ mol\ S_2O_3^{2-}$ must have reacted with $2.43 \times 10^{-3}\ mol\ I_2$.

Calculations involving unfamiliar redox systems

Sometimes you will be required to interpret the results from redox systems you are not familiar with. You will be given any relevant equations and experimental results that you need. The procedures will be the same; some general steps you may need to use include:

- determining the numbers of moles used for any of the substances you have concentrations and volumes for, using $n = cV$ and number of moles $= \dfrac{\text{mass}}{\text{molar mass}}$, where necessary

- identifying the reaction stoichiometry using the balanced equations

- deciding on the amounts that have reacted for known substances, and deducing the amounts of unknowns

- using the equations discussed above to calculate unknown quantities or concentrations.

An example is given in the next worked example.

WORKED EXAMPLE 3

A 25.0 cm^3 portion of hydrogen peroxide was poured into a 250 cm^3 volumetric flask and made up to the mark with distilled water.

A 25.0 cm^3 portion of this solution was acidified and titrated against 0.0200 mol dm^{-3} potassium permanganate solution. 38.00 cm^3 of the MnO_4^- solution was required for complete oxidation.

Calculate the concentration of the original hydrogen peroxide solution.

The following equation represents the oxidation of hydrogen peroxide:

$$H_2O_2(aq) \rightarrow O_2(g) + 2H^+(aq) + 2e^-$$

Step 1: You will need to write the fully balanced equation for the reaction to find the stoichiometry between the two reactants.

The equation for MnO_4^- ions is:

$$MnO_4^-(aq) + 8H^+(aq) + 5e^- \rightarrow Mn^{2+}(aq) + 4H_2O(l)$$

The fully balanced equation is:

$$2MnO_4^-(aq) + 6H^+(aq) + 5H_2O_2(aq) \rightarrow 2Mn^{2+}(aq) + 8H_2O(l) + 5O_2(g)$$

Step 2: Calculate the amount of MnO_4^- that reacted.

$$n(MnO_4^-) = c \times \frac{V}{1000} = 0.0200 \times \frac{38.00}{1000} = 7.60 \times 10^{-4} \text{ mol}$$

Step 3: Deduce the amount of H_2O_2 that was used in the titration.

equation	$2MnO_4^-(aq)$ reacts with	$5 H_2O_2(aq)$
moles from equation	2 mol	5 mol
actual moles	7.60×10^{-4} mol	1.90×10^{-3} mol

Step 4: Calculate the concentration, in mol dm^{-3}, of the hydrogen peroxide solution.

$$c = \frac{n}{V} = \frac{1.90 \times 10^{-3}}{25.0/1000} = 0.0760 \text{ mol dm}^{-3}$$

Step 5: The first sentence of this worked example states that the solution was diluted by a factor of 10 at the start of the experiment. Therefore, the concentration of the original hydrogen peroxide solution is 0.760 mol dm^{-3}.

LEARNING TIP

You will need to be able to perform structured titration calculations *and* unstructured titration calculations. Structured calculations will have clear steps shown. For unstructured calculations, you will be given the experimental results only and any equations required – you will need to decide the steps to carry out the calculation for yourself.

Estimating the copper content of solutions and alloys

Redox titrations can be used to estimate the concentration of a solution containing $Cu^{2+}(aq)$ ions. The copper(II) solution is mixed with aqueous iodide ions, $I^-(aq)$, and a redox reaction occurs. This can be shown as two half-equations:

- iodide ions are oxidised to iodine, $2I^- \rightarrow I_2 + 2e^-$

- copper(II) ions are reduced to copper(I) ions, $Cu^{2+} + e^- \rightarrow Cu^+$

So, the overall ionic equation is:

$$2Cu^{2+}(aq) + 4I^-(aq) \rightarrow 2CuI(s) + I_2(aq)$$

The reaction produces a light brown/yellow solution and a white precipitate of copper(I) iodide, but the precipitate appears light brown due to the iodine in the solution.

The mixture of copper iodide and iodine can be titrated against sodium thiosulfate of known concentration. As the iodine reacts, the iodine colour gets paler during the titration. When the colour is a pale straw, a small amount of starch is added to help with the identification of the end point. A blue–black colour forms. The blue–black colour disappears sharply at the end point because all the iodine has reacted.

You can then use the amount of thiosulfate used to determine the concentration of iodine in the titration mixture, and hence the concentration of copper(II) ions in the original solution.

One application for this method is to calculate (within experimental error) the concentration of copper(II) ions in an unknown solution, or the percentage of copper in alloys such as brass and bronze.

To obtain the copper(II) in solution, the alloy is first reacted with nitric acid. Brass and bronze can both be reacted to produce a solution containing copper(II) ions, which can then be reacted with potassium iodide solution. The iodine that forms is then titrated against sodium thiosulfate.

LEARNING TIP

In Worked example 4, the rounding of the answer was not carried out until the last step in order to give as accurate an answer as possible. Early rounding can lead to incorrect (inaccurate) answers.

The value is given to 3 significant figures as this matches the lowest number of significant figures given in the information provided in the question. Remember, you are only justified to quote to the lowest number of significant figures given.

WORKED EXAMPLE 4

A sample of bronze was analysed to find the proportion of copper it contained.

0.500 g of the bronze was reacted with nitric acid to give a solution containing Cu^{2+} ions. The solution was neutralised and reacted with iodide ions to produce iodine.

The iodine was titrated with 0.200 mol dm^{-3} sodium thiosulfate and 22.40 cm^3 was needed to reach the end point.

Calculate the percentage of copper in the bronze.

The relevant equations are:

$$Cu(s) \rightarrow Cu^{2+}(aq) + 2e^-$$
$$2Cu^{2+}(aq) + 4I^-(aq) \rightarrow 2CuI(s) + I_2(aq)$$
$$2S_2O_3^{2-}(aq) + I_2(aq) \rightarrow 2I^-(aq) + S_4O_6^{2-}(aq)$$

Step 1: Calculate the amount of thiosulfate used.

$$n(S_2O_3^{2-}) = c \times \frac{V}{1000} = 0.200 \times \left(\frac{22.40}{1000}\right) = 4.48 \times 10^{-3} \, mol$$

Step 2: Calculate the reacting amounts.
- 2 mol Cu^{2+} produces 1 mol I_2, which reacts with 2 mol $S_2O_3^{2-}$
- 1 mol Cu reacted with nitric acid to produce 1 mol Cu^{2+}
- So 1 mol Cu is equivalent to 1 mol $S_2O_3^{2-}$

Step 3: Calculate the % copper.

$$n(Cu) = n(S_2O_3^{2-}) = 4.48 \times 10^{-3} \, mol$$

Mass of copper $= n \times M = 4.48 \times 10^{-3} \times 63.5 = 0.28448 \, g$

$$\% \text{ copper in bronze} = \frac{0.28448}{0.500} \times 100 = 56.9\%$$

Questions

1. 25.0 cm^3 of 0.150 mol dm^{-3} Fe^{2+} required 32.50 cm^3 of potassium permanganate for complete oxidation in acid solution.

   Calculate the concentration of the potassium permanganate solution used in the reaction.

2. A sample of metal of mass 2.50 g was thought to contain iron. The metal was reacted with sulfuric acid to form iron(II) sulfate and the resulting solution made up to 250 cm^3.

   25.0 cm^3 of this solution was titrated against 0.0180 mol dm^{-3} potassium permanganate solution and 24.80 cm^3 of the solution was required.

   Calculate the percentage by mass of iron in the metal sample.

(9) Standard electrode potentials

By the end of this topic, you should be able to demonstrate and apply your knowledge and understanding of:

* use of the term *standard electrode (redox) potential, E$^\ominus$*, including its measurement using a hydrogen electrode

* the techniques and procedures used for the measurement of cell potentials of:
   (i) metals or non-metals in contact with their ions in aqueous solution
   (ii) ions of the same element in different oxidation states in contact with a Pt electrode

Electricity from redox reactions

During a redox reaction, electrons will be transferred. The flow of electrons involves electrical energy, so redox reactions are of huge importance in the utilisation of electrical energy – in fact batteries and cells rely on redox reactions.

In topic 5.2.7 we considered half-equations. When we consider redox reactions that are part of a cell, each half-equation involved is known as a half cell. Different half cells can be connected together to make a cell. This is because, together, the half cells cause electrons to flow between each other, releasing electrical energy. This will be covered in more detail in the next topic, 5.2.10.

Half cells

A half cell comprises an element in two oxidation states. The simplest half cell has a metal placed in an aqueous solution of its ions.

For example, a copper half cell comprises a solution containing $Cu^{2+}(aq)$ ions (oxidation state +2) into which a strip or rod of copper metal (oxidation state 0) is placed.

An equilibrium exists at the surface of the copper between these oxidation states of copper:

$$Cu^{2+}(aq) + 2e^- \rightleftharpoons Cu(s)$$

* The forward reaction involves electron gain (reduction).

* The reverse reaction involves electron loss (oxidation).

By convention, the equilibrium is written with the electrons on the left-hand side. The electrode potential of the half cell indicates its tendency to lose or gain electrons in this equilibrium.

The hydrogen half cell

Most half cells comprise metals in equilibrium with their ions, as discussed above for copper. This is achieved, in practice, by placing the metal into a solution of the metal ions (i.e. a compound of the metal in aqueous solution). The piece of solid metal acts as an electrode when the half cell is connected to another half cell to form a cell.

Half cells can also be made from non-metals in equilibrium with non-metal ions.

For example:

* a half cell of bromine and its ions

$$Br_2(g) + 2e^- \rightleftharpoons 2Br^-(aq)$$

The most common example of a half cell comprising a non-metal and non-metal ions is the hydrogen half cell, comprising hydrogen gas, H_2, in contact with hydrogen ions, H^+

$$2H^+(aq) + 2e^- \rightleftharpoons H_2(g)$$

If a hydrogen half cell were to be connected to another half cell, to form a cell, there is no solid piece of metal that can act as the electrode. This is overcome by the use of a platinum electrode.

> ### KEY DEFINITION
>
> The **standard electrode potential of a half cell, E$^\ominus$,** is the e.m.f. of a half cell compared with a standard hydrogen half cell, measured at 298 K with solution concentrations of $1 \, mol \, dm^{-3}$ and a gas pressure of 100 kPa.

Figure 1 Copper half cell – copper metal is placed in an aqueous solution containing $Cu^{2+}(aq)$ ions. An equilibrium is set up at the surface of the solid Cu with aqueous Cu^{2+} ions:
$$Cu^{2+}(aq) + 2e^- \rightleftharpoons Cu(s)$$

The platinum is inert and does not react at all – its sole purpose is to be in contact with both the $H_2(g)$ and the H^+ ions and to allow the transfer of electrons into, and out of, the half cell via a connecting wire. The surface of the platinum electrode is coated with platinum black, a spongy coating in which electrons can be transferred between the non-metal and its ions.

A standard hydrogen half cell is shown in Figure 2 and comprises:

- hydrochloric acid, HCl(aq), of concentration $1\,mol\,dm^{-3}$, as the source of $H^+(aq)$

- hydrogen gas, $H_2(g)$, at 100 kPa pressure

- an inert platinum electrode to allow electrons to pass into or out of the half cell via a connecting wire.

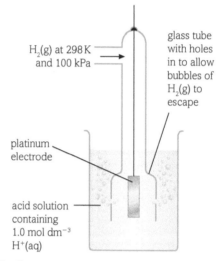

Figure 2 A standard hydrogen half cell.

Metal ion/metal ion half cells

This type of half cell contains ions of the same element in different oxidation states. For example, a half cell can contain $Fe^{3+}(aq)$ and $Fe^{2+}(aq)$ ions:

$$Fe^{3+}(aq) + e^- \rightleftharpoons Fe^{2+}(aq)$$

This type of half cell would again need to involve a platinum electrode as there is no solid piece of metal that could act as an electrode.

A standard $Fe^{3+}(aq)/Fe^{2+}(aq)$ half cell is made up of:

- a solution containing $Fe^{2+}(aq)$ and $Fe^{3+}(aq)$ ions with the same concentrations ('equimolar')

- an inert platinum electrode to allow electrons to pass into or out of the half cell via a connecting wire.

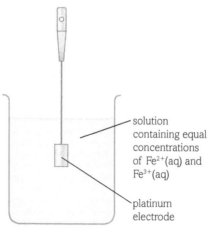

Figure 3 A standard Fe^{3+}/Fe^{2+} half cell.

Determining standard electrode potentials

Different half cells have different electrode potentials. When two half cells are connected together to form a cell, they will have an overall cell potential. This is a measure of how well electrons can be 'pushed around' the cell. The larger the overall cell potential, the more the electrons are 'pushed around'. The actual value of the overall cell potential will depend on the electrode potential of the half cells involved.

Measuring standard electrode potentials using a hydrogen half cell

We can determine the **standard electrode potential of a half cell** by connecting it to a hydrogen half cell. The tendency for different half cells to accept or release electrons is measured as an electromotive force (e.m.f.), or voltage, measured in volts, V. The hydrogen half cell has an e.m.f. value of 0 V, so it can be used as a reference to measure other half cells against.

Figure 4 shows how a hydrogen half cell can be used to measure the standard electrode potential of a Zn^{2+}/Zn half cell. The reading on the voltmeter gives the standard electrode potential of the zinc half cell.

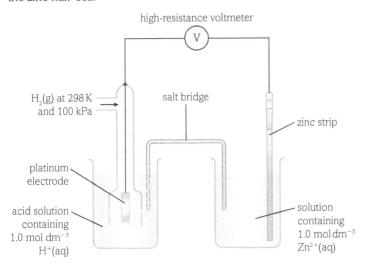

Figure 4 Measuring the standard electrode potential of a Zn^{2+}/Zn half cell.

Notice that in Figure 4 the two cells are joined by the following:

- a wire – this allows the electrons carrying charge to flow through it

- a salt bridge – this connects the two solutions and allows ions carrying charge to be transferred between the half cells. Salt bridges are usually made from a piece of filter paper soaked in an aqueous solution of an ionic substance, usually $KNO_3(aq)$ or $NH_4NO_3(aq)$.

The electrochemical series

We can list half cells in order of their standard electrode potentials – this is known as the electrochemical series. A small portion of this is shown below. The horizontal arrows show each reaction as a reduction reaction as electrons are gained. If the electrode potential is a negative value, the backward reaction will occur when compared with the standard hydrogen electrode. If the electrode potential is a positive value the forward reaction will occur when compared to the standard hydrogen electrode.

← preferred direction of reaction

$Fe^{2+}(aq) + 2e^- \rightleftharpoons Fe(s)$	$E^\ominus = -0.44\,V$
$2H^+(aq) + 2e^- \rightleftharpoons H_2(g)$	$E^\ominus = 0\,V$
$Cu^{2+}(aq) + 2e^- \rightleftharpoons Cu(s)$	$E^\ominus = +0.34\,V$
$Br_2(g) + 2e^- \rightleftharpoons 2Br^-(aq)$	$E^\ominus = +1.07\,V$
$Cl_2(g) + 2e^- \rightleftharpoons 2Cl^-(aq)$	$E^\ominus = +1.36\,V$

→ preferred direction of reaction

The more negative the $E^\ominus$ value, the greater the tendency towards the half cell undergoing oxidation; the more positive the $E^\ominus$ value, the greater the tendency towards the half cell undergoing reduction, when connected in a cell.

From this small series we could say that the half cell $Fe^{2+} + 2e^- \rightleftharpoons Fe$

- has the most negative $E^\ominus$ value

- therefore has the greatest tendency to release electrons and shift the equilibrium to the left.

$E^\ominus$ values for standard electrode potentials can be used to predict what will happen when half cells are connected together to form a cell. This will be covered in the next topic, 5.2.10.

Questions

1 (a) State what a standard hydrogen half cell consists of.

(b) When two different half cells are connected, state how the electric current flows:
   (i) between the two electrodes
   (ii) between the two solutions.

2 A standard electrode potential for the half cell
$Cu^{2+}(aq) + 2e^- \rightleftharpoons Cu(s)$ was measured as +0.34 V.
Explain how this value would have been determined.

(10) Standard cell potentials

By the end of this topic, you should be able to demonstrate and apply your knowledge and understanding of:

* calculation of a standard cell potential by combining two standard electrode potentials

* prediction of the feasibility of a reaction using standard cell potentials and the limitations of such predictions in terms of kinetics and concentration

* application of principles of electrode potentials to modern storage cells

* the explanation that a fuel cell uses the energy from the reaction of a fuel with oxygen to create a voltage and the changes that take place at each electrode.

Making cells from half cells

A simple electrochemical cell can be made by connecting together two half cells with different electrode potentials:

* one half cell releases electrons

* the other half cell gains electrons.

The difference in electrode potential is measured with a voltmeter.

A copper–zinc cell is an example of this and is shown in Figure 1.

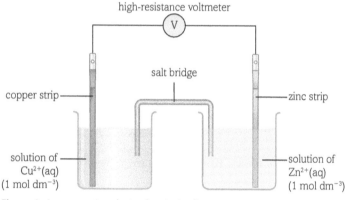

Figure 1 A copper–zinc electrochemical cell.

The two half cells are joined using a wire and a salt bridge to allow the charge to be carried between each cell via the electrons and the ions.

The redox equilibrium occurring, and the standard electrode potential for each half cell, are shown below:

$$Zn^{2+}(aq) + 2e^- \rightleftharpoons Zn(s) \qquad E^\ominus = -0.76\,V$$
$$Cu^{2+}(aq) + 2e^- \rightleftharpoons Cu(s) \qquad E^\ominus = +0.34\,V$$

The Zn^{2+}/Zn equilibrium releases electrons more readily than the Cu^{2+}/Cu equilibrium. We know this because the half cell has the more negative value, so has a greater tendency towards the equilibrium shifting left.

* The Zn^{2+}/Zn equilibrium releases electrons into the wire, making zinc the negative electrode.

* Electrons flow along the wire to the Cu electrode of the Cu^{2+}/Cu half cell.

* The Zn^{2+}/Zn equilibrium loses electrons and moves to the left:

$$Zn^{2+} + 2e^- \rightleftharpoons Zn$$

* The Cu^{2+}/Cu equilibrium gains electrons and moves to the right:

$$Cu^{2+}(aq) + 2e^- \rightleftharpoons Cu(s)$$

The reading on the voltmeter measures the potential difference of the cell – the difference between the electrode potentials of the half cells. The bigger the value, the further away from the equilibrium position the reaction moves. The reading on the voltmeter can be taken as the cell potential as long as any ions of the same element have a concentration of $1\,mol\,dm^{-3}$ or are equimolar.

The standard cell potential can be also calculated using the following equation:

$$E^\ominus_{cell} = E^\ominus(\text{positive terminal}) - E^\ominus(\text{negative terminal})$$

The cell reaction is the overall chemical reaction taking place in the cell – the sum of the reduction and oxidation half reactions taking place in each half cell.

Examples of using electrode potentials to predict a reaction and then writing the cell reaction are shown in Worked examples 1 and 2.

WORKED EXAMPLE 1

The standard cell potential of a silver–copper cell
A silver–copper cell is made by connecting together two half cells:

* an Ag^+/Ag half cell
* a Cu^{2+}/Cu half cell.

Step 1: Identify the two relevant redox equilibria and the sign of each electrode. The more negative of the two systems is the negative terminal of the cell.

$$Ag^+(aq) + e^- \rightleftharpoons Ag(s) \qquad E^\ominus = +0.80\,V \quad \text{positive terminal}$$
$$Cu^{2+}(aq) + 2e^- \rightleftharpoons Cu(s) \qquad E^\ominus = +0.34\,V \quad \text{negative terminal}$$

Step 2: Calculate the standard cell potential.
The standard cell potential, $E^\ominus_{cell}$, is simply the difference between the standard electrode potentials of the half cells.

$$E^\ominus_{cell} = E^\ominus(\text{positive terminal}) - E^\ominus(\text{negative terminal})$$
$$E^\ominus_{cell} = +0.80 - (+0.34)$$
$$= 0.46\,V$$

WORKED EXAMPLE 2

The cell reaction in a silver–copper cell

Step 1: Work out the direction of electron flow in the redox equilibria.

- From Worked example 1, we know the Cu^{2+}/Cu equilibrium has the more negative $E^\ominus$ value. This shows that the Cu^{2+}/Cu redox equilibrium has a greater tendency to lose electrons than the Ag^+/Ag half cell.
- The Cu^{2+}/Cu equilibrium releases electrons into the wire. Electrons flow along the wire to the Ag electrode of the Ag^+/Ag half cell.
- The Ag^+/Ag equilibrium gains the electrons.

Step 2: Combine the half-equations (remembering to show them in the direction the reaction moves for each) to give the cell reaction. The equations taking place at each electrode are:

$Cu(s) \rightarrow Cu^{2+}(aq) + 2e^-$ oxidation
$Ag^+(aq) + e^- \rightarrow Ag(s)$ reduction

The Ag^+/Ag half-equation must be multiplied by 2 to balance the electrons:

$Cu(s) \rightarrow Cu^{2+}(aq) + 2e^-$
$2Ag+(aq) + 2e^- \rightarrow 2Ag(s)$

The two half-equations are then added together and electrons are cancelled:

$Cu(s) + 2Ag^+(aq) + 2e^- \rightarrow 2Ag(s) + Cu^{2+}(aq) + 2e^-$

This gives the equation for the overall cell reaction:

$Cu(s) + 2Ag^+(aq) \rightarrow 2Ag(s) + Cu^{2+}(aq)$

Using standard cell potentials to predict the feasibility of reactions

By calculating the cell potential for a reaction, using the standard electrode potentials for each half cell, we can determine whether electrons are likely to flow and hence the feasibility of any reaction.

The electrochemical series can give a quick indication of how species will react. The same, brief version that you met in topic 5.2.9 is shown below.

$Fe^{2+}(aq) + 2e^- \rightleftharpoons Fe(s)$ $E^\ominus = -0.44\,V$
$2H^+(aq) + 2e^- \rightleftharpoons H_2(g)$ $E^\ominus = 0\,V$
$Cu^{2+}(aq) + 2e^- \rightleftharpoons Cu(s)$ $E^\ominus = +0.34\,V$
$Br_2(g) + 2e^- \rightleftharpoons 2Br^-(aq)$ $E^\ominus = +1.07\,V$
$Cl_2(g) + 2e^- \rightleftharpoons 2Cl^-(aq)$ $E^\ominus = +1.36\,V$

The half cells at the top of the list have more negative values and therefore have a higher tendency towards being oxidised and losing their electrons. When half cells are in combination, the one with the more negative value (this can also mean less positive, in the case of two cells with positive $E^\ominus$ values) will be the species to follow the oxidation reaction and the other half cell will follow the reduction reaction and gain the electrons. We can summarise this by saying that a species will react with the species below it, and to the left of the equation.

For example we could predict that:

- iron would react with $H^+(aq)$ to form $H_2(g)$ – i.e. iron would react with acids
- copper would not react with $H^+(aq)$ – i.e. copper would not react with acids
- copper would react with $Cl_2(g)$.

An example of how likely reactions are determined from half-cell equations is shown in Worked example 3.

WORKED EXAMPLE 3

Two redox equilibria are shown below:

$Cu^{2+}(aq) + 2e^- \rightleftharpoons Cu(s)$ $E^\ominus = +0.34\,V$
$NO_3^-(aq) + 2H^+(aq) + e^- \rightleftharpoons NO_2(g) + H_2O(l)$ $E^\ominus = +0.80\,V$

We can predict the redox reaction that may take place between the species present in these redox equilibria by treating the redox systems as if they are half cells.

Step 1: Work out the direction of electron flow in the redox equilibria.

- The Cu^{2+}/Cu equilibrium has the more negative (less positive) $E^\ominus$ value – this shows that the Cu^{2+}/Cu redox equilibrium supplies electrons and moves to the left.
- The NO_3^-/NO_2 equilibrium has the more positive (less negative) $E^\ominus$ value – this shows that the NO_3^-/NO_2 redox equilibrium accepts electrons and moves to the right.

$Cu^{2+}(aq) + 2e^- \rightleftharpoons Cu(s)$ $E^\ominus = +0.34\,V$ less positive

$NO_3^-(aq) + 2H^+(aq) + e^- \rightleftharpoons NO_2(g) + H_2O(l)$ $E^\ominus = +0.80\,V$ more positive

We can therefore predict that Cu will react with NO_3^- and H^+.

Step 2: Write down the half-equations for the oxidation and reduction. The half-equations taking place are:

$Cu(s) \rightarrow Cu^{2+}(aq) + 2e^-$ oxidation
$NO_3^-(aq) + 2H^+(aq) + e^- \rightarrow NO_2(g) + H_2O(l)$ reduction

Step 3: Check the balance of electrons. The second half-equation must be multiplied by 2 to balance the electrons:

$Cu(s) \rightarrow Cu^{2+}(aq) + 2e^-$
$2NO_3^-(aq) + 4H^+(aq) + 2e^- \rightarrow 2NO_2(g) + 2H_2O(l)$

Step 4: Construct the overall equation. The two half-equations are added together and the electrons cancelled:

$Cu(s) + 2NO_3^-(aq) + 4H^+(aq) + 2e^- \rightarrow Cu^{2+}(aq) + 2NO_2(g) + 2H_2O(l) + 2e^-$

This gives the equation for the overall reaction:

$Cu(s) + 2NO_3^-(aq) + 4H^+(aq) \rightarrow Cu^{2+}(aq) + 2NO_2(g) + 2H_2O(l)$

Positive cell potentials indicate reactions will be spontaneous.

Limitations of predictions of feasibility from cell potentials

Electrode potentials and concentration

Non-standard conditions alter the value of an electrode potential. The half-equation for the copper half cell is:

$$Cu^{2+}(aq) + 2e^- \rightleftharpoons Cu(s)$$

From Le Chatelier's principle, on increasing the concentration of $Cu^{2+}(aq)$:

- the equilibrium opposes the change by moving to the right
- electrons are removed from the equilibrium
- the electrode potential becomes less negative, or more positive.

A change in electrode potential resulting from concentration changes means that predictions made on the basis of the *standard* value may not be valid.

Will a reaction actually take place?

Remember that these are equilibrium processes in aqueous conditions.

- Predictions can be made about the equilibrium position but not about the reaction rate, which may be extremely slow because of a high activation energy.
- The actual conditions used for a reaction may be different from the standard conditions used to measure $E^\ominus$ values. This will affect the value of the electrode potential (see above).
- Standard electrode potentials apply to aqueous equilibria – many reactions take place under very different conditions.

As a general working rule:

- The larger the difference between $E^\ominus$ values, the more likely it is that a reaction will take place.
- If the difference between $E^\ominus$ values is less than 0.4 V, then a reaction is unlikely to take place.

Storage and fuel cells

Electrochemical cells are used as our modern-day cells and batteries. Cells can be divided into three main types.

- Non-rechargeable cells – provide electrical energy until the chemicals have reacted to such an extent that the voltage falls. The cell is then 'flat' and is discarded.
- Rechargeable cells – the chemicals in the cell react, providing electrical energy. The cell reaction can be reversed during recharging. The chemicals in the cell are regenerated and the cell can be used again. Common examples include:
  - nickel and cadmium (Ni-Cad) batteries, used in rechargeable batteries
  - lithium-ion and lithium-polymer batteries, used in laptops.
- Fuel cells – the cell reaction uses external supplies of a fuel and an oxidant, which are consumed and need to be continuously supplied. The cell will continue to provide electrical energy so long as there is a supply of fuel and oxidant.

The risks and benefits of using electrochemical cells must be weighed up when cells are designed. For example, lithium cells, which are often found in the home, can provide high levels of battery life but have been found to have several drawbacks. These include toxicity on being ingested and rapid discharge of current, which can cause fires and even explosions. This has led to some controls on such batteries, including restrictions on the transport of lithium-based batteries and limited sales of the batteries to individual consumers in some countries.

Fuel cells

Modern fuel cells are based on hydrogen, or hydrogen-rich fuels such as methanol, CH_3OH.

A fuel cell uses energy from the reaction of a fuel with oxygen to create a voltage.

- The reactants flow in and products flow out while the electrolyte remains in the cell.

- Fuel cells can operate virtually continuously so long as the fuel and oxygen continue to flow into the cell. Fuel cells do not have to be recharged.

Figure 2 shows a simple hydrogen–oxygen fuel cell with an alkaline electrolyte.

The redox equilibria are shown below. The more negative of the two systems is the negative terminal of the cell.

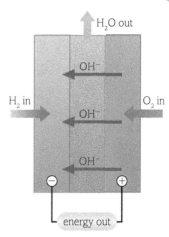

Figure 2 Hydrogen–oxygen fuel cell.

$$2H_2O(l) + 2e^- \rightleftharpoons H_2(g) + 2OH^-(aq) \qquad E^\ominus = -0.83\ V \quad \text{negative terminal}$$

$$\tfrac{1}{2}O_2(g) + H_2O(l) + 2e^- \rightleftharpoons 2OH^-(aq) \qquad E^\ominus = +0.40\ V \quad \textbf{positive terminal}$$

The more negative hydrogen system provides the electrons. This equilibrium is reversed when writing the half-equations at each electrode. The half-equations are added together and electrons cancelled to give the equation for the overall cell reaction:

$$H_2(g) + 2OH^-(aq) \rightarrow 2H_2O(l) + 2e^-$$

$$\tfrac{1}{2}O_2(g) + H_2O(l) + 2e^- \rightarrow 2OH^-(aq)$$

Overall:

$$H_2(g) + \tfrac{1}{2}O_2(g) \rightarrow H_2O(l)$$

$$E^\ominus_{cell} = E^\ominus(\textbf{positive terminal}) - E^\ominus(\text{negative terminal})$$

$$= 0.40 - (-0.83)$$

$$= 1.23\ V$$

Other fuel cells have been developed – these are based on hydrogen-rich fuels such as methanol, natural gas and petrol.

Questions

1 You are provided with the following standard electrode potentials:

$$\tfrac{1}{2}Cl_2(g) + e^- \rightleftharpoons Cl^-(aq) \qquad E^\ominus = +1.36\ V$$
$$Cu^{2+}(aq) + 2e^- \rightleftharpoons Cu(s) \qquad E^\ominus = +0.34\ V$$
$$Fe^{2+}(aq) + 2e^- \rightleftharpoons Fe(s) \qquad E^\ominus = -0.44\ V$$
$$Cr^{3+}(aq) + 3e^- \rightleftharpoons Cr(s) \qquad E^\ominus = -0.74\ V$$

(a) Calculate the standard cell potential for the following cells, prepared under standard conditions:
 (i) Cl_2/Cl^- and Cu^{2+}/Cu
 (ii) Fe^{2+}/Fe and Cr^{3+}/Cr
 (iii) Cr^{3+}/Cr and Cl_2/Cl^-

(b) For each cell in (a), determine the overall cell reaction.

2 Use the standard electrode potentials below to answer the questions that follow.

$$Fe^{3+}(aq) + e^- \rightleftharpoons Fe^{2+}(aq) \qquad E^\ominus = +0.77\ V$$
$$Br_2(aq) + 2e^- \rightleftharpoons 2Br^-(aq) \qquad E^\ominus = +1.07\ V$$
$$Ni^{2+}(aq) + 2e^- \rightleftharpoons Ni(s) \qquad E^\ominus = -0.25\ V$$
$$O_2(g) + 4H^+(aq) + 4e^- \rightleftharpoons 2H_2O(l) \qquad E^\ominus = +1.23\ V$$

(a) Predict the reactions that could take place under standard conditions.

(b) Write an overall equation for each reaction in (a).

THINKING BIGGER

HYDROGEN FUEL CELLS

Hydrogen is the simplest element. An atom of hydrogen consists of only one proton and one electron (although heavier isotopes do occur naturally). It is also the most abundant element in the universe but hydrogen doesn't occur naturally as a gas on the Earth – it's always combined with other elements in compounds such as water. How, then, might hydrogen power the car of the future? Read the following article and answer the questions below.

HYDROGEN GETS ONBOARD

Figure 1 A hydrogen powered car.

[In 2005], Honda proudly launched its new hydrogen powered fuel cell car – the FCX – which boasts a maximum speed of 93 mph and 'zero emissions'. The car, the second generation FCX, runs on compressed hydrogen and is Honda's 26th fuel cell powered vehicle, built at a cost of around $1.5 million (£0.8 million). It is about the same size as a Honda Civic but weighs about 1000 lbs more and will only travel about 190 miles on a full tank.

Many of the other automobile manufacturers are also in the fuel cell building game; General Motors, US, is one of the most aggressive and has pledged to have a fuel cell car ready to go into mass production by 2010. Given the number of technological hurdles still to be overcome, this could be quite a challenge.

One of the biggest hurdles remains finding a material capable of storing enough on-board hydrogen for a vehicle to cover over 300 miles on a full tank without adding significant weight or volume relative to a petrol car. Specifically, 5–13 kg H_2 is required to propel a

highly fuel efficient automobile for 300 miles and must fit into a space comparable to a conventional petrol tank.

Because of volumetric constraints the hydrogen must be stored as a solid or liquid, either by physisorption to high surface area materials or by chemical bonding (covalent or ionic) to lightweight elements. Hydrogen stored on high surface area materials is weakly bound and generally requires low temperatures to stabilise it. Meanwhile, hydrogen covalently bound to light metals requires high temperatures to release it.

If one conservatively assumes that a vehicle's physical system – everything needed to deliver hydrogen to the fuel cell: tanks, heaters, chillers, pumps, tubes and hoses – weighs about one third to half the weight of the storage material then the target material should approach a minimum of 12 weight per cent of hydrogen. This implies that the elements that make up materials for on-board hydrogen storage can only come from the first two rows of the periodic table. However, small amounts of higher weight elements could be tolerated at low concentrations for catalytic purposes.

Unfortunately neither highly compressed hydrogen gas nor liquefied hydrogen is likely to be capable of sufficient volumetric density to meet the mileage target. The limitation is due to simple physics, specifically the very weak intermolecular interactions between hydrogen molecules. H_2's small quadrupole and low polarisability do not provide significant binding through electrostatic, induction or dispersion interactions. Consequently, the van der Waals forces between two hydrogen molecules are negligible compared with the strength of a hydrogen bond or a chemical bond. Because of the weak intermolecular interactions, the volumes of both compressed H_2 (40 g l^{-1} at 700 bar) and liquid H_2 (70 g l^{-1} at 20K) fall far short of the automotive industry's volumetric target.

Source

- http://www.rsc.org/chemistryworld/Issues/2006/March/HydrogenOnBoard.asp

Where else will I encounter these themes?

Book 1   5.1   5.2   YOU ARE HERE

Let us start by considering the nature of the writing in the article.

1. This article is written from a fairly neutral standpoint. How might the article be rewritten, *without* changing the facts, to convince you to buy a hydrogen-powered car?

Now we will look at the chemistry in, or connected to, this article. Do not worry if you are not ready to give answers to these questions yet. You may like to return to the questions once you have covered other topics later in the book.

2. a. Write an equation for the complete combustion, under standard conditions, of hydrogen.

b. Given that the standard enthalpy of combustion of hydrogen is $-242\,kJ\,mol^{-1}$, calculate the amount of energy released when 13 kg of hydrogen is completely combusted under standard conditions.

c. At a hydrogen density of $40\,g\,dm^{-3}$ at a pressure of 700 bar, what volume would the fuel tank have to be to contain 13 kg of hydrogen?

3. a. Explain why hydrogen molecules have such weak intermolecular forces.

b. Use the kinetic model to explain why these forces become more significant at very low temperatures.

4. Use the data in Table 1 to answer the questions.

Substance	Enthalpy of formation /kJ mol^{-1}	Standard entropy content at 298 K /K^{-1} mol^{-1}
H_2		131
O_2		205
H_2O	-242	189

Table 1

a. Calculate the standard entropy change for the complete combustion of hydrogen to give water.

b. Use the data in the table and your answer to part (a) to calculate the standard change in free energy ($\Delta G^{\ominus}$) for the combustion of hydrogen.

c. Using the data in the table, calculate the temperature, in K, at which the combustion of hydrogen to produce water is no longer feasible. Explain your answer.

> **DID YOU KNOW?**
> Although predicted as early as 1935 it was only in 2011 that the creation of the metallic state of hydrogen was finally reported. At a pressure of 260 billion Pa, scientists at the Max Planck Institute in Germany reported that this form of hydrogen had finally been made. Metallic hydrogen helps explain, amongst other things, the strong magnetic field generated by the 'gas giant' Jupiter, which, unlike Earth, has no iron core.

> **LEARNING TIP**
> Remember that $g\,l^{-1} = g\,dm^{-3}$ and 1 bar = 100 kPa.

Activity

The concept of free energy $\Delta G^{\ominus}$ is a central one to chemistry in that it links a number of areas of the specification:

1. $\Delta G = -RT\ln K$
2. $\Delta G = -nFE$
3. $\Delta G = \Delta H - T\Delta S$

Choose one of the three equations above and give a short (5–8 slides) presentation on the application of the equation in an aspect of the chemistry you have covered to date. You should also highlight the limitations of using the equation in predicting chemical reactions.

Practice questions

1. Which of the following represents the lattice enthalpy of magnesium chloride? [1]

    A. $Mg(s) + Cl_2(g) \rightarrow MgCl_2(s)$

    B. $Mg(g) + 2Cl(g) \rightarrow MgCl_2(s)$

    C. $Mg^{2+}(g) + 2Cl^-(g) \rightarrow MgCl_2(g)$

    D. $Mg^{2+}(g) + 2Cl^-(g) \rightarrow MgCl_2(s)$

2. Consider the following reaction:

    $CaCO_3(s) + 2HCl(aq) \rightarrow CaCl_2(aq) + H_2O(l) + CO_2(g)$

    Which of the following statements is true? [1]

    A. The reaction will have a negative enthalpy change and a positive entropy change.

    B. The reaction will have both a positive enthalpy and entropy change.

    C. The reaction will have both a negative enthalpy and entropy change.

    D. There will be a very small entropy change because the number of moles of reactants and products are the same.

3. Look at the following redox half-equations:

    $Fe^{3+}(aq) + e^- \rightarrow Fe^{2+}(aq)$

    $Cr_2O_7{}^{2-}(aq) + 14H^+(aq) + 6e^- \rightarrow 2Cr^{3+}(aq) + 7H_2O(l)$

    What volume of 0.01 M acidified potassium dichromate is required to exactly reduce 23.00 cm^3 of 0.05 M iron(II) sulfate solution? [1]

    A. 9.58 cm^3

    B. 19.17 cm^3

    C. 18.39 cm^3

    D. 19.00 cm^3

4. If a reaction is exothermic and results in a positive change in entropy, which of the following statements is true? [1]

    A. The reaction is likely to only be feasible at high temperatures.

    B. The reaction will only be feasible if there is sufficient activation energy.

    C. The reaction will always be feasible.

    D. The reaction will never be feasible.

    [Total: 4]

5. Redox reactions are used to generate electrical energy from electrochemical cells.

    (a) **Table 1** shows three redox systems, and their standard redox potentials.

redox system	$E^{\ominus}$/V
$Cu^+(aq) + e^- \rightleftharpoons Cu(s)$	+0.52
$Cr^{3+}(aq) + 3e^- \rightleftharpoons Cr(s)$	−0.74
$Sn^{4+}(aq) + 2e^- \rightleftharpoons Sn^{2+}(aq)$	+0.15

    Table 1

    (i) Draw a labelled diagram to show how the standard electrode potential of a Sn^{4+}/Sn^{2+} redox system could be measured. [3]

    (ii) Using the information in **Table 1**, write equations for the reactions that are feasible. Suggest **two** reasons why these reactions may **not** actually take place. [5]

    (b) Modern fuel cells are being developed as an alternative to the direct use of fossil fuels. The 'fuel' can be hydrogen but many other substances are being considered.

    In a methanol fuel cell, the overall reaction is the combustion of methanol.

    As with all fuel cells, the fuel (methanol) is supplied at one electrode and the oxidant (oxygen) at the other electrode.

    Oxygen reacts at the negative electrode of a methanol fuel cell:

    $$O_2 + 4H^+ + 4e^- \rightarrow 2H_2O$$

    (i) Write an equation for the complete combustion of methanol. [1]

    (ii) Deduce the half-equation for the reaction that takes place at the positive electrode in a methanol fuel cell. [1]

    (iii) State **two** advantages of vehicles using fuel cells compared with the combustion of conventional fossil fuels. [2]

    (iv) Suggest **one** advantage of using methanol, rather than hydrogen, in a fuel cell for vehicles. Justify your answer. [1]

    [Total: 13]

    [Q4, F325 Jan 2011]

6. Lattice enthalpy can be used as a measure of ionic bond strength. Lattice enthalpies are determined indirectly using an enthalpy cycle called a Born–Haber cycle.

Table 2 shows the enthalpy changes that are needed to determine the lattice enthalpy of magnesium chloride, $MgCl_2$.

letter	enthalpy change	energy/ kJ mol^{-1}
A	1st electron affinity of chlorine	−349
B	1st ionisation energy of magnesium	+736
C	atomisation of chlorine	+150
D	formation of magnesium chloride	−642
E	atomisation of magnesium	+76
F	2nd ionisation energy of magnesium	+1450
G	lattice enthalpy of magnesium chloride	

Table 2

(a) On the cycle below, write the correct letter in each empty box. [3]

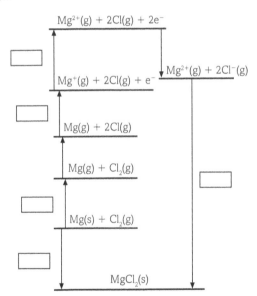

(b) Use the Born–Haber cycle to calculate the lattice enthalpy of magnesium chloride. [2]

(c) Magnesium chloride has stronger ionic bonds than sodium chloride. Explain why. [3]

[Total: 8]

[Q1, F325 June 2010]

7. (a) Define the terms in the equation
$$\Delta G = \Delta H - T\Delta S$$
and give the units for each term. [3]

(b) Using a specific example in each case:

(i) explain how some endothermic reactions can become energetically feasible under certain conditions. [4]

(ii) explain why some reactions are never energetically feasible. [4]

(c) Under standard conditions an equimolar mixture of nitrogen and oxygen is kinetically stable but thermodynamically unstable. Explain this statement. [3]

[Total: 14]

Physical chemistry and transition elements

TRANSITION METALS

Introduction

Colourful precious stones owe their beauty to the transition metals. Emeralds are mainly made of the mineral beryl, which is also known as beryllium aluminium silicate. These precious stones also contain chromium impurities that give them their characteristic light-green transparent appearance. The chromium impurities in the mineral corundum make rubies red. The amazing array of colours generated by transition metals are also used to make pottery glazes and fabric dyes.

Many industrial processes rely on transition metals to make the process economical. These elements have the ability to make multiple stable oxidation states and it is this property that makes them useful as catalysts. Without iron as a catalyst in the Haber process, for example, there would not be enough synthetic fertiliser for crops, seriously diminishing the production of food for the rising population.

The vast majority of colour in nature and chemistry is thanks to the reactions of transition metals.

All the maths you need

To unlock the puzzles of this chapter you need the following maths:

* Determining the oxidation state of species
* Completing titration calculations
* Writing balanced ionic and half-equations
* Representing complex ions in 3D and 2D models

What have I studied before?

- How to determine oxidation states
- Acid–base titration calculations
- How to use VSEPR theory to determine shapes of molecules
- How to determine the electronic structure of elements and ions using sub-shell notation
- Catalysts

What will I study later?

- How entropy and Gibbs' free energy can be used to explain why ligand substitution reactions occur

What will I study in this chapter?

- How a transition metal is different from a d-block element
- Why transition metal compounds are coloured
- How a complex ion is formed and its shape
- Examples of uses of transition metal complexes
- How to safely carry out a redox titration and perform calculations related to the titre
- How transition metals can act as catalysts and industrial examples
- Ligand substitution reactions

Transition metals

By the end of this topic, you should be able to demonstrate and apply your knowledge and understanding of:

* the electron configuration of atoms and ions of the d-block elements of Period 4 (Sc–Zn), given the atomic number and charge

* the elements Ti–Cu as transition elements, i.e. d-block elements that have an ion with an incomplete d-sub-shell

---

KEY DEFINITION

A **transition element** (transition metal) is a d-block element that has an incomplete d-sub-shell as a stable ion.

What are transition metals?

The d-block

The d-block elements are found between Group 2 and Group 3 in the periodic table. For these elements, the highest energy sub-shell is a d-sub-shell. To understand the electron configuration we will consider the Period 4 d-block elements.

1	2											3	4	5	6	7	0
1.0 **H** hydrogen 1																	4.0 **He** helium 2
6.9 **Li** lithium 3	9.0 **Be** beryllium 4											10.8 **B** boron 5	12.0 **C** carbon 6	14.0 **N** nitrogen 7	16.0 **O** oxygen 8	19.0 **F** fluorine 9	20.2 **Ne** neon 10
23.0 **Na** sodium 11	24.3 **Mg** magnesium 12											27.0 **Al** aluminium 13	28.1 **Si** silicon 14	31.0 **P** phosphorus 15	32.1 **S** sulfur 16	35.5 **Cl** chlorine 17	39.9 **Ar** argon 18
39.1 **K** potassium 19	40.1 **Ca** calcium 20	45.0 **Sc** scandium 21	47.9 **Ti** titanium 22	50.9 **V** vanadium 23	52.0 **Cr** chromium 24	54.9 **Mn** manganese 25	55.8 **Fe** iron 26	58.9 **Co** cobalt 27	58.7 **Ni** nickel 28	63.5 **Cu** copper 29	65.4 **Zn** zinc 30	69.7 **Ga** gallium 31	72.6 **Ge** germanium 32	74.9 **As** arsenic 33	79.0 **Se** selenium 34	79.9 **Br** bromine 35	83.8 **Kr** krypton 36

Figure 1 Period 4 d-block elements.

Transition elements

A **transition element** is a d-block element that has an incomplete d-sub-shell as a stable ion.

Scandium and zinc, the first and last members of the Period 4 d-block elements, are *not* classed as transition metals because they do not have any ions with partially filled d-orbitals.

* Scandium forms only the Sc^{3+} ion, in which the d-orbitals are empty.

* Zinc forms only the Zn^{2+} ion, in which the d-orbitals are full.

Writing electron configurations

The electron configuration of an element is the arrangement of the electrons in an atom of the element. Electrons occupy orbitals in order of energy level. Figure 2 shows the order of energy levels for the sub-shells making up the first four shells.

* The sub-shell energy levels in the third and fourth energy levels overlap.

* The 4s-sub-shell fills before the 3d-sub-shell.

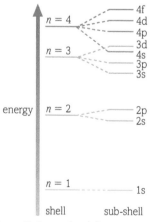

Figure 2 Energy-level diagram showing the overlap of the 3d and 4s sub-shells.

LEARNING TIP

Remember that the 4s-sub-shell has a lower energy than the 3d-sub-shell. This means that the 4s-orbital fills before the orbitals in the 3d-sub-shell.

Across the Period 4 d-block elements, from scandium to zinc, the 3d-orbitals are being filled. The pattern is regular except for chromium and copper, which do not follow the principle of completely filling the lowest energy levels first.

- Chromium – the five 3d-orbitals and the 4s-orbital all contain one electron, with no orbital being completely filled.

- Copper – the five 3d-orbitals are full, but there is only one electron in the 4s-orbital.

In these two elements, the electron repulsions between the outer electrons are minimised, resulting in an increased stability of the chromium and copper atoms.

- In chromium atoms, the 4s-orbital and the five 3d-orbitals are all half-filled.

- In copper atoms, the five 3d-orbitals are all filled and the 4s-orbital is half-filled.

The electron configurations of the elements from scandium to zinc are shown in Figure 3.

> **LEARNING TIP**
>
> When writing the electron configurations, the official way is to show all the $n = 3$ sub-shells before the $n = 4$ energy level. In this book, the energy levels have been shown in order of sub-shell filling with the 4s before the 3d. Either response is perfectly acceptable in exams, so cobalt can be shown as:
> $1s^2 2s^2 2p^6 3s^2 3p^6 4s^2 3d^7$
> or $1s^2 2s^2 2p^6 3s^2 3p^6 3d^7 4s^2$

Element	Z	Electron configuration	Noble gas configuration	Electron in box diagram
Scandium	21	$1s^2\ 2s^2\ 2p^6\ 3s^2\ 3p^6\ 4s^2\ 3d^1$	$[Ar]\ 4s^2\ 3d^1$	↑↓ ↑ ▢ ▢ ▢ ▢
Titanium	22	$1s^2\ 2s^2\ 2p^6\ 3s^2\ 3p^6\ 4s^2\ 3d^2$	$[Ar]\ 4s^2\ 3d^2$	↑↓ ↑ ↑ ▢ ▢ ▢
Vanadium	23	$1s^2\ 2s^2\ 2p^6\ 3s^2\ 3p^6\ 4s^2\ 3d^3$	$[Ar]\ 4s^2\ 3d^3$	↑↓ ↑ ↑ ↑ ▢ ▢
Chromium	24	$1s^2\ 2s^2\ 2p^6\ 3s^2\ 3p^6\ 4s^1\ 3d^5$	$[Ar]\ 4s^1\ 3d^5$	↑ ↑ ↑ ↑ ↑ ↑
Manganese	25	$1s^2\ 2s^2\ 2p^6\ 3s^2\ 3p^6\ 4s^2\ 3d^5$	$[Ar]\ 4s^2\ 3d^5$	↑↓ ↑ ↑ ↑ ↑ ↑
Iron	26	$1s^2\ 2s^2\ 2p^6\ 3s^2\ 3p^6\ 4s^2\ 3d^6$	$[Ar]\ 4s^2\ 3d^6$	↑↓ ↑↓ ↑ ↑ ↑ ↑
Cobalt	27	$1s^2\ 2s^2\ 2p^6\ 3s^2\ 3p^6\ 4s^2\ 3d^7$	$[Ar]\ 4s^2\ 3d^7$	↑↓ ↑↓ ↑↓ ↑ ↑ ↑
Nickel	28	$1s^2\ 2s^2\ 2p^6\ 3s^2\ 3p^6\ 4s^2\ 3d^8$	$[Ar]\ 4s^2\ 3d^8$	↑↓ ↑↓ ↑↓ ↑↓ ↑ ↑
Copper	29	$1s^2\ 2s^2\ 2p^6\ 3s^2\ 3p^6\ 4s^1\ 3d^{10}$	$[Ar]\ 4s^1\ 3d^{10}$	↑ ↑↓ ↑↓ ↑↓ ↑↓ ↑↓
Zinc	30	$1s^2\ 2s^2\ 2p^6\ 3s^2\ 3p^6\ 4s^2\ 3d^{10}$	$[Ar]\ 4s^2\ 3d^{10}$	↑↓ ↑↓ ↑↓ ↑↓ ↑↓ ↑↓

Figure 3 Electron configurations for the d-block elements of Period 4.

The electron configurations of d-block ions

In their reactions, transition element atoms lose electrons to form positive ions. Transition metals lose their 4s electrons before the 3d electrons. This seems surprising because the 4s orbitals are filled first. The 3d and 4s energy levels are very close together and, once electrons occupy the orbitals, the 4s electrons have a higher energy and are lost first.

WORKED EXAMPLE

An iron atom has 26 electrons and has an electronic configuration of $1s^2 2s^2 2p^6 3s^2 3p^6 4s^2 3d^6$.

- To form the 3+ ion, an iron atom has to lose three electrons.
- Fe^{3+} will have an electron configuration of $1s^2 2s^2 2p^6 3s^2 3p^6 3d^5$.
- Fe has lost two electrons from the 4s-sub-shell and one electron from the 3d-sub-shell.
  A copper atom has 29 electrons and its electron configuration is $1s^2 2s^2 2p^6 3s^2 3p^6 4s^1 3d^{10}$.
- To form the 2+ ion a Cu atom has to lose two electrons – one from the 4s-sub-shell and one from the 3d-sub-shell.
- Cu^{2+} has an electron configuration of $1s^2 2s^2 2p^6 3s^2 3p^6 3d^9$.

LEARNING TIP

When writing the electron configurations of the ions, it is easier to start with the electron configuration of the elements and then remove the required number of electrons.

When forming ions in transition element chemistry, the 4s electrons are always removed before any electrons are taken from the 3d-sub-shells.

Questions

1 State the difference between a d-block element and a transition element.

2 Write the electronic configurations of the following atoms and ions.
  (a) Cr
  (b) Mn^{2+}
  (c) Sc^{3+}
  (d) Ti

By the end of this topic, you should be able to demonstrate and apply your knowledge and understanding of:

* illustration, using at least two transition elements, of:
  (i) the existence of more than one oxidation state for each element in its compounds
  (ii) the formation of coloured ions
  (iii) the catalytic behaviour of the elements and their compounds and their importance in the manufacture of chemicals by industry

Physical properties

The transition elements are all metals. They are lustrous in appearance and have high densities, melting points and boiling points. When solid, transition metals exist as giant metallic lattices containing delocalised electrons, which move freely to conduct electricity.

Figure 1 The Statue of Liberty is made of pure copper sheeting, hung on a framework of steel. Steel is an alloy made mainly of the transition metal iron. The exception is the flame of the torch, which is coated with gold leaf. The green appearance of the copper comes from copper(II) carbonate, formed by reaction of copper with atmospheric gases.

Figure 2 Transition metals are used in a wide variety of applications.

Chemical properties

Transition elements can have different oxidation states. The compounds of transition metals form coloured solutions when dissolved in water. Transition elements often catalyse chemical reactions. These properties are a result of the electron configurations of the transition elements – in particular, the partially filled d-orbitals.

Variable oxidation states

The transition elements, from titanium to copper, all form ions with more than one stable oxidation state. These metals also all form compounds with metal ions in the +2 oxidation state. Often, this results from losing the two electrons in the 4s-orbital. The 4s electrons are lost first because they are in the highest occupied energy level. However, because the 3d and 4s energy levels are so close in energy, the 3d electrons can also be lost when an atom forms a stable ion.

Figure 3 Coloured compounds of iron in different oxidation states.

You need to be able to work out the oxidation numbers of the transition element ions found in compounds. To do this you need to use the rules for determining oxidation number. The main oxidation numbers of the first row (Period 4) of the d-block elements are shown in Figure 4; the colours of the most common ions, formed with six water ligands, are also shown.

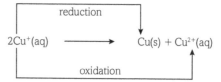

Sc	Ti	V	Cr	Mn	Fe	Co	Ni	Cu	Zn
	+2	+2	+2	+2	+2	+2	+2	+2	+2
+3	+3	+3	+3	+3	+3	+3	+3	+3	
	+4	+4	+4	+4	+4	+4	+4		
	+5	+5	+5	+5	+5	+5			
			+6	+6	+6				
				+7					

Figure 4 Oxidation numbers and colours of the common d-block metal ions.

The highest oxidation state of a transition element is often found in a strong oxidising agent. Manganese forms a compound called potassium manganate, $KMnO_4$, a purple solid used as an oxidising agent in redox titrations. Chromium is found in potassium dichromate, $K_2Cr_2O_7$, an orange crystalline solid that acts as an oxidising agent in the preparation of aldehydes and ketones from alcohols. In these compounds, manganese and chromium have their maximum oxidation states of +7 and +6, respectively.

The systematic names of these compounds show the oxidation states:

- potassium manganate(VII), $KMnO_4$: oxidation number of Mn = +7

- potassium dichromate(VI), $K_2Cr_2O_7$: oxidation number of Cr = +6.

Figure 5 Crystals of potassium dichromate(VI) and potassium manganate(VII).

Disproportionation

Disproportionation is where one species is both oxidised and reduced in the same chemical reaction. For example, when copper(I) oxide is reacted with hot dilute H_2SO_4, a brown precipitate of copper and a blue solution of copper(II) sulfate are formed.

The chemical equation for this is:

$$Cu_2O + H_2SO_4 \rightarrow Cu + CuSO_4 + H_2O$$

In this reaction, the Cu^+ is both oxidised to Cu^{2+} in $CuSO_4$ and reduced to Cu. This disproportionation reaction can be shown in the following ionic equation:

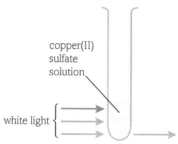

Figure 6 Red copper(I) oxide will undergo disproportionation to produce copper and copper(II) sulfate.

Coloured compounds

When white light passes through a solution containing transition metal ions, some of the wavelengths of visible light are absorbed. The colour that we observe is a mixture of the wavelengths of light that have *not* been absorbed. For example, a solution of copper(II) sulfate appears pale blue because the solution absorbs the red/orange region of the electromagnetic spectrum and reflects or transmits the blue (Figure 7).

copper(II) sulfate solution

white light

Figure 7 How light is absorbed and transmitted by transition metal compound solutions.

Many coloured inorganic compounds contain transition metal ions. In fact, it is quite difficult to find coloured inorganic compounds that do not contain transition metal ions. Colour in inorganic chemistry is linked to the partially filled d-orbitals of transition metal ions.

Scandium(III) has the electron configuration $1s^2 2s^2 2p^6 3s^2 3p^6$ and is colourless in aqueous solution. Scandium(III) is formed from scandium ($1s^2 2s^2 2p^6 3s^2 3p^6 4s^2 3d^1$) by the loss of two 4s electrons and one 3d electron. There is no partially filled d-orbital and there is no colour.

Transition metals as catalysts

A catalyst is a substance that changes the rate of a chemical reaction by providing an alternative reaction pathway, with a lower activation energy. You learned about catalysts and catalysis in Book 1, topic 3.2.8.

Transition metals and their compounds are very effective catalysts. There are two main ways in which this catalysis takes place:

- Transition metals provide a surface on which a reaction can take place. Reactants are adsorbed onto the surface of the metal and held in place while a reaction occurs. After the reaction, the products are desorbed and the metal remains unchanged.

- Transition metal ions have the ability to change their oxidation states by gaining or losing electrons. They then bind to reactants, forming intermediates as part of a chemical pathway, often with a lower activation energy, which speeds up the reaction.

Figure 8 View through the element of a catalytic converter from a car exhaust. The inner surface is coated with an alloy containing platinum, rhodium and palladium.

Transition metals as industrial catalysts

Transition metal catalysts are used in industry to improve the profits of an industrial process. They can reduce the time it takes to make a product and reduce the amount of energy needed to make the reaction occur.

However, as transition metal compounds can be toxic, they must be handled with care. When substances bind to the active site and stop the transition metal catalysts from working, the catalysts can no longer be used. These catalysts must be disposed of in such a way as not to cause harm or pollution.

Haber process

The Haber process is used to make ammonia, NH_3, from the reaction of nitrogen and hydrogen.

$$N_2(g) + 3H_2(g) \rightleftharpoons 2NH_3(g)$$

- The catalyst is iron metal – it is used to increase the rate of reaction and to lower the temperature at which the reaction takes place.

Much of the ammonia produced in the Haber process is used in manufacturing agricultural fertilisers.

Contact process

The Contact process is used to convert sulfur dioxide, SO_2, into sulfur trioxide, SO_3, which is used to manufacture sulfuric acid, H_2SO_4.

$$2SO_2(g) + O_2(g) \rightleftharpoons 2SO_3(g)$$

- The catalyst used is vanadium(V) oxide, V_2O_5, in which vanadium has the +5 oxidation state.

Sulfuric acid is an important inorganic chemical with many uses including the production of fertilisers, detergents, adhesives and explosives, and also as the electrolyte in car batteries.

Figure 9 Sample of vanadium(V) oxide – used as the catalyst in the Contact process.

Hydrogenation of alkenes

Hydrogen can be added across the C=C double bonds in unsaturated compounds to saturate them – the process is called hydrogenation. You learned about the hydrogenation of alkenes in Book 1, topic 4.1.9.

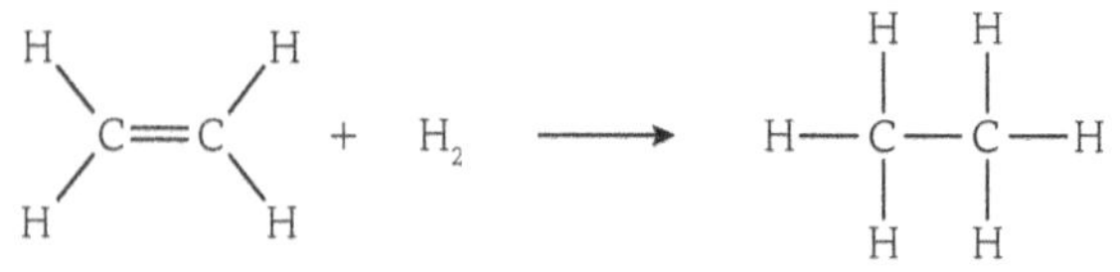

Figure 10 Hydrogenation of the C=C bond in alkenes.

- The catalyst is nickel metal – it is used to lower the temperature and pressure needed to carry out the reaction.

This process is used in the hydrogenation of unsaturated vegetable oils to make spreadable margarines.

Decomposition of hydrogen peroxide

Hydrogen peroxide decomposes slowly at room temperature and pressure into water and oxygen. A catalyst is added to increase the reaction rate.

$$2H_2O_2(aq) \rightarrow 2H_2O(l) + O_2(g)$$

- A suitable catalyst is manganese(IV) oxide, MnO_2, in which manganese has the +4 oxidation state. Manganese(IV) oxide is commonly called manganese dioxide.

The catalytic decomposition of hydrogen peroxide is often used in the laboratory as a simple and convenient preparation of oxygen gas.

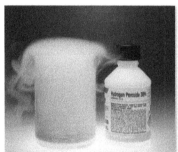

Figure 11 Hydrogen peroxide is rapidly decomposed into water and oxygen in the presence of manganese dioxide catalyst.

Hydrogen production

Zinc granules will react with dilute sulfuric acid to form hydrogen and zinc sulfate. When copper(II) sulfate is added, the rate of reaction increases. The transition metal ion acts as a catalyst.

$$Zn(s) + H_2SO_4(aq) \xrightarrow{CuSO_4} ZnSO_4(aq) + H_2(g)$$

This reaction is often used in a laboratory as a simple and convenient preparation of hydrogen gas.

Questions

1. State three properties of the transition elements.

2. Zinc is a d-block element and forms an ion Zn^{2+}. Suggest and explain the colour of solutions containing this ion.

3. What is the oxidation state of vanadium in V_2O_5 and in VO^{2+}?

4. Explain what you understand by the term *catalyst*.

5. Give an example of a chemical reaction that uses a transition metal compound as a catalyst.

By the end of this topic, you should be able to demonstrate and apply your knowledge and understanding of:

* explanation and use of the term *ligand* in terms of coordinate (dative covalent) bonding to a metal ion or metal, including bidentate ligands

* use of the terms *complex ion* and *coordination number* and examples of complexes with:
  (i) six-fold coordination with an octahedral shape
  (ii) four-fold coordination with either a planar or tetrahedral shape

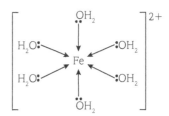

Figure 1 Complex ion $[Fe(H_2O)_6]^{2+}$ – each of the six H_2O ligands forms one coordinate bond (dative covalent bond) to the central metal ion. The shapes of six-coordinate complex ions are discussed later in this topic.

KEY DEFINITIONS

A **complex ion** is a transition metal ion bonded to one or more ligands by coordinate bonds (dative covalent bonds).
A **ligand** is a molecule or ion that can donate a pair of electrons to the transition metal ion to form a coordinate bond.
The **coordination number** is the total number of coordinate bonds formed between a central metal ion and its ligands.

Complex ions

In solution, transition metal compounds form **complex ions**. A complex ion consists of a central metal ion surrounded by ligands.

* A **ligand** is a molecule or ion that donates a pair of electrons to the central metal ion to form a coordinate, or dative covalent, bond.

* A coordinate bond is a bond in which one of the atoms provides both electrons for the covalent bond. A coordinate bond is the name commonly used in transition metal chemistry for a dative covalent bond.

* The transition metal ion accepts the pair of electrons from the ligand in forming the coordinate bond.

DID YOU KNOW?

Ligands are Lewis bases as they are electron donors. See topic 5.1.9.

An example of a complex ion is $[Cu(H_2O)_6]^{2+}$. The central metal ion is Cu^{2+} and the ligands are molecules of water, H_2O. Each water molecule donates a lone pair of electrons from its oxygen atom to the Cu^{2+} ion to form a coordinate bond.

In a complex ion, the **coordination number** indicates the number of coordinate bonds to the central metal ion. In the complex ion $[Cu(H_2O)_6]^{2+}$, the coordination number is 6 because there are six coordinate bonds in total from the ligands to the Cu^{2+} ion.

In the formula of the complex ion:

* square brackets group together the species making up the complex ion

* the overall charge is shown outside the brackets.

The overall charge is the sum of the individual charges of the transition metal ion and those of the ligands present in the complex. The six coordinate bonds in the complex ion $[Fe(H_2O)_6]^{2+}$ are shown in Figure 1.

Common ligands

All ligands have one or more lone pairs of electrons in their outer energy level. Some ligands are neutral and carry no charge – other ligands are negatively charged. In a monodentate ligand, the ligand donates just one pair of electrons to the central metal ion to form one coordinate bond.

Name of ligand	Formula	Charge
Water	$:OH_2$	None – neutral ligand
Ammonia	$:NH_3$	None – neutral ligand
Thiocyanate	$:SCN^-$	−1
Cyanide	$:CN^-$	−1
Chloride	$:Cl^-$	−1
Hydroxide	$:OH^-$	−1

Table 1 Some common ligands, their formulae and charges.

Bidentate ligands have two pairs of electrons from different atoms to donate to the central metal. This means two coordinate bonds can be formed. The most common bidentate ligand is ethane-1,2-diamine, $NH_2CH_2CH_2NH_2$, often shortened to 'en'. Each nitrogen atom in a molecule of $NH_2CH_2CH_2NH_2$ donates its lone pair to the metal ion.

WORKED EXAMPLE 1

State the formula and charge of the complex ion made from one titanium(III) ion and six water molecules.
$$[Ti(H_2O)_6]^{3+}$$
The formula shows that six H_2O ligands are bonded to one titanium(III) ion. As water is a neutral ligand, the overall charge on the ion is the same as the transition metal ion, which is +3.

WORKED EXAMPLE 2

What is the oxidation number of the transition metal in the complex ion $[Co(H_2O)_5Cl]^+$?
This complex ion has five neutral water ligands with no charge, and one chloride ligand with a charge of −1. Overall the complex ion has a 1+ positive charge.

Co	$(H_2O)_5$	Cl^-	Overall charge
?	5 × 0	−1	+1

In $[Co(H_2O)_5Cl]^+$, cobalt must have the oxidation state +2 to give the overall charge of +1.

A multidentate ligand

A hexadentate ligand has six lone pairs of electrons available to form coordinate bonds. One example of a hexadentate ligand is ethylenediaminetetraacetic acid, shortened to EDTA. This ligand exists in complexes as the ion $EDTA^{4-}$ (see Figure 2).

EDTA is used to bind metal ions and is known as a chelating agent. This means that EDTA decreases the concentration of metal ions in solutions by binding them into a complex.

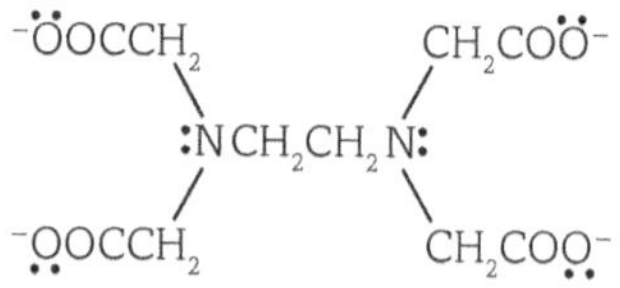

Figure 2 $EDTA^{4-}$ – note the six pairs of electrons, each of which can form a coordinate bond.

DID YOU KNOW?

Uses of EDTA

The ability of EDTA to bind to metals is exploited in industry, and EDTA is used in many different consumer products:

- in detergents – it binds to calcium and magnesium ions to reduce hardness in water
- in some foods – as a stabiliser to remove metal ions that might catalyse the oxidation of the product
- in medical applications – it is added to blood samples to prevent clotting and used to treat patients suffering from lead and mercury poisoning.

Shapes of complex ions with six-fold coordination

Complex ions exist in a number of different shapes, with the most common being the octahedral shape. Octahedral complexes have six coordinate bonds attached to the central ion. The outer face of the shape is an eight-sided octahedron. In an octahedral complex, four of the ligands are in the same plane, one ligand is above the plane and the remaining ligand below the plane. All bond angles are 90° or 180°.

When cobalt(II) salts are dissolved in water, a pale pink solution is formed containing the octahedral complex ion $[Co(H_2O)_6]^{2+}$ – Figure 3 shows its shape.

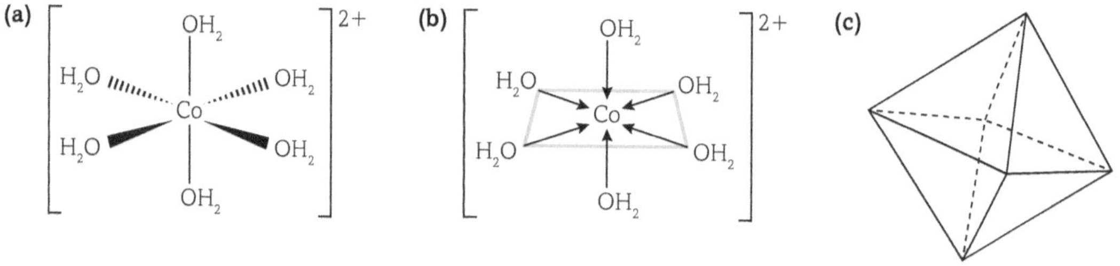

Figure 3 (a) $[Co(H_2O)_6]^{2+}$ is an octahedral complex ion. (b) $[Co(H_2O)_6]^{2+}$ showing shape. (c) The 3D octahedral shape.

In Figure 3:

- the solid wedge bonds come out of the plane of the paper, towards you
- the hatched wedge bonds go into the plane of the paper, away from you
- the ligands attached by the wedge bonds are at the corners of a square, with each bond separated by 90°

LEARNING TIP

Transition metal complex ions are drawn using wedge-shaped bonds, which indicate bonds coming out of the plane of the paper. Hatched lines show bonds going into the paper. Solid lines are in the plane of the paper.

- the solid lines represent bonds in the plane of the paper
- the 'solid bonds' are separated from each wedge bond by 90°.

Octahedral complexes can also occur with multidentate ligands. Figure 4 shows $[Ni(en)_3]^{2+}$, an example of an octahedral transition metal ion containing a bidentate ligand.

- Each 'en' ligand forms two coordinate bonds to the central metal ion.
- There are three 'en' ligands, giving a total of $3 \times 2 = 6$ coordinate bonds.
- The coordination number is 6.

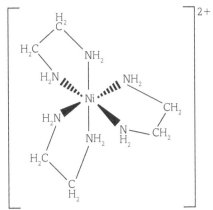

Figure 4 Complex ion $[Ni(en)_3]^{2+}$ (or $[Ni(NH_2CH_2CH_2NH_2)_3]^{2+}$).

Shapes of complex ions with four-fold coordination

Chloride ligands are so large that only four can fit around a transition metal ion. So, chloride complexes form a tetrahedral shape (see Figure 5).

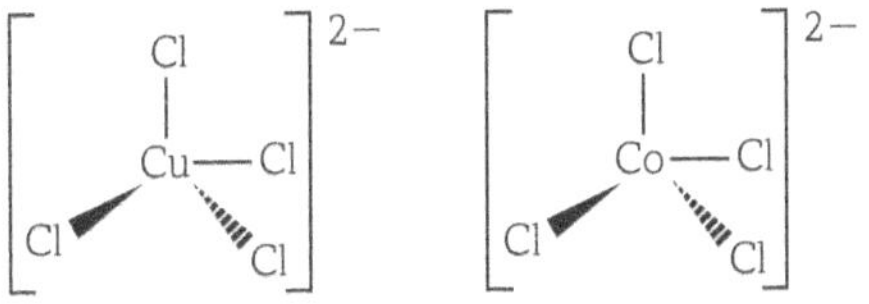

Figure 5 $[CuCl_4]^{2-}$ and $[CoCl_4]^{2-}$ complex ions are a tetrahedral shape.

A rare configuration is a square planar complex. This is when the four monodentate ligands arrange themselves in the same plane as the metal atom, as shown in Figure 6.

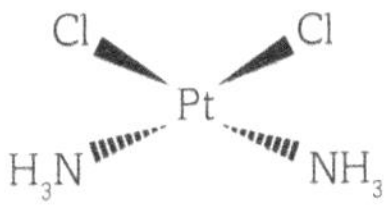

Figure 6 *Cis*-platin is a square planar complex that is used as a chemotherapy drug. See the next topic, 5.2.4.

Questions

1. State the formula and charge of each of these complex ions:
   (a) one cobalt(II) ion and four chloride ions
   (b) one iron(III) ion, five water molecules and a chloride ion.

2. In the complex ion $[Cu(NH_3)_4(H_2O)_2]^{2+}$, ammonia molecules and water molecules are both ligands. Explain why water can behave as a ligand.

3. Using any stated complex ion, explain what is meant by the term *coordination number*.

4 Stereoisomerism in complex ions

By the end of this topic, you should be able to demonstrate and apply your knowledge and understanding of:

* types of stereoisomerism shown by metal complexes, including those associated with bidentate and multidentate ligands:
  (i) *cis–trans* isomerism e.g. $Pt(NH_3)_2Cl_2$
  (ii) optical isomerism e.g. $[Ni(NH_2CH_2CH_2NH_2)_3]^{2+}$

* use of *cis*-platin as an anti-cancer drug and its action by binding to DNA preventing cell division

---

KEY DEFINITION

Stereoisomers are species with the same structural formula but with a different arrangement of the atoms in space.

What is a stereoisomer?

LEARNING TIP

You learned about stereoisomerism in Book 1, topic 4.1.4 when looking at organic chemistry, but isomerism is also found among the complexes of the transition elements.

There are two types of stereoisomerism in transition element chemistry:

* *cis–trans* isomerism
* optical isomerism.

The definition for stereoisomerism must be broadened to accommodate both organic molecules and transition metal complexes:

* **stereoisomers** are molecules or complexes with the same structural formula but with a different spatial arrangement of these atoms.

In transition element chemistry, it is possible to have stereoisomers from complexes containing all types of ligands – monodentate and multidentate. You met multidentate ligands in the previous topic, 5.3.3.

Cis–trans isomerism

Some octahedral complex ions contain two different ligands, for example:

* four monodentate ligands and two different monodentate ligands
* two bidentate ligands and two different bidentate ligands.

These complex ions can exist as *cis* and *trans* isomers.

The earliest examples of stereoisomerism involved Co^{3+} complex ions. In 1889, Sophus Mads Jørgensen, a Danish chemist, observed that some complex ions could form salts that had different colours.

However it was Alfred Werner, a Swiss chemist, who succeeded in explaining the existence of two isomers of $[Co(NH_3)_4Cl_2]^+$. Werner proposed that there were two isomers of this ion, one of which was purple and the other green. He suggested that the cobalt(III) ion is surrounded by four NH_3 ligands and two Cl^- ligands at the corners of an octahedron.

He called the purple isomer '*cis*':

* the two Cl^- ligands are at adjacent corners of the octahedron
* the two Cl^- ligands are at 90° to one another.

He called the green isomer '*trans*':

* the two Cl^- ligands are at opposite corners of the octahedron
* the two Cl^- ligands are at 180° to one another.

The *cis* and *trans* isomers of $[Co(NH_3)_4Cl_2]^+$ are shown in Figure 2.

Figure 1 Alfred Werner, one of the pioneers of transition element chemistry.

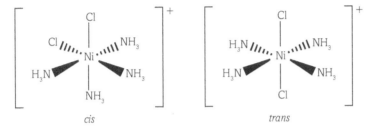

Figure 2 *Cis–trans* isomerism in $[Co(NH_3)_4Cl_2]^+$.

Cis–trans isomerism in four-coordinate complexes

Cis–trans isomerism is also possible in some four-coordinate complexes with a square planar shape. In this shape, the ligands are arranged at the corners of a square. This structure is rather like an octahedral complex, but without the ligands above and below the plane. For *cis–trans* isomerism, the complex must contain two different ligands, with two of one ligand and two of another. Figure 3 shows the structures of the *cis* and *trans* isomers of $[NiCl_2(NH_3)_2]$.

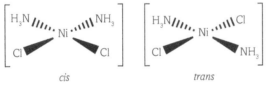

Figure 3 *Cis–trans* isomerism in $[NiCl_2(NH_3)_2]$.

Transition metal complexes in medicine

Cis-platin is one of the most effective drugs against many forms of cancer. *Cis*-platin is the *cis*-isomer of a platinum complex, $[PtCl_2(NH_3)_2]$.

Cells divide to make more cells of the same type. Before cell division, a copy is made of the DNA present in the cell. After cell division, two new cells are formed, each containing the same DNA. Cancer is spread throughout an organism by cell division of a cancerous cell that can reproduce itself.

The current explanation of *cis*-platin's effect is that it acts by binding to the DNA of fast-growing cancer cells. This prevents cell division, so the cells are prevented from reproducing by these changes to the DNA structure. Activation of a cell's own repair mechanism eventually leads to the death of the cancer-containing cell. However, there are also several other theories for the action of *cis*-platin – showing that the science involved is currently uncertain.

Figure 4 Structure of *cis*-platin.

The use of *cis*-platin is the basis for modern chemotherapy treatment of cancer. Unfortunately, treatment is associated with some unpleasant side effects. These include severe sickness, hair loss and fatigue.

Cis-platin is a very effective drug, but scientists have now developed a new generation of compounds with lower doses and fewer side effects. The most common of these is carboplatin, which was developed in the late 1980s to treat advanced ovarian cancer. It slows cancer growth by reacting with a cell's DNA.

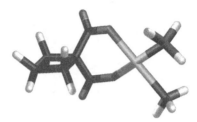

Figure 5 Structure of carboplatin. Can you spot where the platinum is? Do some research to find out what the other atoms are in the structure.

Bidentate and multidentate ligands

Cis–trans isomerism from bidentate ligands

Octahedral complexes containing bidentate ligands can also show *cis–trans* isomerism. The complex ion $[Cr(C_2O_4)_2(H_2O)_2]^-$ has *cis* and *trans* isomers.

Figure 6 shows the ethanedioate ion, $C_2O_4^{2-}$, which is the bidentate ligand present in this complex ion.

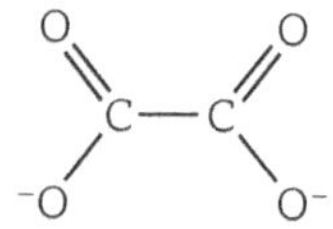

Figure 6 Structure of the ethanedioate ligand, $C_2O_4^{2-}$. Each O^- in the ion can form a coordinate bond with a transition metal ion.

The *cis* and *trans* isomers of $[Cr(C_2O_4)_2(H_2O)_2]^-$ are shown in Figure 7.

LEARNING TIP

The ethanedioate ion can come from ethanedioic acid. When the acid is put into water it will partially dissociate to form the bidentate ligand.

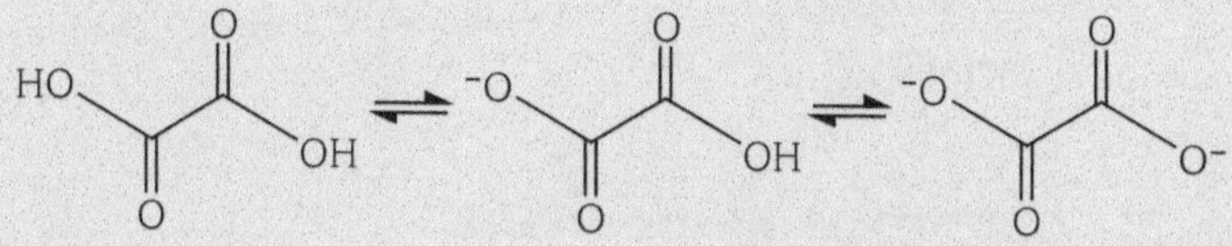

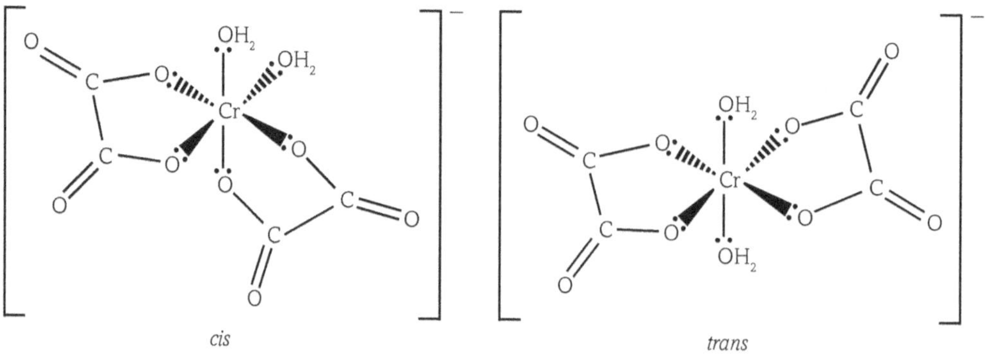

cis   *trans*

Figure 7 *Cis* and *trans* isomers of $[Cr(C_2O_4)_2(H_2O)_2]^-$.

Optical isomers

Optical isomerism in transition element chemistry is associated with octahedral complexes that contain multidentate ligands.

The requirements for optical isomerism are:

- a complex with three molecules or ions of a bidentate ligand – e.g. $[Ni(en)_3]^{2+}$

- a complex with two molecules or ions of a bidentate ligand, and two molecules or ions of a monodentate ligand – e.g. $[Co(en)_2Cl_2]$

- a complex with one hexadentate ligand – e.g. $[Cu(EDTA)]^{2-}$.

Optical isomers are non-superimposable mirror images of each other. Optical isomers rotate plane-polarised light differently – one of the isomers rotates the light clockwise and the other rotates the light anticlockwise.

DID YOU KNOW?

A mixture containing equal amounts of two optical isomers has no effect on plane-polarised light because the rotations cancel out. This is known as a racemic mixture.

Optical isomerism can be seen in the structures of the two isomers of $[Ni(NH_2CH_2CH_2NH_2)_3]^{2+}$, shown in Figure 8.

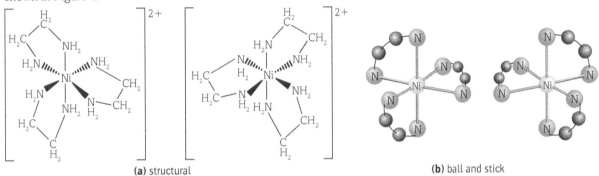

(a) structural   (b) ball and stick

Figure 8 Optical isomers of $[Ni(NH_2CH_2CH_2NH_2)_3]^{2+}$.

Questions

1 What kind of stereoisomers can be formed by the complex ion $[Ru(H_2O)_4Cl_2]^+$? Draw diagrams to illustrate your answer.

2 Define the term *stereoisomers*.

3 The compound $[Co(H_2O)_2(NH_2CH_2CH_2NH_2)_2]^{2+}$ can form a pair of optical isomers. Draw diagrams to show the two isomers.

(5) Ligand substitution in complexes

By the end of this topic, you should be able to demonstrate and apply your knowledge and understanding of:

* ligand substitution reactions and the accompanying colour changes in the formation of:
   (i) $[Cu(NH_3)_4(H_2O)_2]^{2+}$ and $[CuCl_4]^{2-}$ from $[Cu(H_2O)_6]^{2+}$
   (ii) $[Cr(NH_3)_6]^{3+}$ from $[Cr(H_2O)_6]^{3+}$

KEY DEFINITION

Ligand substitution is a reaction in which one ligand in a complex ion is replaced by another ligand.

Ligand substitution reactions

A **ligand substitution** reaction is one in which a ligand in a complex ion is replaced by another ligand. In many ligand substitution reactions, water molecules in an aqueous solution of a complex ion are replaced by another ligand.

The reaction of aqueous copper(II) ions and ammonia

An aqueous solution of copper(II) ions contains $[Cu(H_2O)_6]^{2+}$ complex ions, which have a characteristic pale blue colour. When an excess of aqueous ammonia solution is added, the pale blue solution changes colour and a deep blue solution forms. This can be represented by the equilibrium equation:

$$[Cu(H_2O)_6]^{2+}(aq) + 4NH_3(aq) \rightleftharpoons [Cu(NH_3)_4(H_2O)_2]^{2+}(aq) + 4H_2O(l)$$

pale blue solution   deep blue solution

In the reaction, four of the water ligands are replaced by four ammonia ligands. Ligand substitution has taken place because one type of ligand has been replaced by another ligand.

The product, $[Cu(NH_3)_4(H_2O)_2]^{2+}$, has six ligands is an octahedral complex ion. The shape of the product is shown in Figure 1. The copper–oxygen bonds are longer than the copper–nitrogen bonds, so the product is strictly described as having a distorted octahedral shape.

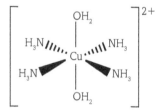

Figure 1 Octahedral shape of $[Cu(NH_3)_4(H_2O)_2]^{2+}$.

If you carry out this experiment in the laboratory you will observe two changes. On addition of a small amount of ammonia, a pale blue precipitate of copper(II) hydroxide, $Cu(OH)_2$, forms. This is because the ammonia acts as an alkali. When ammonia is put into water it reacts to form ammonium hydroxide, which can partially ionise, releasing the hydroxide ion.

On addition of an excess of aqueous ammonia, the pale blue precipitate dissolves and a deep blue solution is formed containing $[Cu(NH_3)_4(H_2O)_2]^{2+}$ ions. These colour changes are shown in Figure 2.

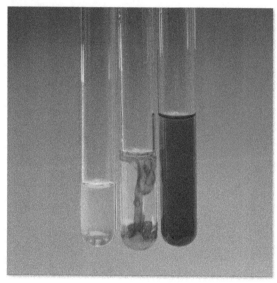

Figure 2 Stepwise reaction of aqueous copper(II) ions with aqueous ammonia.

The reaction of copper(II) ions and hydrochloric acid

When concentrated hydrochloric acid is added to an aqueous solution containing copper(II) ions, the pale blue solution initially forms a green solution before finally turning yellow. This reaction exists in equilibrium and can be reversed by adding water to the yellow solution to return it to the original blue colour.

This can be represented by the equilibrium equation:

$$[Cu(H_2O)_6]^{2+}(aq) + 4Cl^-(aq) \rightleftharpoons [CuCl_4]^{2-}(aq) + 6H_2O(l)$$

pale blue solution   yellow solution

Note that the $[Cu(H_2O)_6]^{2+}$ complex ion has six ligands, but the $[CuCl_4]^{2-}$ complex ion has only four ligands. This occurs because the chloride ligands are larger than the water ligands and have stronger repulsions, so fewer chloride ligands can fit around the central metal ion. $[CuCl_4]^{2-}$ has a tetrahedral shape, like a CH_4 molecule.

As the reaction is an equilibrium system, you can apply Le Chatelier's principle (Book 1, topic 3.2.10) to predict how the colour of the solution changes when the reaction conditions are modified.

The colours in this equilibrium are shown in Figure 3.

Figure 3 Equilibrium reaction of aqueous copper(II) ions with chloride ions.

The reaction of chromium(III) with ammonia

An aqueous solution of chromium(III) contains $[Cr(H_2O)_6]^{3+}$ complex ions and has a grey–green or dark green colour. When a small amount of ammonia is added, hydrogen ions are pulled off the water ligands. This produces a green precipitate of $[Cr(H_2O)_3(OH)_3]$.

The balanced chemical equation is:

$$[Cr(H_2O)_6]^{3+} + 3NH_3 \rightarrow [Cr(H_2O)_3(OH)_3] + 3NH_4^+$$

When excess ammonia is added, some of the green precipitate re-dissolves to form a dark green solution. This is more noticeable when concentrated ammonia is used. This occurs as the ligands are replaced by ammonia.

The balanced chemical equation starting from the aqueous solution is:

$$[Cr(H_2O)_6]^{3+} + 6NH_3 \rightleftharpoons [Cr(NH_3)_6]^{3+} + 6H_2O$$

> **LEARNING TIP**
>
> When concentrated ammonia is used, it is more likely that the hexaamminechromium(III) ion is made rather than the precipitate of chromium(III) hydroxide.

Questions

1 What do you understand by the term *ligand substitution*?

2 A solution of ammonia was slowly added to an aqueous solution containing copper(II) ions until the ammonia was in excess. Initially a pale blue precipitate formed, followed by the formation of a deep blue solution.

(a) Identify the pale blue precipitate and write an equation for its formation.

(b) Write the formula of the complex ion in the deep blue solution.

(6) Ligand substitution and precipitation reactions

By the end of this topic, you should be able to demonstrate and apply your knowledge and understanding of:

* explanation of the biochemical importance of iron in haemoglobin, including ligand substitution involving O_2 and CO

* reactions, including ionic equations, and the accompanying colour changes of aqueous Cu^{2+}, Fe^{2+}, Fe^{3+}, Mn^{2+} and Cr^{3+} with aqueous sodium hydroxide and aqueous ammonia, including:
  (i) precipitation reactions
  (ii) complex formation with excess aqueous sodium hydroxide and aqueous ammonia

Haemoglobin and ligand substitution

A tiny drop of blood contains millions of red blood cells, which carry oxygen around the body. These cells are efficient because they contain haemoglobin, a complex protein composed of four polypeptide chains.

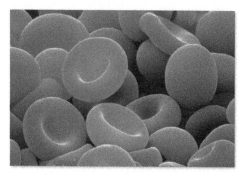

Figure 1 Scanning electron micrograph of red blood cells – these biconcave, disc-shaped cells transport oxygen from the lungs to all the cells in the body.

Each protein chain contains four non-protein components called haem groups. Each haem group has an Fe^{2+} ion at its centre – oxygen can reversibly bind to the Fe^{2+} ion. This allows haemoglobin to carry out its function of transporting oxygen around the body.

When the Fe^{2+} ion in each haem group is bound to oxygen, the haem group is red in colour. As blood passes through the lungs, the haemoglobin picks up oxygen and carries it to every living cell, where it can be released.

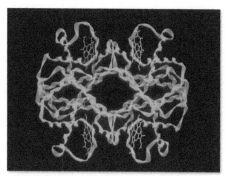

Figure 2 Haemoglobin is an oxygen-carrying macromolecule found in red blood cells. It consists of four polypeptide chains (blue); each chain carries a haem complex (purple) capable of reversibly binding oxygen. At its centre, the haem group contains an iron(II) ion, which binds one oxygen molecule.

LEARNING TIP

Carbon dioxide is not transported by haemoglobin. This waste gas diffuses into plasma.

Figure 3 shows the coordinate bonds formed around the Fe^{2+} ions in a haem group.

- There are four coordinate bonds between the Fe^{2+} ion and the nitrogen atoms in the haem structure.

- A further coordinate bond is formed to the protein globin.

- A final coordinate bond can form to an oxygen molecule, which is then transported.

Carbon monoxide – the silent killer

Carbon monoxide and oxygen can both bind to haemoglobin at the same place. Carbon monoxide binds more strongly to the haemoglobin than oxygen. If both carbon monoxide and oxygen are present in the lungs, carbon monoxide will bind to the haemoglobin – leaving fewer haemoglobin molecules to bind to oxygen molecules. Tissues can then be starved of oxygen because less oxygen is carried around the body. The reaction is an example of ligand substitution because carbon monoxide molecules can replace oxygen molecules in haemoglobin.

This reaction is not reversible for carbon monoxide. Once on a haem group, a carbon monoxide molecule cannot be removed and the haemoglobin can no longer perform its function of carrying oxygen around the body. Low levels of carbon monoxide in the blood can cause headaches, nausea and potential suffocation. High levels of carbon monoxide in the blood can be fatal.

Carbon monoxide is formed during the incomplete combustion of carbon-containing fuels. It is a colourless and odourless gas, so we have no warning if it is present in the air around us. Sensors, rather like smoke detectors, are available to give some warning of this silent killer.

Burning tobacco also releases carbon monoxide and this is one of the reasons why long-term smokers become short of breath.

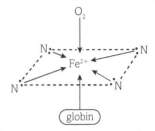

Figure 3 Part of the structure of haemoglobin showing the coordinate bonds formed to the central metal ion.

Precipitation reactions

A precipitation reaction is one in which soluble ions, in separate solutions, are mixed together to produce an insoluble compound – this settles out of solution as a solid. The insoluble compound is called a precipitate.

Transition metal ions in aqueous solution react with aqueous sodium hydroxide, NaOH(aq), to form coloured precipitates. A blue solution of copper(II) ions reacts with aqueous sodium hydroxide to form a pale blue precipitate of copper(II) hydroxide, as shown in Figures 4 and 5. The equation for the reaction is:

$$Cu^{2+}(aq) + 2OH^-(aq) \rightarrow Cu(OH)_2(s)$$

      pale blue   pale blue
        solution   precipitate

Cobalt(II), iron(II) and iron(III) ions also form precipitates with sodium hydroxide and ammonia, and these reactions are detailed in Table 1 and in Figures 4 and 5.

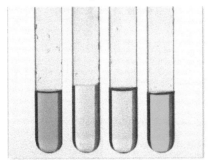

Figure 4 Solutions of $CO^{2+}(aq)$, $Fe^{2+}(aq)$, $Fe^{3+}(aq)$ and $Cu^{2+}(aq)$.

LEARNING TIP

$[Cu(H_2O)_6]^{2+}$ is often summarised to Cu^{2+}. Copper(II) hydroxide can be represented as $Cu(OH)_2$ or $Cu(OH)_2(H_2O)_4$. Ammonia can act as a base providing the OH^- ligand but it can also be a ligand in its own right.

Ion	Observation with NaOH	Observation with ammonia
$Cu^{2+}(aq)$	Blue gelatinous precipitate formed $[Cu(H_2O)_6]^{2+}(aq) + 2OH^-(aq) \rightarrow [Cu(OH)_2(H_2O)_4](s) + 2H_2O$	Blue gelatinous precipitate formed, which re-dissolves to a blue solution $[Cu(H_2O)_6]^{2+}(aq) + 2NH_3(aq) \rightarrow [Cu(OH)_2(H_2O)_4](s) + 2NH_4^+(aq)$
$Fe^{2+}(aq)$	Pale green solution containing $Fe^{2+}(aq)$ forms a green precipitate, turning a rusty brown at its surface on standing in air* $Fe^{2+}(aq) + 2OH^-(aq) \rightarrow Fe(OH)_2(s)$	Dark green precipitate formed $[Fe(H_2O)_6]^{2+}(aq) + 2NH_3(aq) \rightarrow [Fe(OH)_2(H_2O)_4](s) + 2NH_4^+(aq)$
$Fe^{3+}(aq)$	Pale yellow solution containing $Fe^{3+}(aq)$ forms a rusty brown precipitate (iron(III) hydroxide) $Fe^{3+}(aq) + 3OH^-(aq) \rightarrow Fe(OH)_3(s)$	Rusty brown precipitate formed $[Fe(H_2O)_6]^{3+}(aq) + 3NH_3(aq) \rightarrow [Fe(OH)_3(H_2O)_3](s) + 3NH_4^+(aq)$
$Mn^{2+}(aq)$	Brown precipitate forms, which darkens in air as it oxidises $[Mn(H_2O)_6]^{2+}(aq) + 2OH^-(aq) \rightarrow [Mn(OH)_2(H_2O)_4](s) + 2H_2O$ The manganese ion can be summarised to $Mn(OH)_2$. Oxidation can be summarised by: $4Mn(OH)_2 + O_2 \rightarrow 2Mn_2O_3 + 4H_2O$	Brown precipitate forms, which darkens in air as it oxidises $[Mn(H_2O)_6]^{2+}(aq) + 2NH_3(aq) \rightarrow [Mn(OH)_2(H_2O)_4](s) + 2NH_4^+(aq)$
$Cr^{3+}(aq)$	Green precipitate forms, which will then re-dissolve in excess hydroxide $[Cr(H_2O)_6]^{3+}(aq) + 3OH^-(aq) \rightarrow [Cr(OH)_3(H_2O)_3](s) + 3H_2O$ Then $[Cr(OH)_3(H_2O)_3](s) + 3OH^-(aq) \rightarrow [Cr(OH)_6]^{3-}(aq) + 3H_2O$	Green precipitate forms, which will re-dissolve if excess ammonia is added $[Cr(H_2O)_6]^{3+}(aq) + 3NH_3(aq) \rightarrow [Cr(OH)_3(H_2O)_3](s) + 3NH_4^+(aq)$ Then $[Cr(OH)_3(H_2O)_3](s) + 6NH_3(aq) \rightarrow [Cr(NH_3)_6]^{3+}(aq) + 3H_2O + 3OH^-(aq)$

*The precipitate changes colour because green Fe^{2+} ions are readily oxidised to rusty brown Fe^{3+} ions.

Table 1 Reactions of different ions with NaOH and NH_3.

Figure 5 Precipitates of $Co(OH)_2(s)$, $Fe(OH)_2(s)$, $Fe(OH)_3(s)$ and $Cu(OH)_2(s)$.

Questions

1 How many molecules of oxygen can one molecule of haemoglobin carry?

2 A solution of an iron salt reacts with aqueous sodium hydroxide to form a red–brown precipitate. Name the iron complex present and state its oxidation number.

3 Nickel is a transition element that forms an ion with an oxidation state of +2. Write an equation for the reaction of an aqueous solution containing this ion with aqueous sodium hydroxide.

7 Redox reactions

By the end of this topic, you should be able to demonstrate and apply your knowledge and understanding of:

* redox reactions and accompanying colour changes for:
    (i) interconversions between Fe^{2+} and Fe^{3+}
    (ii) interconversions between Cr^{3+} and $Cr_2O_7^{2-}$
    (iii) reduction of Cu^{2+} to Cu^+ and disproportionation of Cu^+ to Cu^{2+} and Cu

* interpretation and prediction of unfamiliar reactions including ligand substitution, precipitation, redox

KEY DEFINITIONS

Oxidation is loss of electrons, or an increase in oxidation number.
Reduction is gain of electrons, or a decrease in oxidation number.

Oxidation and reduction in transition element chemistry

As transition elements form compounds with variable oxidation states, they can take part in **oxidation** and **reduction** reactions.

MnO_4^- and Fe^{2+}

Iron has two common oxidation states in its compounds, Fe^{2+} and Fe^{3+}. The iron(II) oxidation state is less stable than the iron(III) oxidation state. In the presence of air, or when in contact with another oxidising agent, iron(II) is readily oxidised to iron(III):

$$Fe^{2+}(aq) \rightarrow Fe^{3+}(aq) + e^-$$

Manganese exists in a number of oxidation states, including the +2 and +7 oxidation states. MnO_4^- is a strong oxidising agent. In acidic solution, MnO_4^- can be reduced to form Mn^{2+}:

$$MnO_4^-(aq) + 8H^+(aq) + 5e^- \rightarrow Mn^{2+}(aq) + 4H_2O(l)$$

MnO_4^- is commonly used to oxidise solutions containing iron(II), but is also used as an oxidising agent in many other chemical reactions. Half-equations can be added together to generate an ionic equation. The reaction between iron(II) and acidified manganate(VII) is shown below:

$$MnO_4^-(aq) + 8H^+(aq) + 5Fe^{2+}(aq)$$
$$\rightarrow Mn^{2+}(aq) + 5Fe^{3+}(aq) + 4H_2O(l)$$

We can use these and other redox reactions to calculate (within experimental error) the concentration of a transition metal ion in an unknown solution, or to calculate the percentage composition of a metal in a solid sample of a compound or alloy.

Carrying out redox titrations

Redox titrations are carried out in a similar way to acid–base titrations.

* An acid–base reaction involves the transfer of protons from the acid to the alkali and the formation of water and a salt.

* A redox titration involves the transfer of electrons from one species to another.

Just as an acid can be titrated against an alkali, an oxidising agent can be titrated against a reducing agent.

In an acid–base reaction, the end point is identified using an indicator, which appears as one colour in the acid solution and as a different colour in the alkaline solution. Indicators can also be used in some redox titrations. In some reactions there is a colour change that provides the end point without the need for an indicator.

Aqueous potassium manganate(VII) contains $MnO_4^-(aq)$ and is a common oxidising agent that is self-indicating:

$MnO_4^-(aq)$ becomes $Mn^{2+}(aq)$
purple   almost colourless

$Mn^{2+}(aq)$ ions are very pale pink in colour. In manganate(VII) titrations, the solutions used are very dilute and the $Mn^{2+}(aq)$ ions are present in low concentrations. This means that the very pale pink colour from $Mn^{2+}(aq)$ cannot be seen and the solution appears colourless. The $MnO_4^-(aq)$ ions have such a deep purple colour that this masks any other colour present.

Figure 1 Colours of solutions containing $MnO_4^-(aq)$ (left) and $Mn^{2+}(aq)$ ions.

Redox titrations involving MnO_4^- ions

MnO_4^- is typically used to oxidise solutions containing iron(II) ions. The solution is acidified with sulfuric acid. Hydrochloric acid cannot be used because it reacts with MnO_4^-. This would not only invalidate the titration results but is unsafe as chlorine gas is made.

Typically these reactions are carried out with the potassium manganate(VII) solution in the burette and the iron(II) solution in the conical flask. As the manganate(VII) solution is added to the iron(II) solution, it is decolourised. The end point of the titration is when excess MnO_4^- ions are present. This is observed as the first hint of a permanent pink colour in the solution in the conical flask.

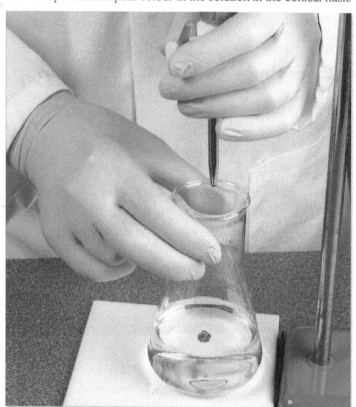

Figure 2 Measuring the concentration of iron(II) ions in a solution of iron(II) sulfate. You can see the purple colour forming as the MnO_4^- runs into the acidified $Fe^{2+}(aq)$ solution in the conical flask.

25.0 cm³ of a solution of an iron(II) salt required 23.00 cm³ of 0.0200 mol dm⁻³ potassium manganate(VII) for complete oxidation in acid solution.
(a) Calculate the amount, in moles, of manganate(VII), MnO_4^-, used in the titration.
(b) Deduce the amount, in moles, of iron(II), Fe^{2+}, in the solution that was titrated.
(c) Calculate the concentration, in mol dm⁻³, of Fe^{2+} in the solution that was titrated.

(a) Calculate the amount, in mol, of MnO_4^- that reacted.
$$n(MnO_4^-) = c \times \frac{V}{1000} = 0.0200 \times \frac{23.00}{1000} = 4.60 \times 10^{-4}\ mol$$
(b) Calculate the amount, in mol, of Fe^{2+}, that was used in the titration.

equation	$MnO_4^-(aq)$ reacts with	$5\ Fe^{2+}(aq)$
moles	1 mol	5 mol
actual moles	4.60×10^{-4} mol	2.30×10^{-3} mol

(c) Calculate the concentration, in mol dm⁻³, of Fe^{2+} in the solution.
$$c(Fe^{2+}) = n \times \frac{1000}{V} = 2.30 \times 10^{-3} \times \frac{1000}{25.0} = 0.0920\ mol\ dm^{-3}$$

For **(a)** we use amount, $n = c \times \dfrac{V}{1000}$

For **(b)** we use the balanced equation (see above) to work out the reacting moles of MnO_4^- and Fe^{2+}:
$$(MnO_4^-(aq)) \equiv (5Fe^{2+}(aq))$$
For **(c)**, we rearrange: $n = c \times \dfrac{V}{1000}$ to get $c = n \times \dfrac{1000}{V}$

Calculating the molar mass and formula of an iron(II) salt

A student weighed out 2.950 g of hydrated iron(II) sulfate, $FeSO_4.xH_2O$, and dissolved it in 50 cm³ of sulfuric acid. This solution was poured into a 250 cm³ volumetric flask and made up to the mark with distilled water. 25.0 cm³ of this solution was titrated with 0.0100 mol dm⁻³ potassium manganate(VII), $KMnO_4$, and 21.20 cm³ of the MnO_4^- solution was used.
(a) Calculate the amount, in moles, of manganate(VII), MnO_4^-, used in the titration.
(b) Deduce the amount, in moles, of iron(II), Fe^{2+}, in the 25.0 cm³ solution used.
(c) Deduce the amount, in moles, of iron(II), Fe^{2+}, in 250 cm³ of the solution.
(d) Determine the molar mass of the iron(II) salt.
(e) Determine the value of x and the formula of the hydrated iron(II) salt.

(a) Calculate the amount, in mol, of MnO_4^- that reacted:
$$n(MnO_4^-) = c \times \frac{V}{1000} = 0.0100 \times \frac{21.20}{1000}$$
$$= 2.12 \times 10^{-4}\ mol$$
(b) Calculate the amount, in mol, of Fe^{2+}, that was used in the titration:

equation	$MnO_4^-(aq)$ reacts with	$5\ Fe^{2+}(aq)$
moles from equation	1 mol	5 mol
actual moles	2.12×10^{-4} mol	1.06×10^{-3} mol

(c) Calculate the amount, in mol, Fe^{2+}, that was used to make up the 250 cm³ solution:
25.0 cm³ $Fe^{2+}(aq)$ contains 1.06×10^{-3} mol
So the 250 cm³ solution contains $10 \times 1.06 \times 10^{-3} = 0.0106$ mol
(d) Calculate the molar mass of the iron(II) salt:
$$M(FeSO_4 \cdot xH_2O) = \frac{mass}{n} = \frac{2.950}{0.0106}$$
$$= 278.3\ g\ mol^{-1}$$
(e) Determine the value of x in the iron(II) salt:
Mass of $FeSO_4$ in 1 mole of $FeSO_4 \cdot xH_2O = 55.8 + 32.1 + (16 \times 4)$
$$= 151.9\ g$$
Mass of H_2O in 1 mole of $FeSO_4 \cdot xH_2O = 278.3 - 151.9 = 126.4$ g
Therefore the value of $x = \dfrac{126.4}{18.0} = 7$.
So the formula of the hydrated salt is $FeSO_4 \cdot 7H_2O$

For **(a)** we use amount, $n = c \times \dfrac{V}{1000}$

For **(b)** we use the balanced equation to work out the reacting amounts of MnO_4^- and Fe^{2+}:
$$(MnO_4^-(aq)) \equiv (5\ Fe^{2+}(aq))$$
For **(c)** the titration used 25.0 cm³ of solution, but the original solid was made up in 250 cm³ of solution. Therefore, we need to multiply the answer to **(b)** by 10.

For **(d)** we rearrange $n = \dfrac{mass}{M}$ to get $M = \dfrac{mass}{n}$

WORKED EXAMPLE 3

Calculate the percentage mass of iron in some tablets

A multi-vitamin tablet contains vitamins A, B, C and D along with some iron. 0.325 g of a powdered tablet was dissolved in water and a little dilute sulfuric acid added.

12.10 cm³ of 0.00200 mol dm⁻³ potassium manganate(VII) solution was titrated until a permanent pale pink colour was observed.

Assume that no vitamin will react with the potassium manganate(VII) solution.

Calculate the percentage mass of iron in the tablets.

Amount of $MnO_4^- = c \times \dfrac{V}{1000} = 0.00200 \times \dfrac{12.10}{1000} = 2.42 \times 10^{-5}$ mol

Amount of $Fe^{2+} = 5 \times 2.42 \times 10^{-5} = 1.21 \times 10^{-4}$ mol

$n = \dfrac{mass}{M}$

Therefore, mass $= n \times M = 1.21 \times 10^{-4} \times 55.8 = 0.00675$ g

Therefore, % iron $= \dfrac{\text{mass of iron}}{\text{mass of tablet}} \times 100 = \dfrac{0.006\,75}{0.325} = 2.08\%$

Figure 3 Some multi-vitamins have traces of iron, which can be analysed by titration.

Sometimes you will be expected to apply your knowledge of titration calculations to unfamiliar situations. You will normally be given some additional information to help you.

WORKED EXAMPLE 4

Oxidising hydrogen peroxide

A 25.0 cm³ portion of hydrogen peroxide was poured into a 250 cm³ volumetric flask and made up to the mark with distilled water.

A 25.0 cm³ portion of this solution was acidified and titrated against 0.0200 mol dm⁻³ potassium manganate(VII) solution. 38.00 cm³ of the MnO_4^- solution was required for complete oxidation.

Calculate the concentration of the original hydrogen peroxide solution.

The following equation represents the oxidation of hydrogen peroxide:

$H_2O_2(aq) \rightarrow O_2(g) + 2H^+(aq) + 2e^-$

Step 1: You will need to write the fully balanced equation for the reaction to find the molar relationship between the two reactants (see topic 5.2.7 on combining half-equations).

The equation for MnO_4^- ions is:

$MnO_4^-(aq) + 8H^+(aq) + 5e^- \rightarrow Mn^{2+}(aq) + 4H_2O(l)$

The fully balanced equation is:

$2MnO_4^-(aq) + 6H^+(aq) + 5H_2O_2(aq) \rightarrow 2Mn^{2+}(aq) + 8H_2O(l) + 5O_2(g)$

Step 2: Calculate the amount of MnO_4^- that reacted.

$n(MnO_4^-) = c \times \dfrac{V}{1000} = 0.0200 \times \dfrac{38.00}{1000} = 7.60 \times 10^{-4}$ mol

Step 3: Deduce the amount of H_2O_2 that was used in the titration.

equation	$2MnO_4^-(aq)$ reacts with	$5H_2O_2(aq)$
moles from equation	2 mol	5 mol
actual moles	7.60×10^{-4} mol	1.90×10^{-3} mol

Step 4: Calculate the concentration, in mol dm⁻³, of the hydrogen peroxide solution.

$c = n \times \dfrac{1000}{V} = 1.90 \times 10^{-3} \times \dfrac{1000}{25.0} = 0.0760$ mol dm⁻³

Step 5: The first sentence of this worked example states that the solution was diluted by a factor of 10 at the start of the experiment. Therefore, the concentration of the original hydrogen peroxide solution is $10 \times 0.0760 = 0.760$ mol dm⁻³.

Potassium dichromate(VI) solution

$Cr_2O_7^{2-}$ ions will oxidise Fe^{2+} ions in solution and so can also be used in redox titrations. The colour change is very subtle so cannot be detected with the naked eye. A redox indicator such as diphenylamine sulfonate is used. This produces a violet–blue coloured end point.

Figure 4 Potassium dichromate titration where this chemical is in the burette.

The two half-equations are:

- Oxidation: $Fe^{2+} \rightarrow Fe^{3+} + e^-$
- Reduction: $Cr_2O_7^{2-} + 14H^+ + 6e^- \rightarrow 2Cr^{3+} + 7H_2O$

The overall ionic equation is:

$$Cr_2O_7^{2-} + 14H^+ + 6Fe^{2+} \rightarrow 2Cr^{3+} + 7H_2O + 6Fe^{3+}$$

Redox titrations – iodine and thiosulfate

The reaction between iodine, I_2, and thiosulfate ions, $S_2O_3^{2-}$, is a redox reaction that is useful in chemical analysis:

$$2S_2O_3^{2-}(aq) + I_2(aq) \rightarrow 2I^-(aq) + S_4O_6^{2-}(aq)$$

In this reaction, aqueous iodine is reduced to iodide ions, I^-, by aqueous sodium thiosulfate, which forms the tetrathionate ion, $S_4O_6^{2-}$.

The concentration of iodine in a solution can be determined by titration with a solution of sodium thiosulfate of known concentration. This titration can be used to determine the concentration of a solution of an oxidising agent that reacts with iodide ions, I^-, to produce iodine, I_2. Examples of oxidising agents that can be analysed using this method include copper(II), Cu^{2+}, dichromate(VI), $Cr_2O_7^{2-}$, and chlorate(I), ClO^-.

The method for analysing an oxidising agent is outlined below.

- Iodide ions are added to the oxidising agent under investigation – a redox reaction takes place, producing iodine.

- The iodine is titrated against a sodium thiosulfate solution of known concentration.

- From the results, you can calculate the amount of iodine, and hence the concentration of the oxidising agent.

Estimating the copper content of solutions and alloys

If a solution containing Cu^{2+}(aq) ions is mixed with aqueous iodide ions, I^- (aq):

- iodide ions are oxidised to iodine, $2I^- \rightarrow I_2 + 2e^-$
- copper(II) ions are reduced to copper(I) ions, $Cu^{2+} + e^- \rightarrow Cu^+$

$$2Cu^{2+}(aq) + 4I^-(aq) \rightarrow 2CuI(s) + I_2(aq)$$

- This produces a light brown/yellow solution and a white precipitate of copper(I) iodide – the precipitate appears to be light brown due to the iodine in the solution.

- This mixture is then titrated against sodium thiosulfate of known concentration. As the iodine reacts, the iodine colour gets paler during the titration. When the solution becomes a pale straw colour, a small amount of starch is added to help with the identification of the end point – a blue–black colour forms.

- The blue–black colour disappears sharply at the end point because all the iodine has reacted.

You can use the results to determine the concentration of iodine in the titration mixture, and then the concentration of copper(II) ions in the original solution can be determined. You can use this method to estimate the concentration of copper(II) ions in an unknown solution or the percentage of copper in alloys such as brass and bronze. The alloy is first reacted with nitric acid. Brass and bronze can both be reacted to produce a solution containing copper(II) ions. This can then be reacted with potassium iodide solution. The iodine that forms is then titrated against sodium thiosulfate.

Worked example 5 is an unstructured titration calculation – you are provided with the titration results and relevant equations, but you need to work out how to tackle the problem yourself.

Figure 5 The famous bronze Champions Statue commemorating four of the players who won the 1966 World Cup.

WORKED EXAMPLE 5

A sample of bronze was analysed to find the proportion of copper it contained. 0.500 g of the bronze was reacted with nitric acid to give a solution containing Cu^{2+} ions. The solution was neutralised and reacted with iodide ions to produce iodine. The iodine was titrated with 0.200 mol dm^{-3} sodium thiosulfate, and 22.40 cm^3 were required. Calculate the percentage of copper in the bronze.

$$Cu(s) \rightarrow Cu^{2+}(aq) + 2e^-$$
$$2Cu^{2+}(aq) + 4I^-(aq) \rightarrow 2CuI(s) + I_2(aq)$$
$$2S_2O_3^{2-}(aq) + I_2(aq) \rightarrow 2I^-(aq) + S_4O_6^{2-}(aq)$$

Step 1: Calculate the amount of thiosulfate used.
$$n(S_2O_3^{2-}) = c \times \frac{V}{1000} = 0.200 \times \frac{22.40}{1000} = 4.48 \times 10^{-3} \, mol$$

Step 2: Calculate the reacting amounts.
2 mol Cu^{2+} produces 1 mol I_2 which reacts with 2 mol $S_2O_3^{2-}$.
1 mol Cu reacted with nitric acid to produce 1 mol Cu^{2+}.
So 1 mol Cu is equivalent to 1 mol $S_2O_3^{2-}$.

Step 3: Calculate the percentage copper.
$$n(Cu) = n(S_2O_3^{2-}) = 4.48 \times 10^{-3} \, mol$$

Mass of copper $= n \times M = 4.48 \times 10^{-3} \times 63.5 = 0.28448 \, g$

% copper in bronze $= \dfrac{0.28448}{0.500} \times 100 = 56.9\%$

LEARNING TIP

Note that we have not rounded anything until the final answer, which is given to three significant figures to match the smallest number of significant figures given in the question. If we had rounded the mass of copper to 0.284 g, we would have found the percentage of copper to be 56.8%.

Questions

1. MnO_4^- is used as an oxidising agent in redox titrations.
   (a) What do you understand by the term *redox reaction*?
   (b) What is the oxidation state of Mn in MnO_4^-?

2. 25.0 cm^3 of 0.150 mol dm^{-3} Fe^{2+} required 32.50 cm^3 of potassium manganate(VII) for complete oxidation in acid solution.
   Calculate the concentration of the potassium manganate(VII) solution used in the reaction.

3. A sample of metal of mass 2.50 g was thought to contain iron. The metal was reacted with sulfuric acid and the resulting solution made up to 250 cm^3. 25.0 cm^3 of this solution was titrated against 0.0180 mol dm^{-3} potassium manganate(VII) solution and 24.80 cm^3 of the solution was required.
   Calculate the percentage by mass of iron in the metal sample.

4. The equation below shows the reduction of potassium dichromate(VI):
   $$Cr_2O_7^{2-}(aq) + 6e^- + 14H^+(aq) \rightarrow Cr^{3+}(aq) + 7H_2O(l)$$

   What volume of 0.00200 mol dm^{-3} iron(II) sulfate will be oxidised by 20.00 cm^3 of 0.0150 mol dm^{-3} potassium dichromate(VI) in acid solution?

8 Testing for ions

By the end of this topic, you should be able to demonstrate and apply your knowledge and understanding of:

* qualitative analysis of ions on a test-tube scale; processes and techniques needed to identify the following ions in an unknown compound:

 (i) anions: CO_3^{2-}, Cl^-, Br^-, I^-, SO_4^{2-}; (ii) cations: NH_4^+, Cu^{2+}, Fe^{2+}, Fe^{3+}, Mn^{2+}, Cr^{3+}

Qualitative analysis

Qualitative tests can tell us which ions or chemicals are present in a sample, but do not indicate how much of these ions or chemicals there are.

Many qualitative tests involve precipitation reactions. In this type of reaction an insoluble solid often forms from two aqueous solutions.

Qualitative tests can be carried out on a small scale, such as in a test tube with only 1 cm³ of the unknown substance.

To identify a particular compound, separate qualitative tests are needed for the positive and negative ions.

> **LEARNING TIP**
> The individual reactions reviewed here have all been studied in your A level Chemistry course so you should be familiar with the full equations, ionic equations and observations.

Positive ions

Positive ions (cations) are identified depending on whether a metal ion or the ammonium ion is present.

Metal ions

Metal ions react with either ammonia or hydroxide ions to produce insoluble coloured compounds.

Method

Half fill a test tube with the sample to be analysed. Add an aqueous solution of ammonia or aqueous sodium hydroxide drop by drop.

* If a blue precipitate is produced, the sample contained copper(II) ions, Cu^{2+}.
* If a brown precipitate is produced, the sample contained manganese(II) ions, Mn^{2+}.

* If a green precipitate is produced, the sample contained either iron(II) ions, Fe^{2+}, or chromium(III) ions, Cr^{3+}, therefore a further test is required.

Sometimes the colour is difficult to see. Add excess ammonia or sodium hydroxide until there is no further change and note the solubility of the precipitate and the colour of the resulting solution if the precipitate dissolved.

> **LEARNING TIP**
> When identifying metal ions the alkali should only be added drop by drop. If too much is added too quickly, the precipitate may re-dissolve before you see it.

> **LEARNING TIP**
> When testing for iron(II) ions, Fe^{2+}, or manganese(II) ions, Mn^{2+} the precipitate sometimes darkens or changes colour at its surface on standing in air. This is the air oxidising the metal ion.

Ammonium ion

The ammonium ion, NH_4^+, reacts with the hydroxide ion to produce ammonia gas and water:

$$NH_4^+(aq) + OH^-(aq) \rightarrow NH_3(aq) + H_2O(aq)$$

Method

Half fill a test tube with the sample to be analysed. Add sodium hydroxide solution and warm gently. If the distinctive smelling gas produced turns damp red litmus paper blue, the sample contained ammonium ions.

> **LEARNING TIP**
> When testing for ammonium ions the red litmus paper must be damp, as the ammonia gas produced needs to dissolve in the water to form an alkali and cause the colour change.

Ion	Name	Compound before addition	With NH_3(aq) or OH^-(aq)	Excess NH_3(aq)	Excess OH^-(aq)
Cu^{2+}	Copper(II) ion	Blue solution	Blue precipitate	Precipitate re-dissolves to give a blue solution	No change
Fe^{2+}	Iron(II) ion	Green solution	Green precipitate	No change	No change
Mn^{2+}	Manganese(II) ion	Pink solution	Brown precipitate	No change	No change
Cr^{3+}	Chromium(III) ion	Violet solution	Green precipitate	Precipitate dissolves to give a purple solution	Precipitate dissolves to give a green solution
Fe^{3+}	Iron(III) ion	Yellow/brown solution	Brown precipitate	No change	No change

Table 1 Results of tests to identify different ions.

Negative ions

Negative ions (anions) can be identified using simple laboratory tests.

Carbonate ions, CO_3^{2-}

Carbonate ions, CO_3^{2-}, react with acids to give bubbles of gas that turn limewater milky, indicating that they are carbon dioxide:

$$CO_3^{2-}(aq) + 2H^+(aq) \rightarrow H_2O(aq) + CO_2(g)$$

Method

Half fill a test tube with the sample to be analysed and add a strong acid. Collect any bubbles and pass them through limewater.

If the gas turns the limewater cloudy, the sample contained carbonate ions, CO_3^{2-}.

Sulfate ion, SO_4^{2-}

Sulfate ions, SO_4^{2-}, react with barium ions to form the insoluble salt, barium sulfate, $BaSO_4$:

$$Ba^{2+}(aq) + SO_4^{2-}(aq) \rightarrow BaSO_4(aq)$$

Method

Half fill a test tube with the sample to be analysed and add dilute hydrochloric acid and barium chloride. If a white precipitate of barium sulfate is produced, the sample contained sulfate ions, SO_4^{2-}.

Halide ions

Halide ions react with silver ions to form different coloured silver halide precipitates:

$$Ag^+(aq) + Cl^-(aq) \rightarrow AgCl(s)$$

$$Ag^+(aq) + Br^-(aq) \rightarrow AgBr(s)$$

$$Ag^+(aq) + I^-(aq) \rightarrow AgI(s)$$

Method

Dissolve the suspected halide in water if it is not already an aqueous solution.

Add some dilute nitric acid and an aqueous solution of silver nitrate.

- If a white precipitate of silver chloride is produced, the sample contained chloride ions, Cl^-.

- If a cream precipitate of silver bromide is produced, the sample contained bromide ions, Br^-.

- If a yellow precipitate of silver iodide is produced, the sample contained iodide ions, I^-.

If the colour is hard to distinguish, first add dilute ammonia and then concentrated ammonia, noting the solubility of the precipitate.

Halide ion	Name	With $AgNO_3(aq)$	Solubility of precipitate formed
Cl^-	Chloride	White precipitate	Soluble in dilute $NH_3(aq)$
Br^-	Bromide	Cream precipitate	Soluble in concentrated $NH_3(aq)$ only
I^-	Iodide	Yellow precipitate	Insoluble in dilute and concentrated $NH_3(aq)$

Table 2 Identification of silver halide precipitates.

DID YOU KNOW?

Sometimes we need to find out which compounds are dissolved in the water people use. An environmental scientist might analyse water for pollutants using the same tests.

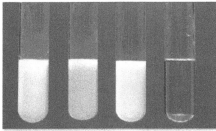

Figure 1 Halide tests using aqueous silver nitrate

Questions

1. Nitric acid was added to sodium carbonate. Write a balanced symbol equation for this reaction.

2. Explain how you would distinguish between the following pairs of compounds:
   (a) potassium chloride and potassium iodide
   (b) potassium hydroxide and ammonium hydroxide.

3. A solution of sodium hydroxide was added drop by drop to samples of iron(II) nitrate and chromium(III) nitrate. A green precipitate was produced in both cases, therefore the precipitates were indistinguishable. Explain what could then be done to distinguish between the two samples.

OCTOPUS ADAPTATIONS

Most people have heard of haemoglobin: the mammalian oxygen transporter in blood – but evolution has come up with other solutions to the oxygen transport problem! Haemerythrins are a different class of iron-based protein found in certain marine invertebrates and haemocyanins are a third class of proteins found in animals as diverse as spiders and octopuses. These different solutions to the oxygen transport problem may have evolved to enable organisms to overcome specific challenges in their environments, as the following article suggests.

BLUE BLOOD HELPS OCTOPUS SURVIVE BRUTALLY COLD TEMPERATURES

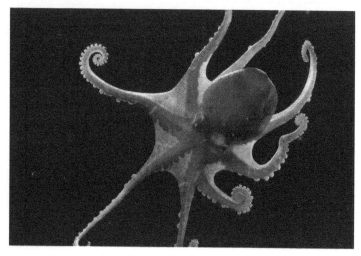

Researchers at the Alfred Wegener Institute for Polar and Marine Research in Germany have found that a specialized pigment in the blood of Antarctic octopods allows them to survive temperatures that often drop below freezing.

It's all down to a respiratory pigment in their blood called haemocyanin that allows the octopus "to maintain an aerobic lifestyle at sub-zero temperatures," said Michael Oellermann, who took part in the research. Haemocyanin also has the effect of making the octopods' blood blue.

"Haemocyanin contains copper, which is the responsible ion for binding oxygen," explained Oellermann in an email. "Haemoglobins contain iron instead, which gives our blood a red color."

Blood chilling

It's not a question of blood freezing. "The reason why octopus blood does not freeze at −1.9 °C [28.58 degrees Fahrenheit] is that they are isosmotic, which means that their blood shares the same salinity as the surrounding seawater," explained Oellermann.

The problem in very cold temperatures is that, without special adaptations, the aerobic process would, in most cases, shut down.

"Cold temperatures increase oxygen affinity to the extent that oxygen cannot be released in the tissue anymore," explained Oellermann, making survival in waters that can drop to as low as 29 degrees Fahrenheit a challenge to say the least.

Comparing octopods

The researchers looked at a particular species of Antarctic octopod called *Pareledone charcoti* and compared it with temperate and warm-adapted octopods to observe the difference in how cold-water octopods transport oxygen in their blood.

"*Pareledone charcoti* decreased the oxygen affinity of its haemocyanin to counter the adverse effect of temperature on oxygen binding accompanied by changes in their protein sequence, to assure sufficient oxygen supply to its tissues and organs," explained Oellermann.

Octopods likely developed the adaptation because they require a lot of oxygen compared to other invertebrates, notes Oellermann, and because they are largely non-migratory and must instead adapt to their environment.

In contrast, said Oellermann, Antarctic icefish survive in the same frigid environment by dint of their lower oxygen demand, rather than through the adapted system of oxygen transport employed by their blue-blooded neighbors.

Interestingly, the same adaptation may be what allows octopods to survive temperatures on the other end of the thermometer, such as the 86-degree F heat often found near thermal vents.

Source

- http://newswatch.nationalgeographic.com/2013/07/10/blue-blood-helps-octopus-survive-brutally-cold-temperatures/

Where else will I encounter these themes?

Let us start by considering the nature of the writing in the article.

1. Do you think this article is aimed at scientists, the general public, or people who are not scientists but have an interest in science? Look back through the extract and find examples to support your answer.

Now we will look at the chemistry in, or connected to, this article. Do not worry if you are not ready to give answers to these questions yet. You may like to return to the questions once you have covered other topics later in the book. Use the timeline at the bottom of the page to help you put this work in context with what you have already learned and what is ahead in your course.

2. Oxygenated haemocyanin is blue in colour whereas deoxygenated haemocyanin is colourless. Using your knowledge of copper chemistry, explain this observation.

3. Describe the type of bond between copper and nitrogen in oxyhaemocyanin as shown below in Figure 1.

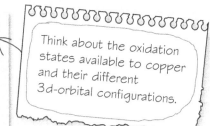

Think about the oxidation states available to copper and their different 3d-orbital configurations.

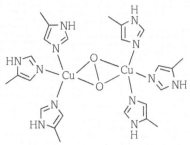

Figure 1

4. Suggest why zinc is unlikely to be able to form a protein complex that is capable of reversibly binding oxygen.

5. **a.** Use the two half-equations below to show that the reaction between oxygen and copper(I) ions under standard conditions is feasible:

$$Cu^{2+}(aq) + e^- \rightleftharpoons Cu^+(aq) \qquad E^\ominus = +0.15 \ e/V$$
$$O_2(g) + 2H_2O(l) + 4e^- \rightleftharpoons 4OH^-(aq) \qquad E^\ominus = +0.40 \ e/V$$

b. Calculate the total entropy change for this reaction under standard conditions (298 K). (F = Faraday's constant, 96 500 C).

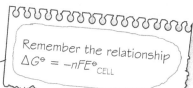

Remember the relationship $\Delta G^\ominus = -nFE^\ominus_{CELL}$

Activity

Design a poster to illustrate the role of transition metals and their compounds as catalysts. You could choose to focus on their role in biological systems (enzymes) or their role in industrial chemistry. In each case choose a specific example to illustrate the general principles.

DID YOU KNOW?

There is evidence that some of the earliest life forms on the planet used a diverse range of transition metals in order to survive and flourish. Metabolism in Methanogens seems to involve Ni, Co and even W. Williams and Frausto da Silva suggest that the central use of Co, Ni and W has practically disappeared from more recent life forms. Quite why this is can only be a matter of speculation.

Williams, R. J. P and Frausto da Silva, J. J. R. (1996) *The Natural Selection of the Chemical Elements*, Clarendon Press, Oxford.

1. Which of the following reactions **does not** involve a change in oxidation state? [1]
   A. $2Cu^+(aq) \rightarrow Cu(s) + Cu^{2+}(aq)$
   B. $Cr_2O_7^{2-}(aq) + 2OH^-(aq) \rightarrow 2CrO_4^{2-}(aq) + H_2O(l)$
   C. $2FeCl_2(s) + Cl_2(g) \rightarrow 2FeCl_3(s)$
   D. $Zn(s) + 2HCl(aq) \rightarrow ZnCl_2(aq) + H_2(g)$

2. Which of the following species can act as a bidendate ligand? [1]
   A. N_2H_4
   B. NH_3
   C. $C_2O_4^{2-}$
   D. $EDTA^{2-}$

3. The ionic equation for the reaction between copper(II) sulfate and potassium iodide solution is shown below.

   $2Cu^{2+}(aq) + 4I^-(aq) \rightarrow 2CuI(s) + I_2(aq)$

   Which of the following statements concerning this reaction is true? [1]
   A. A brown precipitate of copper(I) iodide is formed.
   B. Iodide ions are acting as the reducing agent.
   C. Copper ions are being oxidised.
   D. A purple solution is formed.

4. Cobalt is a transition element. Solid compounds of cobalt are often complexes and in solution, complex ions are formed.

   (a) In its complexes, the common oxidation numbers of cobalt are +2 and +3.
   Complete the electron configurations of cobalt as the element and in the +3 oxidation state:
   cobalt as the element: $1s^2\, 2s^2\, 2p^6\ldots$
   cobalt in the +3 oxidation state: $1s^2\, 2s^2\, 2p^6\ldots$ [2]

   (b) State **one** property of cobalt(II) and cobalt(III), other than their ability to form complex ions, which is typical of ions of a transition element. [1]

   (c) Complex ions contain ligands. State the meaning of the term *ligand*. [1]

   (d) Aqueous cobalt(II) sulfate, $CoSO_4(aq)$, takes part in the following reactions.
   For each reaction, state the formula of the transition element species formed and the type of reaction taking place.
   (i) Aqueous cobalt(II) sulfate, $CoSO_4(aq)$, reacts with aqueous sodium hydroxide. [2]
   (ii) Aqueous cobalt(II) sulfate, $CoSO_4(aq)$, reacts with concentrated hydrochloric acid. [2]

   (e) Cobalt(III) chloride, $CoCl_3$, reacts with ammonia to form a range of complexes. These complexes contain different amounts of ammonia. Information about these complexes is summarised in **Table 1**.

The complex ions **C** and **D** are stereoisomers.

complex	formula	formula of complex
A	$CoCl_3(NH_3)_6$	$[Co(NH_3)_6]^{3+}\ 3Cl^-$
B	$CoCl_3(NH_3)_5$	$[Co(NH_3)_5Cl]^{2+}\ 2Cl^-$
C	$CoCl_3(NH_3)_4$	$[Co(NH_3)_4\, Cl_2]^+\ Cl^-$
D	$CoCl_3(NH_3)_4$	$[Co(NH_3)_4Cl_2]^+\ Cl^-$

Table 1

(i) Complete the diagrams below to suggest possible structures for the complex ion in complexes **A** to **D**. [4]

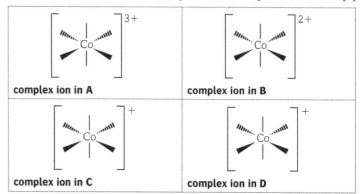

complex ion in A   complex ion in B

complex ion in C   complex ion in D

(ii) Chemists provided evidence for the formulae of these complexes from their reactions with aqueous silver nitrate. Aqueous silver nitrate reacts with aqueous halide ions in a precipitation reaction.

An excess of silver nitrate solution was reacted with 0.0100 mol of one of the complexes **A** to **D**. 2.868 g of a precipitate was formed.

Determine which complex was reacted.

In your answer you should explain how the result of the experiment would allow the formula of the complex to be identified. [3]

[Total: 18]

[Q3, F325 Feb 2012]

5. Redox reactions can be used to generate electrical energy from electrochemical cells.

   (a) A student carries out an investigation based on the redox systems shown in **Table 2** below.

	redox system	$E^{\ominus}/V$
1	$Ni^{2+}(aq) + 2e^- \rightleftharpoons Ni(s)$	-0.25
2	$Fe^{3+}(aq) + e^- \rightleftharpoons Fe^{2+}(aq)$	$+0.77$
3	$Cr^{3+}(aq) + 3e^- \rightleftharpoons Cr(s)$	-0.74

Table 2

The student sets up two standard cells to measure two standard cell potentials.

- **Cell A** is based on redox systems **1** and **2**.
- **Cell B** is based on redox systems **1** and **3**.

(i) Draw a labelled diagram to show how the student could have set up **Cell A**, based on redox systems **1** and **2**, to measure the standard cell potential. [3]

(ii) For each standard cell above:
- what would be the standard cell potential?
- what would be the sign of the Ni electrode? [2]

(b) The student left each cell in **(a)** connected for a length of time.

For each cell, the student weighed the nickel electrode before connecting the cell and after the cell had been disconnected.

The student made the following observations.

- In **Cell A**, the nickel electrode lost mass.
- In **Cell B**, the nickel electrode gained mass.
- In **both** cells, the measured cell potential slowly changed.

Explain these observations. Include equations in your answer. [3]

(c) Nickel metal hydride cells (NiMH cells) are being developed for possible use in cars.

In a NiMH cell, an alloy is used to absorb hydrogen as a metal hydride. For simplicity, the alloy can be represented as M and the metal hydride as MH.

The overall cell reaction in a NiMH cell is shown below.

$$MH + NiO(OH) \rightarrow M + Ni(OH)_2$$

The half-equation at one electrode is shown below.

$$NiO(OH) + H_2O + e^- \rightarrow Ni(OH)_2 + OH^-$$

(i) Deduce the half-equation at the other electrode. [1]

(ii) State a method, other than absorption, that is being developed to store hydrogen for possible use as a fuel in cars. [1]

[Total: 10]

[Q5, F325 Feb 2012]

6. Chromium shows typical properties of a transition element. The element's name comes from the Greek word 'Chroma' meaning colour because of its many colourful compounds.

(a) Write down the electron configuration of:
(i) a Cr atom [1]
(ii) a Cr^{3+} ion. [1]

(b) An acidified solution containing orange $Cr_2O_7^{2-}$ ions reacts with zinc in a redox reaction to form a solution containing Zn^{2+} ions and blue Cr^{2+} ions.

The unbalanced half-equations are shown below.

$$Zn \rightarrow Zn^{2+} + e^-$$
$$Cr_2O_7^{2-} + H^+ + e^- \rightarrow Cr^{2+} + H_2O$$

Balance these equations and construct an overall equation for this reaction. [3]

(c) Aqueous solutions of Cr^{3+} ions contain ruby-coloured $[Cr(H_2O)_6]^{3+}$ complex ions. If an excess of concentrated ammonia solution is added, the solution changes to a violet colour as the hexaammine chromium(III) complex ion forms.

(i) What type of reaction has taken place? [1]

(ii) Suggest an equation for this reaction. [2]

(d) Chromium picolinate, $Cr(C_6H_4NO_2)_3$, is a bright red complex, used as a nutritional supplement to prevent or treat chromium deficiency in the human body.

In this complex,
- chromium has the +3 oxidation state,
- picolinate ions, $C_6H_4NO_2^-$, act as bidentate ligands.

The structure of the picolinate ion is shown below.

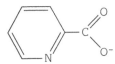

$Cr(C_6H_4NO_2)_3$ exists as a mixture of stereoisomers.

(i) What is meant by the term *ligand*? [1]

(ii) How is the picolinate ion able to act as a **bidentate** ligand? [2]

(iii) Why does $Cr(C_6H_4NO_2)_3$ exist as a mixture of stereoisomers? Draw diagrams of the stereoisomers as part of your answer. [3]

(e) Compound **A** is an orange ionic compound of chromium with the percentage composition by mass N, 11.11%; H, 3.17%; Cr, 41.27%; O, 44.45%. Compound **A** does **not** have water of crystallisation.

On gentle heating, compound **A** decomposes to form three products, **B**, **C** and water.

B is a green oxide of chromium with a molar mass of 152.0 g mol^{-1}.

C is a gas. At RTP, each cubic decimetre of **C** has a mass of 1.17 g.

In the steps below, show all your working.

- Calculate the empirical formula of compound A.
- Deduce the ions that make up the ionic compound A.
- Identify substances B and C.
- Write an equation for the decomposition of compound A by heat. [8]

[Total: 22]

[Q6, F325 June 2010]

119

Organic chemistry and analysis

AROMATIC COMPOUNDS, CARBONYLS AND ACIDS

Introduction

Aromatic compounds were originally thought to be smelly, carbon-based compounds. However, we now classify aromatic compounds as those that have a six-membered carbon ring with delocalised electrons.

Benzene is the simplest aromatic compound and was first isolated by Michael Faraday in 1825. He separated benzene from the liquid residue of whale oil street lamps.

Although aromatic compounds are unsaturated, the delocalising of electrons makes these six-carbon ring structures very stable. By substituting different groups onto the delocalised ring, the reactivity can be controlled.

Research chemists carefully design multi-stage reactions, using different reagents and conditions, to change one functional group into another. This is the beginning of organic synthesis and how we make many of our pharmaceuticals.

All the maths you need

To unlock the puzzles of this chapter you need the following maths:

- Use thermodynamic data to explain why the model of benzene has changed over time

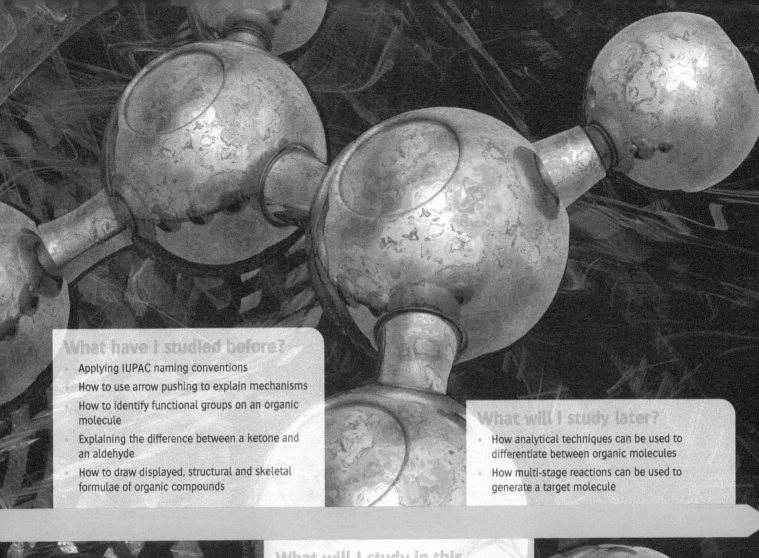

What have I studied before?

- Applying IUPAC naming conventions
- How to use arrow pushing to explain mechanisms
- How to identify functional groups on an organic molecule
- Explaining the difference between a ketone and an aldehyde
- How to draw displayed, structural and skeletal formulae of organic compounds

What will I study later?

- How analytical techniques can be used to differentiate between organic molecules
- How multi-stage reactions can be used to generate a target molecule

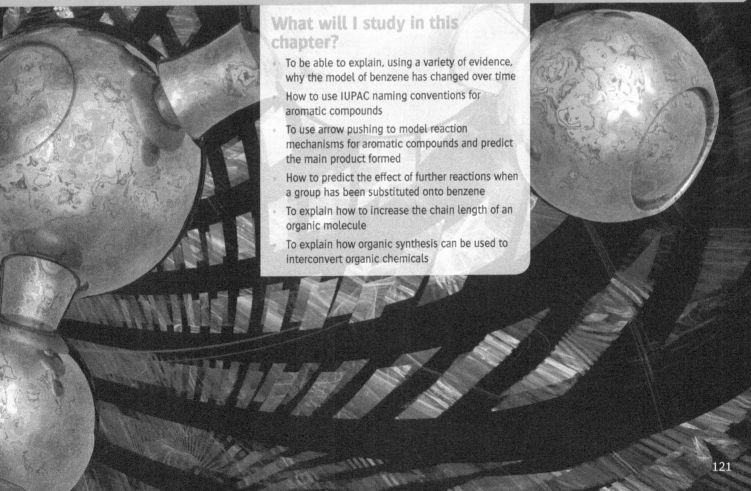

What will I study in this chapter?

- To be able to explain, using a variety of evidence, why the model of benzene has changed over time
- How to use IUPAC naming conventions for aromatic compounds
- To use arrow pushing to model reaction mechanisms for aromatic compounds and predict the main product formed
- How to predict the effect of further reactions when a group has been substituted onto benzene
- To explain how to increase the chain length of an organic molecule
- To explain how organic synthesis can be used to interconvert organic chemicals

(1) Benzene and its structure

By the end of this topic, you should be able to demonstrate and apply your knowledge and understanding of:

* the comparison of the Kekulé model of benzene with the subsequent delocalised models for benzene in terms of p-orbital overlap forming a delocalised π-system

* the experimental evidence for a delocalised, rather than Kekulé, model for benzene in terms of bond lengths, enthalpy change of hydrogenation and resistance to reaction

---

KEY DEFINITIONS

Benzene is a naturally occurring aromatic compound, which is a very stable planar ring structure with delocalised electrons.

A **model** is a simplified version that allows us to make predictions and understand observations more easily.

Benzene is a naturally occurring aromatic hydrocarbon with a ring structure. You learned about aromatic hydrocarbons in Book 1, topic 4.1.1.

Benzene is the simplest member of the arene homologous series. It has the molecular formula C_6H_6 and an empirical formula of CH. It is a liquid at room temperature and is a key ingredient added to gasoline as it increases the efficiency of the car engine.

Kekulé's model of benzene

Although the empirical and molecular formula of benzene has been known since the 1850s, the structure was very difficult to determine. Experimental evidence collected over many years has seen a development in the **model** of the structure of benzene.

Friedrich August Kekulé (later known as August Kekulé von Stradonitz) was a German chemist who, in 1865, suggested that benzene was a six-membered ring with alternating single and double bonds between the carbon atoms (Figure 1). He discovered that when one group was added to benzene, only one isomer was ever made; but when two groups were added, there were always three structural isomers produced. He used this experimental evidence about the types of isomers produced when benzene reacted, to support his theory.

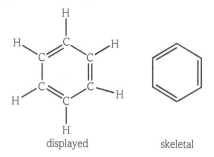

displayed   skeletal

Figure 1 Kekulé's model of benzene.

Problems with the Kekulé model

Although the Kekulé model is still used today to draw chemical reaction mechanisms, the structure was proved incorrect. There are three pieces of experimental evidence that do not support the Kekulé model.

1. Unlike alkenes, benzene is resistant to addition reactions.

2. Enthalpy of hydrogenation of benzene shows that benzene is much more stable than was predicted.

3. All six carbon bonds in benzene are the same length.

Resistance to reaction

Using the Kekulé model, you would expect benzene to undergo similar reactions to alkenes. Ethene will readily undergo addition reactions but benzene tends to undergo substitution of a hydrogen atom rather than addition reactions. Kekulé tried to explain this by saying the double and single bonds changed position in a very fast equilibrium (Figure 2).

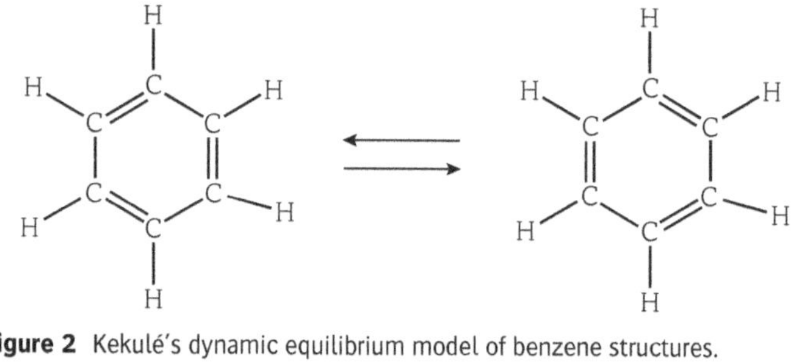

Figure 2 Kekulé's dynamic equilibrium model of benzene structures.

Hydrogenation

Hydrogenation is the addition of hydrogen to an unsaturated chemical. Using bond enthalpy data we can calculate the enthalpy change for the complete hydrogenation of cyclohexene and cyclo-1,3,5-hexatriene. Cyclo-1,3,5-hexatriene is the Kekulé model of benzene (Figure 3). However, experimentally it is found that the enthalpy change for hydrogenation of benzene is $-208\,kJ\,mol^{-1}$, which shows it is $152\,kJ\,mol^{-1}$ more energetically stable than predicted (Figure 4).

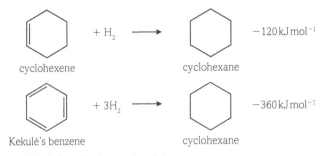

Figure 3 Enthalpy of hydrogenation data.

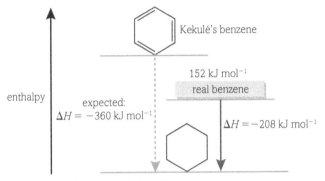

Figure 4 Chart to show the enthalpy of hydrogenation.

Bond lengths

X-ray diffraction techniques have shown that all six carbon bonds in benzene are 0.140 nm, which is between a C–C single bond at 0.147 nm and a C=C double bond at 0.135 nm. Kekulé's structure suggests that there should be three shorter C=C double bonds and three longer C–C single bonds. This evidence disproved the Kekulé structure as all six bonds are the same length.

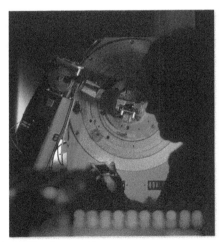

Figure 5 An X-ray diffraction machine.

Delocalised structure of benzene

It is now thought that benzene has a delocalised electron structure. The delocalised model can explain all three pieces of experimental evidence that do not support Kekulé's structure.

In the delocalised structure, each of the six carbon atoms donates one electron from its p-orbital. These electrons combine to form a ring of delocalised electrons above and below the plane of the molecule. The electrons in the rings are said to be delocalised as they are able to move freely within the ring and do not belong to a single atom. Therefore, unlike Kekulé's structure, all bonds in this ring are identical, so they are the same length.

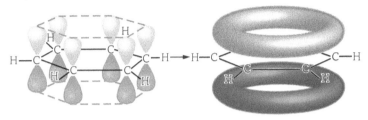

Figure 6 Delocalised electron model of benzene.

This delocalisation of electrons leads to benzene being about 152 kJ mol^{-1} more stable than expected when using the Kekulé model. Because so much energy is needed to disrupt this delocalisation, benzene is very stable and resistant to addition reaction.

When representing benzene in equations or mechanisms you can use either the Kekulé structure or the delocalised electron structure.

Questions

1 Describe how Kekulé came up with his model for benzene.

2 Explain why the Kekulé model has changed over time.

2 Naming aromatic compounds

By the end of this topic, you should be able to demonstrate and apply your knowledge and understanding of:

* use of IUPAC rules of nomenclature for systematically naming substituted aromatic compounds

---

KEY DEFINITIONS

A **substitution reaction** is where a group or atom is exchanged for another group or atom in a chemical reaction.
A **benzene derivative** is a benzene ring that has undergone a substitution reaction.

Aromatic compounds contain one or more benzene rings. This class of compounds is very important in biochemistry as they are found in many proteins, including the nucleotides in DNA.

Structure of benzene derivatives

As benzene has delocalised electrons it is energetically more stable than initially calculated. So, it rarely undergoes addition reactions but it will undergo **substitution reactions**. In these reactions, a hydrogen atom is substituted for a different group and a **benzene derivative** is formed.

Table 1 shows the common groups that will substitute a hydrogen atom on a benzene ring.

Substitute	Formula	Naming prefix
Chlorine atom	$-Cl$	chloro
Bromine atom	$-Br$	bromo
Nitro functional group	$-NO_2$	nitro
Alkyl chains	e.g. $-CH_3$	e.g. methyl

Table 1 Common groups that will substitute a hydrogen atom on a benzene ring.

Single-substituted benzene derivative

When one hydrogen is replaced by another group, a single-substitution reaction has occurred. Using the steps outlined in Book 1, topic 4.1.1, the compound can be named.

* Stem – the longest chain of carbon atoms is the aromatic ring and has the stem benzene.

* Prefix – there is one ethyl group on the ring so there is no need to number. So, the prefix is ethyl.

The name of this compound is ethylbenzene.

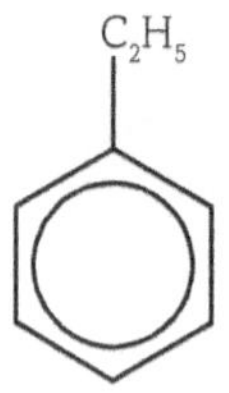

C_2H_5

$C_6H_5C_2H_5$

Figure 1 Ethylbenzene.

Double-substituted benzene derivative

Two hydrogen atoms on different carbon atoms of benzene can be replaced by two groups. If the groups are different, the name is written as a prefix in alphabetical order. One group will be added first and be given carbon number 1, the second group will then be given a number to state which carbon atom it is on with respect to the first substituted group. When naming double-substituted benzene derivatives, the smallest possible numbers should be used.

WORKED EXAMPLE

Benzene can undergo a reaction to make methylbenzene. This can then react to substitute another hydrogen atom from the aromatic ring for a bromine atom.

What is the name of the compound formed?

- Stem – the longest chain of carbon atoms is the aromatic ring and has the stem benzene.
- Prefix – there is one methyl group on the ring and one bromine group. As 'b' comes before 'm' in the alphabet, bromo is the first prefix. The usual priority is given, so bromine is given carbon 1 and the methyl is numbered with respect to the bromine group.

So, the name of this compound is 1-bromo-2-methylbenzene.

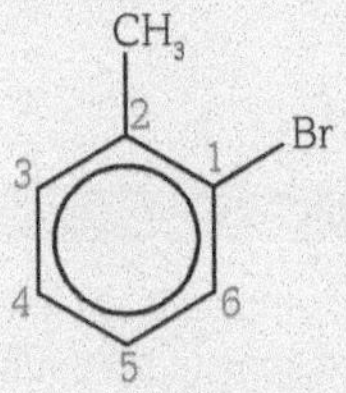

Figure 2 1-Bromo-2-methylbenzene.

LEARNING TIP

There are three structural isomers of bromomethylbenzene because the bromine atom can also take places on carbon atoms 3 and 4.

Remember that 1-bromo-6-methylbenzene is actually the same structure as 1-bromo-2-methylbenzene. The latter name is used as it contains the *lowest* number. So, there are only three ways to place the methyl and bromine groups and hence three structural isomers.

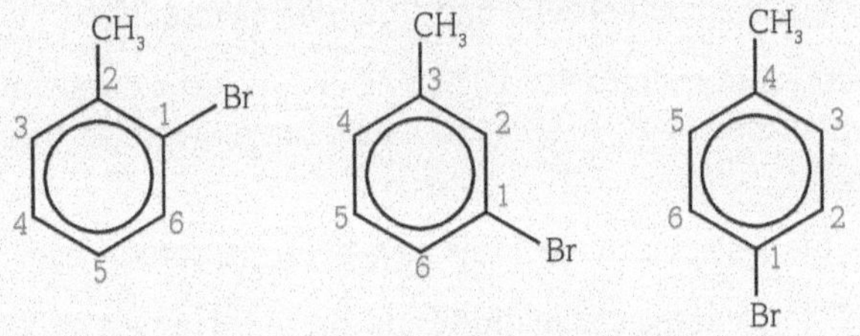

1-bromo-2-methylbenzene   1-bromo-3-methylbenzene   1-bromo-4-methylbenzene

Figure 3 Structural isomers of bromomethylbenzene.

Multiple substitutions

Further hydrogen atoms on the aromatic ring can be substituted. This may take a number of stages in an organic synthesis. For example, benzene can react to make methylbenzene and then two nitro functional groups can substitute a further two hydrogen atoms. To show there is more than one of the same functional group, prefixes such as *di* and *tri* are used. In this case it is 4-methyl-1, 3-dinitrobenzene.

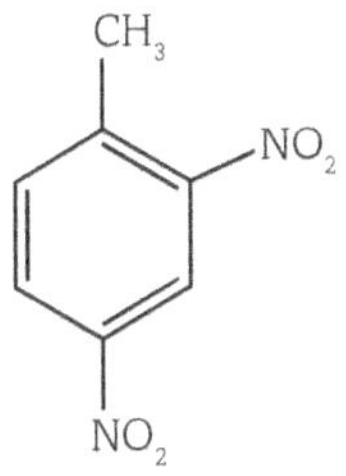

Figure 4 4,methyl-1,3-dinitrobenzene.

LEARNING TIP

It is important that everyone uses the same naming convention so that every organic chemical can have a unique name. This allows the structural formula to be generated.

The IUPAC nomenclature system has provided a logical set of rules for generating unique and universal names with information about structure and functional groups. In this way, scientists from all over the world can communicate, despite language barriers, and be sure that they are discussing the same chemical.

Prefix

In some organic chemicals the main functional group is not the aromatic ring. If a hydrogen atom is removed from the benzene ring it becomes a phenyl group. This is the same as when one hydrogen is removed from methane to form a methyl group.

DID YOU KNOW?

When making the plastic poly(phenylethene), the monomer is phenylethene. The common name previously used for the monomer was styrene, and so the plastic is commonly called polystyrene. This addition polymer has many uses, including in packaging.

You learned about polymers in Book 1, topic 4.1.10. You will find out more about polymers in topics 6.2.5 and 6.2.6.

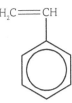

Figure 5 Phenylethene.

Questions

1 Name the following chemicals:

(a)   (b)

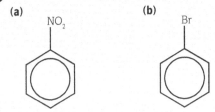

2 Draw the structure of chlorobenzene.

3 How many isomers are there of ethylbenzene?

4 How many isomers are there of chloroethylbenzene?

③ Electrophilic substitution

By the end of this topic, you should be able to demonstrate and apply your knowledge and understanding of:

* the electrophilic substitution of aromatic compounds with:
  (i) concentrated nitric acid in the presence of concentrated sulfuric acid
  (ii) a halogen in the presence of a halogen carrier

* the interpretation of unfamiliar electrophilic substitution reactions of aromatic compounds, including prediction of mechanisms

---

KEY DEFINITIONS

Electrophilic substitution is a substitution reaction where an electrophile is attracted to an electron-rich atom or part of a molecule and a new covalent bond is formed by the electrophile accepting an electron pair.

A **reaction mechanism** is a model with steps to explain and predict a chemical reaction.

As we have seen, electrons are delocalised on the benzene ring. This makes the structure stable and unlikely to undergo addition reactions. As the ring is so electron-dense it is susceptible to electrophilic attack and a hydrogen atom can be substituted by another group.

Electrophilic substitution

Electrophilic substitution can be considered as a series of reaction steps:

1. Electrons above and below the plane of atoms in the benzene ring attract an electrophile.

2. The electrophile accepts a pair of π electrons from the delocalised ring and makes a covalent bond. This is the slowest step and known as the *rate-determining step*. Find out more about this in topic 5.1.4.

3. A reactive intermediate is formed where the delocalised electrons have been disrupted.

4. The unstable intermediate releases an H^+ ion and the stable product is formed. This is a very fast step.

This **reaction mechanism** can be summarised using curly arrows. See Figure 1, where X^+ represents a general electrophile.

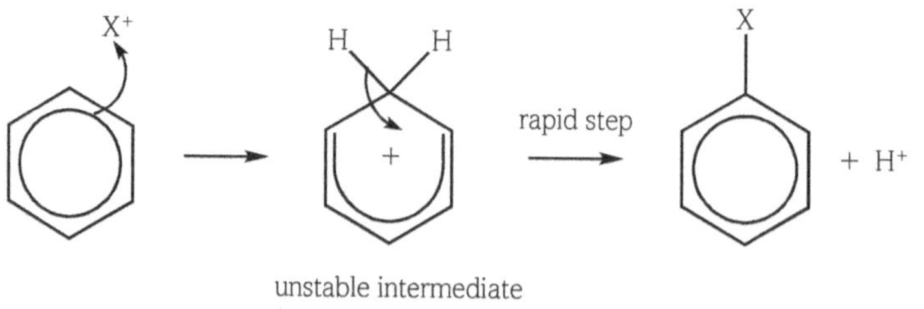

Figure 1 Electrophilic substitution in benzene.

LEARNING TIP

The arrows show the movement of electrons. Each half of an arrow head (shows the movement of one electron. In this mechanism, all the arrow heads should be (as two electrons are moving to the same place in each step. It is important that the arrow in the first step starts at the delocalised electrons and finishes on the electrophile. In the second step, the arrow should start in the middle of the C–H bond and end on the + section of the delocalised electrons.

Reactions of benzene

DID YOU KNOW?

Benzene is a carcinogen, which means it is likely to cause cancer. So, experiments in schools and colleges do not use benzene but use safer aromatic compounds like methyl benzoate. But the theory of aromatic chemistry is focused on benzene as this simplifies the mechanisms.

Nitration

This is an electrophilic **substitution reaction** where a hydrogen atom is exchanged for a nitro group ($-NO_2$). For the nitration of benzene the reagent is concentrated nitric acid, with concentrated sulfuric acid as a catalyst.

The following balanced chemical equation summarises the reaction:

$$C_6H_6 + HNO_3 \rightarrow C_6H_5NO_2 + H_2O$$

Initially the concentrated nitric acid and concentrated sulfuric acid are mixed together in a flask held in an ice bath. Then benzene is added and a reflux condenser is set up, keeping the mixture at 50 °C to prevent further substitution reactions occurring (Figure 2).

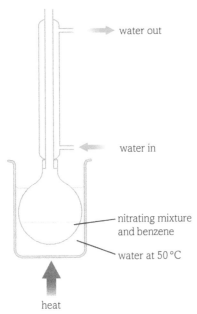

Figure 2 Quickfit set-up for nitration of benzene.

In the reaction, the sulfuric acid is needed to generate the NO_2^+ electrophile from the nitric acid. The sulfuric acid is regenerated after the nitration and is therefore a catalyst. The steps of the mechanism are shown below:

$$HNO_3 + H_2SO_4 \rightarrow NO_2^+ + HSO_4^- + H_2O$$

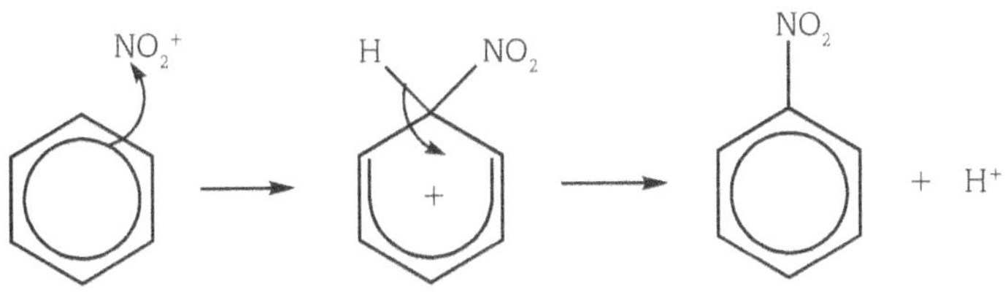

$$H^+ + HSO_4^- \rightarrow H_2SO_4$$

Figure 3 Mechanism for the nitration of benzene.

Halogenation

Benzene does not directly react with halogens as the aromatic ring is too stable. A halogen carrier such as iron (which forms an iron halide in situ), iron halides or aluminium halides are used. The halogen carrier will generate a positive halogen ion.

Table 1 shows the halogen carriers that can be used for halogenation of benzene.

Type of reaction	Halogen carrier
Chlorination	$AlCl_3$, $FeCl_3$ or Fe
Bromination	$AlBr_3$, $FeBr_3$ or Fe

Table 1 Halogen carriers that can be used for halogenation of benzene.

For example, bromine can react with iron(III) bromide to form a positive bromine ion that can act as an electrophile. This can be represented by the following balanced chemical equation:

$$Br_2 + FeBr_3 \rightarrow Br^+ + FeBr_4^-$$

The Br^+ is generated in situ. It can then attack the benzene ring and electrophilic substitution occurs:

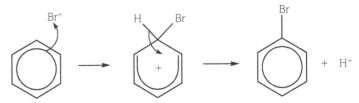

Figure 4 Mechanism for the bromination of benzene.

The halogen carrier is a catalyst and regenerated at the end of the halogenation, as the released H^+ from the benzene ring forms HBr. The following balanced chemical equation illustrates this:

$$FeBr_4^- + H^+ \rightarrow HBr + FeBr_3$$

Questions

1. State the conditions for the nitration of benzene.
2. State the reagents for the chlorination of benzene.
3. Write a mechanism to show the chlorination of benzene.

(4) Halogenation and Friedel–Crafts

By the end of this topic, you should be able to demonstrate and apply your knowledge and understanding of:

* the explanation of the relative resistance to bromination of benzene, compared with alkenes, in terms of the delocalised electron density of the π-system in benzene compared with the localised electron density of the π-bond in alkenes

* the electrophilic substitution of aromatic compounds with a haloalkane or acyl chloride in the presence of a halogen carrier (Friedel–Crafts reaction) and its importance to synthesis by formation of a C–C bond to an aromatic ring

* the mechanism of electrophilic substitution in arenes for nitration and halogenation

Halogenation

> **KEY DEFINITION**
>
> A **Friedel–Crafts reaction** is a substitution reaction where hydrogen is exchanged for an alkyl, or acyl, chain.

Although benzene and aromatic compounds are unsaturated they are not as reactive as alkenes. In Book 1, topic 4.1.8, we saw that ethene, the simplest alkene, has a π-bond between the two carbon atoms. This is a region of high electron density and will readily undergo addition reactions to become saturated.

Shaking ethene with bromine water will cause decolourisation as the coloured bromine is used to form the colourless 1,2-dibromoethane.

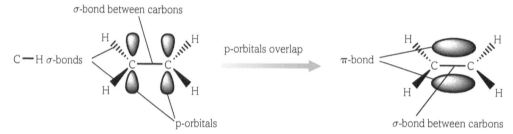

Figure 1 Bonding in ethene.

If cyclohexene is mixed with bromine water, an addition reaction occurs. The first part of the mechanism is the bromine molecule having an induced dipole due to the interaction of the π-bond of the cyclohexene.

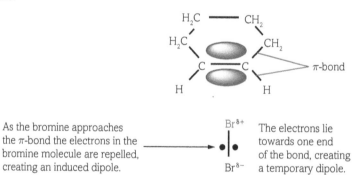

As the bromine approaches the π-bond the electrons in the bromine molecule are repelled, creating an induced dipole.

$Br^{\delta+}$ The electrons lie towards one end of the bond, creating
$Br^{\delta-}$ a temporary dipole.

Figure 2 Electrophilic substitution in benzene.

When benzene is shaken with bromine water, no reaction occurs. This surprising result occurs even though the electrons in benzene are delocalised in π-bonds. So benzene must have a lower electron density between the carbon atoms than an alkene.

When non-polar molecules like bromine approach the benzene ring, there is not enough electron density between the carbon atoms to induce a dipole and start the reaction. This is also the case when attempting to substitute alkyl halides like haloalkanes. By using a halogen carrier, a stronger electrophile can be generated and alkylation can occur.

Friedel–Crafts

In Paris during the nineteenth century, French chemist Charles Friedel worked with James Crafts, an American chemist, to develop a technique for aromatic electrophilic substitutions where hydrogen is substituted for an alkyl chain. This occurs by breaking a C–H bond and forming a C–C bond and is called alkylation. It is very difficult to add alkyl groups to benzene and this was a significant breakthrough in organic synthesis. In all these reactions a strong Lewis acid (which accepts a pair of electrons; see topic 5.1.9) is used as a catalyst.

Haloalkanes

Haloalkanes like chloromethane are mixed with a halogen carrier such as iron (III) chloride. The halogen carrier acts as a catalyst and is regenerated at the end of the reaction. A reactive carbocation is made which undergoes electrophilic substitution with the benzene ring.

The general mechanism in Figure 3 shows a chloroalkane undergoing a **Friedel–Crafts reaction,** where R is any alkyl group. This reaction will occur at room temperature.

$$R—Cl + FeCl_3 \longrightarrow R^+ + FeCl_4^-$$

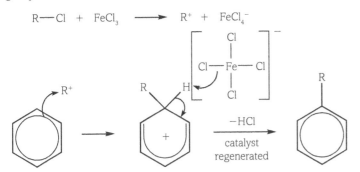

Figure 3 Friedel–Crafts reaction. You may need to apply this in unfamiliar situations.

Multiple substitutions are likely and therefore a mixture of products is made. The products can be separated using fractional distillation or chromatography. The actual yield of a singly-substituted product can be improved by adding excess benzene.

The mixture of products is caused as each successive substitution makes the delocalised π-electrons more nucleophilic and therefore more susceptible to electrophilic attack. This increase in reactivity is due to the alkyl chain donating electrons to the aromatic ring.

Acyl chloride

An acyl chloride has the functional group RCOCl (see Figure 4) and is very reactive. This can be used in a Friedel–Crafts reaction as the halogen carrier to substitute just one hydrogen atom. As the carbonyl group withdraws electrons from the aromatic ring, a less reactive ketone is made. So, only one substitution can occur.

Figure 4 Bonding in an acyl chloride.

The reaction is called acylation. The reaction mixture is held at about 60 °C for 30 minutes under reflux, for the reaction to occur.

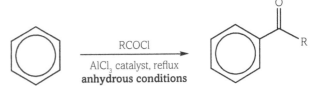

Figure 5 Friedel–Crafts reaction.

Questions

1 State and explain the observations when bromine water is added to benzene.

2 Explain why the Friedel–Crafts reaction was seen as such a breakthrough in organic chemistry.

3 Explain why only one substitution occurs in acylation but multiple substitutions can occur in alkylation.

⑤ Phenols

By the end of this topic, you should be able to demonstrate and apply your knowledge and understanding of:

* the weak acidity of phenols shown by the neutralisation reaction with NaOH but absence of reaction with carbonates

* the relative ease of electrophilic substitution of phenol compared with benzene, in terms of electron pair donation to the π-system from an oxygen p-orbital in phenol

> **KEY DEFINITION**
>
> **Phenols** are a class of aromatic compounds where a hydroxyl group is directly attached to the aromatic ring.

Phenol is the first member of a type of aromatic compounds where a hydroxyl group (–OH) is attached directly to an aromatic ring.

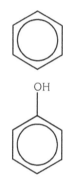

Figure 1 Benzene and phenol.

If the hydroxyl group is attached to an alkyl chain on the aromatic ring then the compound is no longer a phenol derivative. It would be described as an aromatic alcohol.

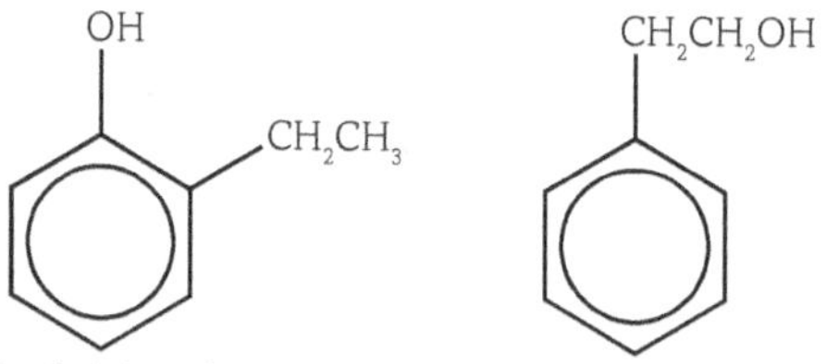

OH
CH₂CH₃

2-ethylphenol
(a phenol)

CH₂CH₂OH

phenylethanol
(not a phenol)

Figure 2 2-ethylphenol and phenylethanol.

Acidity

Phenol is a weak acid, as it partially dissociates in water. The chemical equilibrium can be shown in a balanced chemical equation:

$$C_6H_5OH + H_2O \rightleftharpoons H_3O^+ + C_6H_5O^-$$

As a weak acid, phenol will react with strong bases to form a salt and water. You can find out more about reactions of weak acids in topic 5.1.10. A balanced chemical equation illustrates the reaction between phenol and sodium hydroxide:

$$C_6H_5OH + NaOH \rightarrow C_6H_5O^-Na^+ + H_2O$$

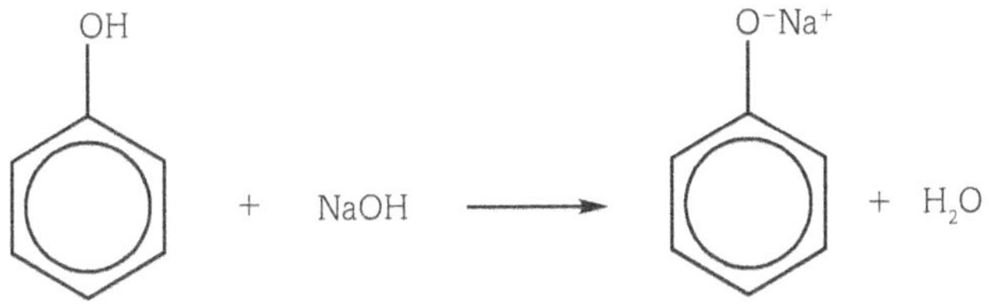

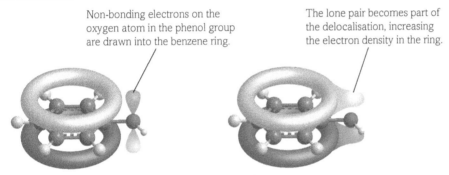

Figure 3 Phenol forms the salt sodium phenoxide when reacted with sodium hydroxide.

Phenol is an acid because it reacts with strong bases such as NaOH. However, it is a weak acid because it does not react with carbonates. Phenol will not react with weak bases such as sodium carbonate.

Reactivity

Phenol is more reactive than benzene. This is due to the p-orbital electrons from the oxygen of the hydroxyl group adding to the delocalised electrons of the aromatic ring (Figure 4). So the π-system of the aromatic ring becomes more nucleophilic. The increase in electron density allows the aromatic ring in phenol to be more susceptible to electrophilic attack as it can induce a dipole in non-polar molecules.

Non-bonding electrons on the oxygen atom in the phenol group are drawn into the benzene ring.

The lone pair becomes part of the delocalisation, increasing the electron density in the ring.

Figure 4 Overlap of a lone pair of electrons from oxygen in phenol.

For example, as the aromatic ring in phenol is more electron-dense, it can induce a dipole in the non-polar bromine molecule. So phenol can undergo direct halogenation, unlike benzene.

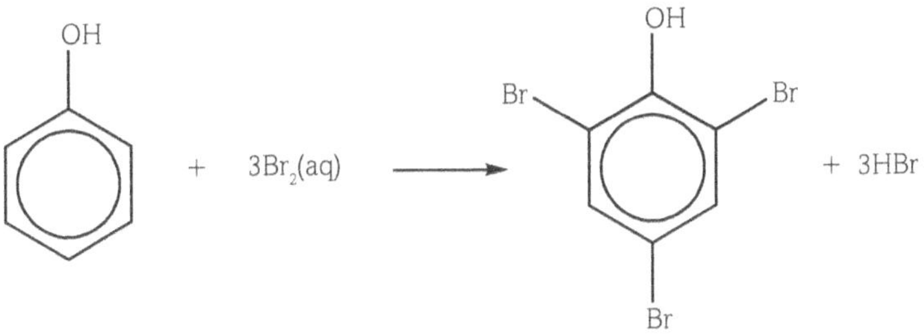

Figure 5 Bromination of phenols.

Questions

1. Explain why phenylpropanol is *not* a phenol.

2. Explain why phenol can undergo direct bromination but benzene cannot.

3. Write a balanced chemical equation for the reaction of phenol with potassium hydroxide.

(6) Electrophilic substitution in aromatic compounds

By the end of this topic, you should be able to demonstrate and apply your knowledge and understanding of:

* the electrophilic substitution reactions of phenol:
  (i) with bromine to form 2,4,6-tribromophenol
  (ii) with dilute nitric acid to form 2-nitrophenol

* the 2- and 4-directing effect of electron-donating groups (OH, NH₂) and the 3-directing effect of electron-withdrawing groups (NO₂) in electrophilic substitution of aromatic compounds

* the prediction of substitution products of aromatic compounds by directing effects and the importance to organic synthesis

---

KEY DEFINITION

The **directing effect** is how a functional group attached directly to an aromatic ring affects which carbon atoms are more likely to undergo substitution.

Phenol will readily undergo electrophilic substitutions with a variety of reagents without the presence of a catalyst. This enhanced reactivity is due to extra electrons from the oxygen on the hydroxyl group being donated to the π-system of the aromatic ring.

Bromination

Phenol will undergo a triple substitution reaction with bromine water at room temperature. The balanced chemical equation for this reaction is:

$$C_6H_5OH + 3Br_2 \rightarrow C_6H_2Br_3OH + 3HBr$$

The resulting product is a white precipitate of 2,4,6-tribromophenol, which smells of antiseptic.

Figure 1 Bromination of phenol. The white precipitate is 2,4,6-tribromophenol.

Nitration

Phenol will undergo a single substitution reaction with dilute nitric acid (HNO_3) at room temperature. This reaction forms a mixture of 2-nitrophenol and 4-nitrophenol. The balanced chemical equation for this reaction is:

$$C_6H_5OH + HNO_3 \rightarrow C_6H_4(NO_2)OH + H_2O$$

Unlike nitration with benzene, this reaction does not require concentrated nitric acid or a sulfuric acid catalyst.

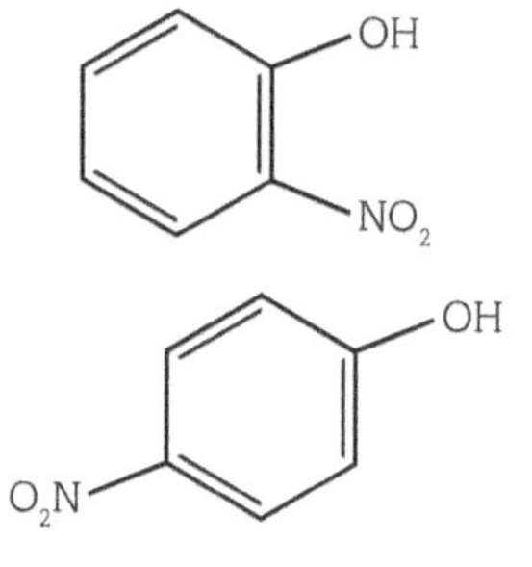

Figure 2 2-nitrophenol and 4-nitrophenol.

However, if concentrated nitric acid *is* used, a triple substitution reaction occurs forming 2,4,6-trinitrophenol.

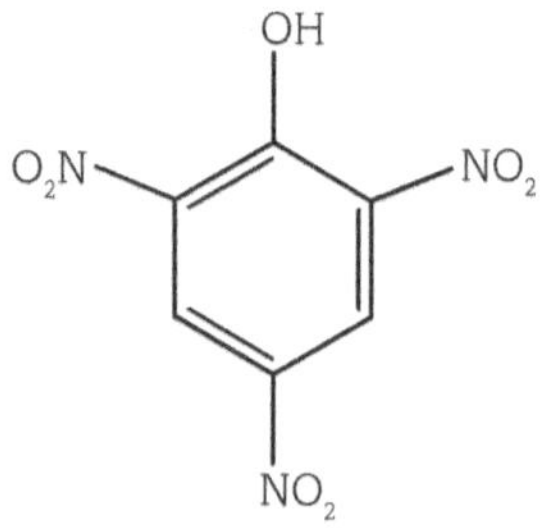

Figure 3 2,4,6-trinitrophenol.

Position of substitution

Substitution reactions can be more favourable for certain carbon atoms when a functional group is directly attached to the aromatic ring.

In phenol, the hydroxyl group pushes additional electrons into the π-system. This makes substitution reactions mainly occur on the 2 and 4 positions of the aromatic ring. The hydroxyl group activates these carbon atoms so that their rate of substitution is faster than the other positions. This is known as the 2- and 4-**directing effect**. This effect is more pronounced in aromatic compounds with an NH_2 group directly attached to the aromatic ring.

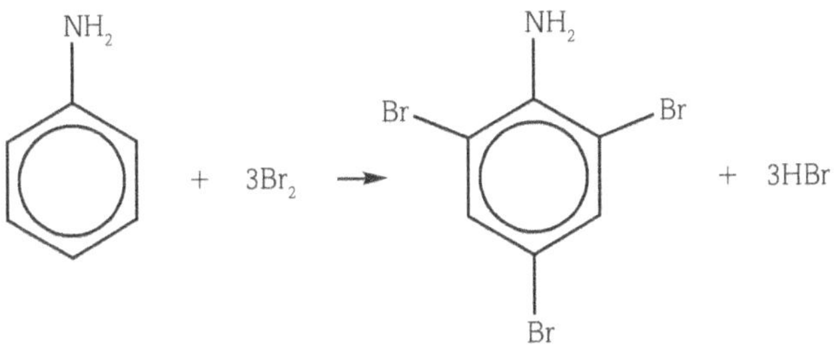

Figure 4 Phenylamine reacting with bromine water demonstrates 2- and 4-directing effect.

When $-NO_2$ groups are directly attached to the aromatic ring, a 3-directing effect is seen. The nitro group withdraws electrons from the π-system and makes the rate of substitution highest on the third carbon atom.

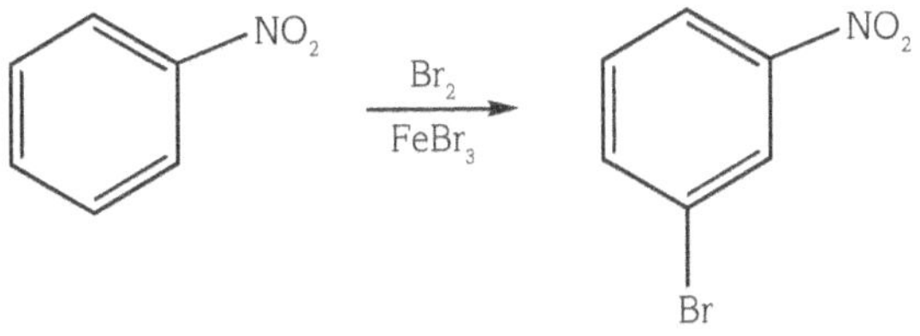

Figure 5 Nitrobenzene reacting with bromine water demonstrates 3-directing effect.

For organic synthesis it is important that a reaction pathway can be designed to maximise the desired product. By considering the electron donating or withdrawing effects of a directly-attached functional group, predictions can be made as to the position(s) in which substitution will take place.

LEARNING TIP

An electron donating group (e.g. OH) on the aromatic ring causes a 2- and 4-directing effect and a 2,4,6 triple-substituted species as a product. An electron withdrawing group (e.g. $-NO_2$) on the aromatic ring causes a 3-directing effect and a single-substituted product.

Questions

1 State the conditions for bromination of phenol.

2 Predict the product produced when nitrobenzene reacts with chlorine water.

3 Predict the product produced when phenol reacts with iodine.

(7) Reactions of carbonyl compounds

By the end of this topic, you should be able to demonstrate and apply your knowledge and understanding of:

* oxidation of aldehydes using $Cr_2O_7^{2-}/H^+$ (i.e. $K_2Cr_2O_7/H_2SO_4$) to form carboxylic acids

* nucleophilic addition reactions of carbonyl compounds with:
  (i) $NaBH_4$ to form alcohols
  (ii) HCN [i.e. NaCN(aq)/H^+(aq)] to form hydroxynitriles

* the mechanism for nucleophilic addition reactions of aldehydes and ketones with $NaBH_4$ and HCN

KEY DEFINITION

A **nucleophile** is a species attracted to an electron-deficient part of a molecule, where it donates a pair of electrons to make a new covalent bond.

A carbonyl functional group is C=O. If the functional group is on the end carbon atom, an *aldehyde* is formed; if the carbonyl is *not* at the end of a carbon chain, then a *ketone* is formed. Each class of compounds has different properties.

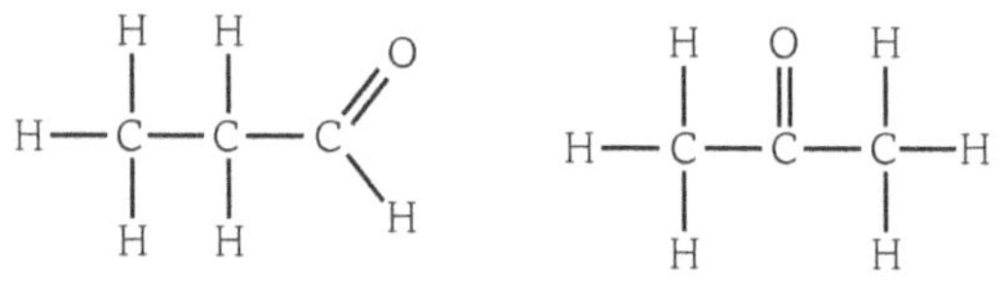

Figure 1 Propanal is an aldehyde and propanone is a ketone.

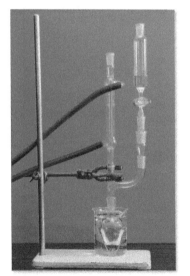

Figure 2 Oxidation of aldehydes is carried out under reflux.

Oxidation

Aldehydes will undergo oxidation to form a carboxylic acid. The reagents are potassium dichromate, $K_2Cr_2O_7$, and sulfuric acid, H_2SO_4. Reagents react in situ to form the oxidising species, $Cr_2O_7^{2-}$, and H^+.

The balanced ionic equation for the oxidation of ethanal to ethanoic acid is:

$$3CH_3CHO(l) + Cr_2O_7^{2-}(aq) + 8H^+(aq) \rightarrow 3CH_3COOH(aq) + 2Cr^{3+}(aq) + 4H_2O(l)$$

Often the oxidising agent is summarised as [O]. So, the simplified balanced symbol equation is:

$$CH_3CHO + [O] \rightarrow CH_3COOH$$

In the laboratory, the reaction mixture is gently heated under reflux. As the reaction proceeds, a colour change is observed, from orange to green, due to the oxidation state of the chromium changing.

Nucleophilic addition reactions

Carbonyl compounds have a dipole in the C=O functional group. This makes them susceptible to nucleophilic attack on the δ^+ carbon atom. A **nucleophile** donates a lone pair of electrons to the electron-deficient carbon. Simultaneously, the π-bond in the C=O breaks and a reactive intermediate is formed. The extra electron pair is quickly donated to the neighbouring hydrogen to form an alcohol group and the stable product.

Nucleophilic addition reactions can be modelled in a general mechanism, where R is an alkyl group or hydrogen, H–A is a small molecule like H_2O or HCN, and Nu^- is a nucleophile.

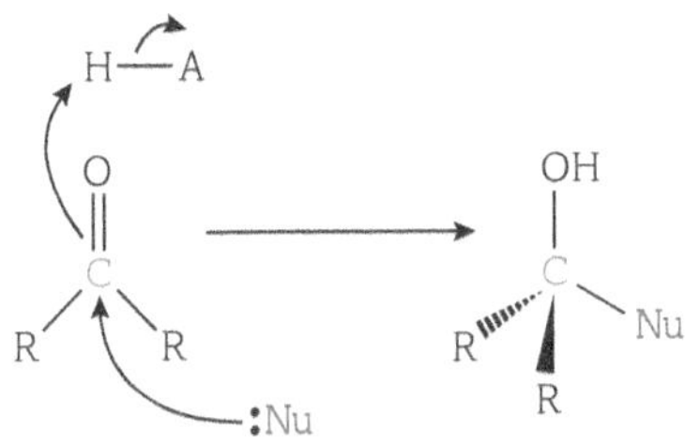

Figure 3 General mechanism for nucleophilic addition on a carbonyl compound.

Sodium tetrahydridoborate(III)

Sodium tetrahydridoborate(III) is more commonly known as sodium borohydride and has the formula $NaBH_4$. It is a reducing agent used commonly in organic synthesis. This compound is made of a BH_4^- ion, which acts as a source of hydride ions, H^-. The hydride ion is the species that is involved in the electrophilic addition and reduction of carbonyl compounds to alcohols.

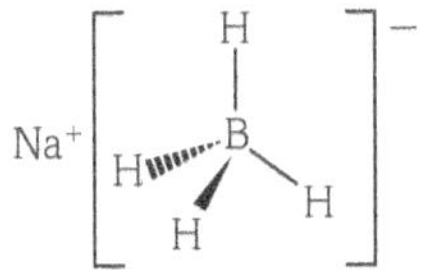

Figure 4 Sodium borohydride.

The reducing agent can be summarised as [H]. The simplified balanced chemical equation for the reduction of pentan-2-one:

$$CH_3COC_3H_7 + 2[H] \rightarrow CH_3CH(OH)C_3H_7$$

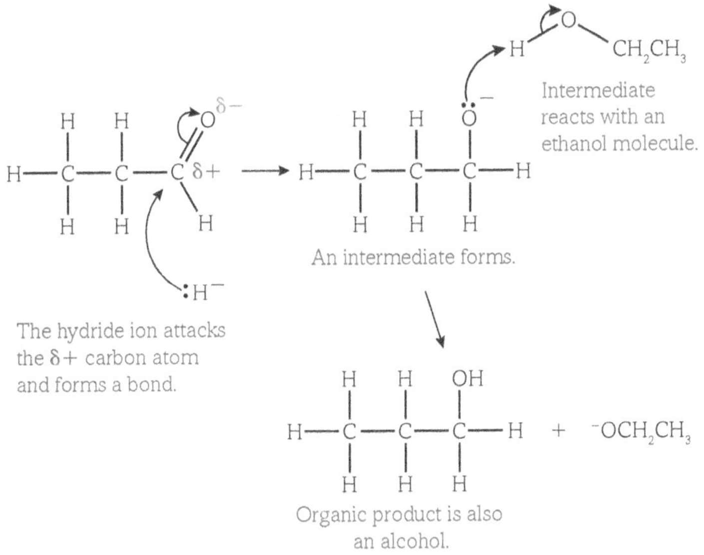

The hydride ion attacks the $\delta+$ carbon atom and forms a bond.

An intermediate forms.

Intermediate reacts with an ethanol molecule.

Organic product is also an alcohol.

Figure 5 Reduction of an aldehyde by nucleophilic addition.

> **LEARNING TIP**
> Aldehydes are reduced to primary alcohols. Ketones are reduced to secondary alcohols.

Hydrogen cyanide

Hydrogen cyanide is a weak acid that will partially ionise in solution. A cyanide nucleophile with a negative charge on the carbon atom is formed. The balanced chemical equation for the reaction is:

$$HCN + H_2O \rightleftharpoons CN^- + H_3O^+$$

Other sources of cyanide include sodium cyanide, NaCN.

The cyanide ion cannot react directly with carbonyl compounds. But, when the reaction is acidified, the carbonyl functional group becomes more reactive as polarity of the $C=O$ bond is increased.

The reaction can be summarised as a mechanism. In the second step, the hydrogen could also be obtained from another small molecule in the reaction mixture such as ethanol.

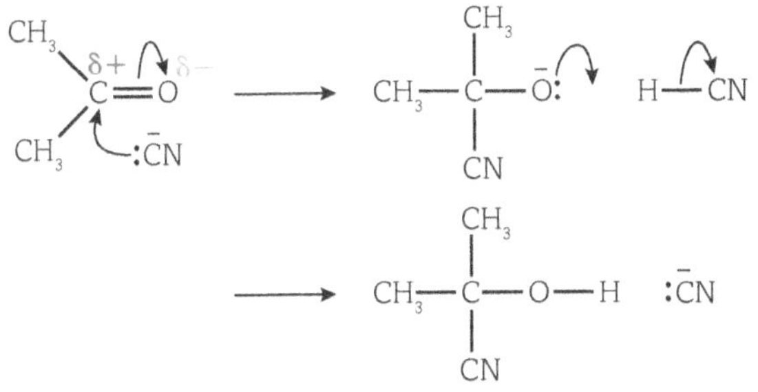

Figure 6 Reaction of hydrogen cyanide with propanone.

Addition of cyanide allows further carbon atoms to be added to the organic molecule. In this reaction a hydroxynitrile is formed, which is an important chemical used in many industrial processes.

> **LEARNING TIP**
> Addition of acid drives the hydrogen cyanide equilibrium to the left and reduces the amount of nucleophiles. There is more about equilibrium systems in topic 5.1.6.
> As a practical compromise, the reaction between propanone and hydrogen cyanide is completed in solutions with an acidity no lower than pH4.

Questions

1. Predict the organic product of the reaction between $NaBH_4$ and propanal.

2. State the reagents needed to oxidise ethanal.

3. Explain the role of an acid in the nucleophilic addition reaction between butanal and hydrogen cyanide.

By the end of this topic, you should be able to demonstrate and apply your knowledge and understanding of:

* use of 2,4-dinitrophenylhydrazine to:
  (i) detect the presence of a carbonyl group in an organic compound
  (ii) identify a carbonyl compound from the melting point of the derivative

* use of Tollens' reagent (ammoniacal silver nitrate) to:
  (i) detect the presence of an aldehyde group
  (ii) distinguish between aldehydes and ketones, explained in terms of the oxidation of aldehydes to carboxylic acids with reduction of silver ions to silver

DID YOU KNOW?

2,4-dinitrophenylhydrazine is sometimes used as a diet pill but is highly toxic. Ingestion can be fatal.

Aldehydes and ketones have the C=O functional group in different places on their carbon chain. This results in them having different chemical properties. Their different reactions with the same reagents can be used to distinguish between these classes of carbonyl compounds.

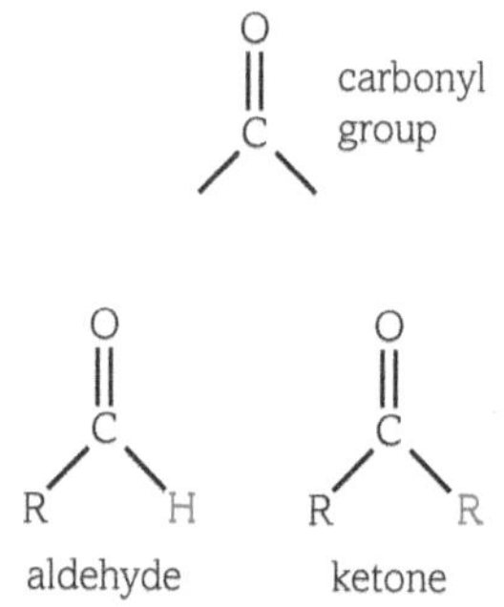

Figure 1 Aldehydes and ketones have different chemical reactions with the same reagent.

2,4-dinitrophenylhydrazine

Testing for a carbonyl functional group

LEARNING TIP

In the examination you will *not* be required to know the equation for this reaction, *nor* the structure of the precipitate.

Brady's reagent is an orange transparent mixture of methanol, sulfuric acid and a solution of 2,4-dinitrophenylhydrazine (2,4-DNP). When this is added to an aldehyde or ketone a yellow/orange precipitate of 2,4-dinitrophenylhydrazone derivative is seen. No precipitation is observed with a carboxylic acid or an ester, despite these compounds also having the C=O bond.

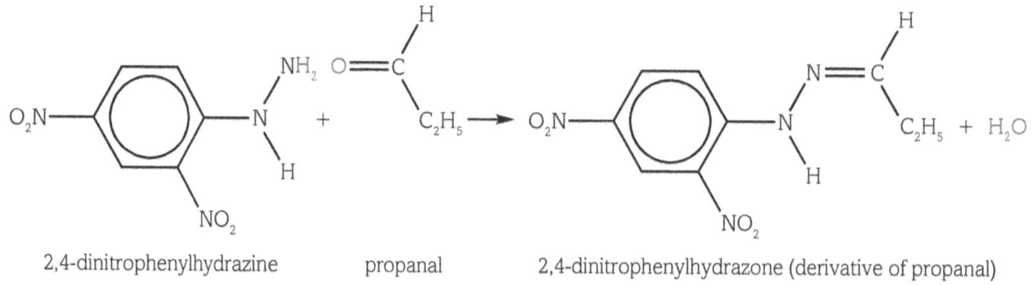

Figure 2 Reaction of propanal with 2,4-dinitrophenylhydrazine.

Figure 3 A 2,4-dinitrophenylhydrazone derivative is made by mixing 2,4-DNP with an aldehyde or a ketone.

DID YOU KNOW?

Qualitative analysis means using experiments to determine the presence or absence of a particular species. All the techniques described in this topic provide qualitative analysis as they allow the identification of a particular functional group.

Identifying a specific aldehyde or ketone

After a positive Brady's reagent test, further processing of the precipitate allows the specific aldehyde or ketone to be identified.

The 2,4-dinitrophenylhydrazone derivative precipitate can be collected by filtration and purified using recrystallisation. After drying, the accurate melting point of the pure product can then be measured through experiment.

The aldehyde or ketone can be identified by comparing the melting point of the 2,4-dinitrophenylhydrazone derivative precipitate with a database. The database is a list of accurately measured melting points of 2,4-dinitrophenylhydrazone derivatives, listed against the aldehyde or ketone that made it.

Although every pure chemical has a specific melting and boiling point that would allow identification, this is experimentally difficult for similar ketones. Ketones with a similar chain length have very similar boiling points, making it challenging to experimentally distinguish between them. However, the 2,4-dinitophenylhydrazone derivatives have very different melting points.

Compound	Boiling point (°C)	Melting point of 2,4-dinitrophenylhydrazone derivative/°C
Heptan-2-one	151	90
Cyclohexanone	156	162
Octan-2-one	173	58

Table 1 Experimental data that can be used to identify ketones.

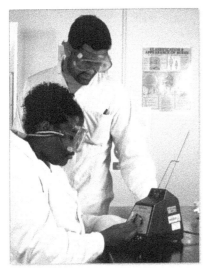

Figure 4 The melting point of the pure 2,4-dinitrophenylhydrazone derivative can be used to identify the aldehyde or ketone present in the sample.

Tollens' reagent

Tollen's reagent, also called ammoniacal silver nitrate, is a colourless chemical that is made in a two-stage process:

1. Sodium hydroxide solution is added to silver nitrate solution until a brown precipitate is formed.

2. Dilute ammonia is added drop-wise until the brown precipitate redissolves.

Tollens' reagent can be used to distinguish between an aldehyde and a ketone. It is a weak oxidising agent and can react with the carbonyl functional group in an aldehyde but not a ketone. When Tollens' reagent is added to a ketone, there is no reaction. This is because ketones cannot be oxidised further.

When Tollens' reagent is added to an aldehyde a silver mirror is observed.

Figure 5 Tollens' reagent does not react with ketones but produces a silver mirror with aldehydes.

When aldehydes react with Tollens' reagent a redox reaction occurs. The silver ions are reduced and the aldehyde functional group is oxidised.

LEARNING TIP

Remember that a redox reaction is a chemical reaction where one species is reduced and other is oxidised.

Refer back to Book 1, topic 2.1.19, and topic 5.3.7 earlier in this book.

The balanced ionic equation for the oxidation of the silver ions is:

$$Ag^+(aq) + e^- \rightarrow Ag(s)$$

This reaction causes silver metal to be precipitated out and this is observed as a silver mirror effect on the inside of the reaction vessel.

Simultaneously, the aldehyde functional group is oxidised to a carboxylic acid. In this equation, the oxidising agent is summarised as [O].

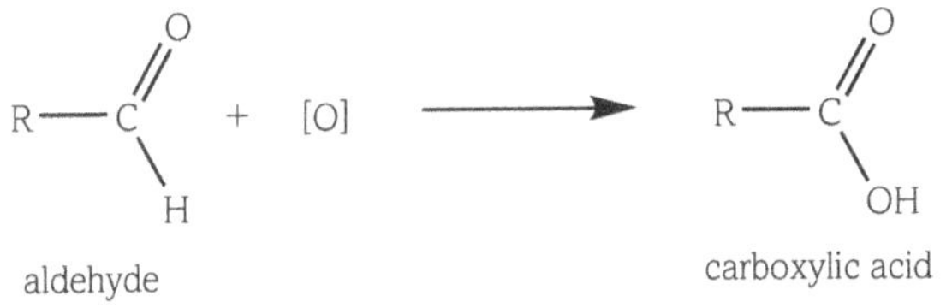

$$R-\overset{\displaystyle O}{\underset{\displaystyle H}{C}} + [O] \longrightarrow R-\overset{\displaystyle O}{C}{-OH}$$

aldehyde   carboxylic acid

Figure 6 Oxidation of an aldehyde using Tollens' reagent.

Questions

1. Two samples of colourless chemicals need to be identified. The chemicals are known to be aldehydes or ketones. List the stages of a qualitative analysis to name these compounds.

2. Explain why a silver mirror is seen when propanal reacts with Tollens' reagent.

3. Describe how propanal, propanone and propanoic acid could be distinguished using a series of chemical tests.

(9) Carboxylic acids

By the end of this topic, you should be able to demonstrate and apply your knowledge and understanding of:

* explanation of the water solubility of carboxylic acids in terms of hydrogen bonding

* reactions in aqueous conditions of carboxylic acids with metals and bases (including carbonates, metal oxides and alkalis)

> **KEY DEFINITION**
> A **base** is a chemical that will react with an acid.

Carboxylic acids contain the functional group –COOH. They are found frequently in nature, from the methanoic acid in ant stings to butanoic acid, which is the smell of rancid butter.

In the IUPAC naming convention, an acid group takes the highest priority. So, the acid functional group generates the suffix of the name. When naming a saturated carboxylic acid, the route of the name is generated by removing the 'e' from the end of the name of the alkane.

Name	Structure
Methanoic acid	HCOOH
Ethanoic acid	CH_3COOH
Propanoic acid	CH_3CH_2COOH
Butanoic acid	$CH_3CH_2CH_2COOH$

Physical properties

Solubility

Figure 1 Ethanoic acid is found in vinegar.

Small carboxylic acids, like those in the table above, are very soluble in polar solvents like water. This is because hydrogen bonds can be formed between the carboxylic acid functional group and water.

hydrogen bonds

Figure 2 Hydrogen bonds between a carboxylic acid functional group in ethanoic acid and water.

As the hydrocarbon chain of a carboxylic acid increases in size, the solubility decreases. This is because only the polar carboxylic acid functional group can form hydrogen bonds with the water. So, as more of the molecule becomes non-polar, then solubility will decrease.

Chemical properties

Carboxylic acids are weak acids as they partially ionise in solution releasing the H^+ ion from the carboxylic acid group, forming a carboxylate ion. So, this class of compounds will undergo typical acid reactions.

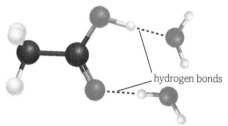

Figure 3 Ethanoic acid is a weak acid.

Carboxylic acid reactions will happen at a slower rate than with a strong acid as the pH is higher and therefore the concentration of $H^+(aq)$ will be lower.

DID YOU KNOW?

When a carboxylic acid reacts with a metal or a metal carbonate, effervescence is observed. This is because a gas is being produced in each case.

Reaction with metals

Carboxylic acids will react with metals above hydrogen in the reactivity series to make hydrogen and a metal salt. The name of the salt is generated from the acid. The suffix of the acid changes from -oic acid to -oate. For example:

sodium + ethanoic acid → sodium ethanoate + hydrogen

$$2Na + 2CH_3COOH \rightarrow 2CH_3COONa + H_2$$

Reaction with metal oxides

Metal oxides react with acids and therefore can be classified as **bases**. Carboxylic acids will react with metal oxides to make water and a metal salt. For example:

magnesium oxide + methanoic acid → magnesium methanoate + water

$$MgO(s) + HCOOH(aq) \rightarrow (HCOO)_2Mg(aq) + H_2O(l)$$

Reaction with metal hydroxides

Group 1 metal hydroxides are soluble bases that release $OH^-(aq)$. Carboxylic acids will react with metal hydroxides to make water and a metal salt. For example:

potassium hydroxide + propanoic acid → potassium propanoate + water

$$KOH(aq) + CH_3CH_2COOH(l) \rightarrow CH_3CH_2COOK(aq) + H_2O(l)$$

Reaction with metal carbonates

Metal carbonates are also bases. Carboxylic acids will react with metal carbonates to make water, carbon dioxide and a metal salt. For example:

sodium carbonate + methanoic acid → sodium methanoate + water + carbon dioxide

$$Na_2CO_3(s) + 2HCOOH(aq) \rightarrow 2HCOONa(aq) + H_2O(l) + CO_2(g)$$

Group 1 metals can also form metal hydrogencarbonates. This is where the carbonic acid (H_2CO_3) has had only one proton exchanged for a metal ion to form the metal hydrogencarbonate ($MHCO_3$ where M is a Group 1 metal). Acids will also react with this class of compound to form a salt, water and carbon dioxide.

Questions

1 Name the following carboxylic acids.

(a)

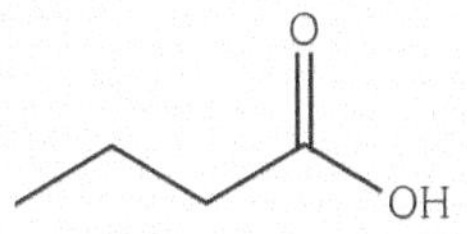

(b)

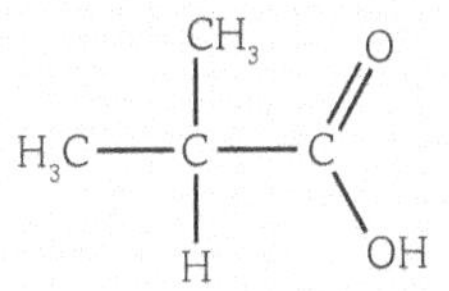

(c)

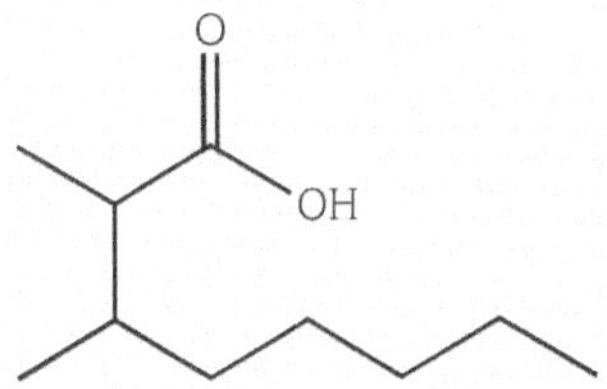

2 Write a balanced symbol equation for the reaction between sodium hydrogencarbonate and ethanoic acid.

By the end of this topic, you should be able to demonstrate and apply your knowledge and understanding of:

* esterification of:
  (i) carboxylic acids with alcohols in the presence of an acid catalyst (e.g. concentrated H_2SO_4)
  (ii) acid anhydrides with alcohols

* hydrolysis of esters:
  (i) in hot aqueous acid to form carboxylic acids and alcohols
  (ii) in hot aqueous alkali to form carboxylate salts and alcohols

Esters contain the functional group R–COO–R' where R and R' are alkyl groups that may be the same or different. This class of compounds are carboxylic acid derivatives, where the H of the acid group has been replaced with an alkyl chain. Esters are found frequently in nature and have a pungent smell. Synthetic esters have many uses including solvents, artificial fruit flavourings and perfumes.

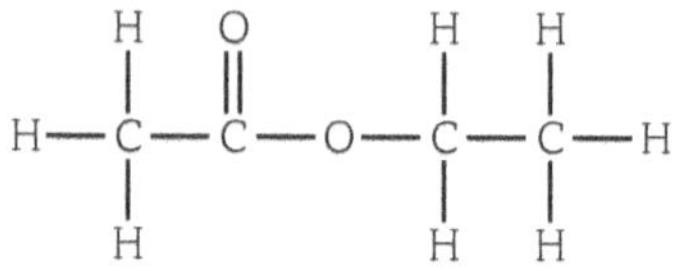

Figure 1 Ethyl ethanoate is a common ester.

Naming esters

To name an ester, look at the structure of the ester – particularly the ester functional group. The first part of the name is the alkyl chain with a carbon atom bonded to only one oxygen atom. The second part of the name is the alkyl chain containing the C=O carbon atom, followed by the suffix -oate.

Structure	Name of ester
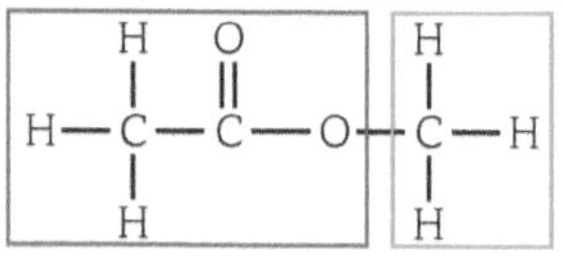 From the acid: ethanoate From the alcohol: methyl	methyl ethanoate
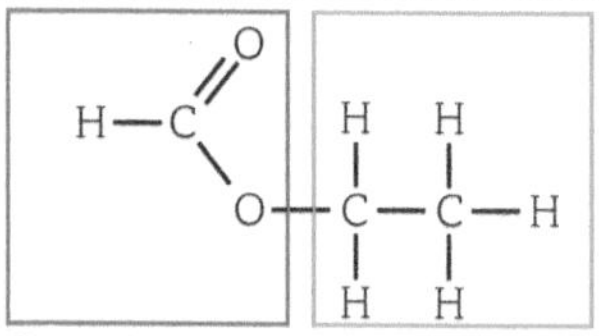 From the acid: methanoate From the alcohol: ethyl	ethyl methanoate
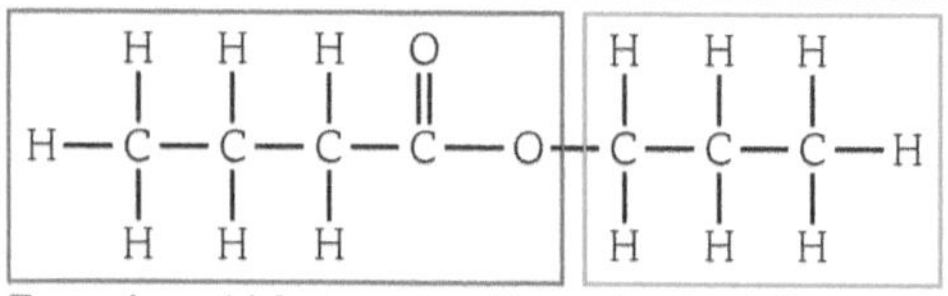 From the acid: butanoate From the alcohol: propyl	propyl butanoate

Esterification

The chemical reaction used to make an ester is known as esterification. There are two main methods for making esters, which are described below.

Carboxylic acids with an alcohol

To prepare a small ester, an alcohol and a carboxylic acid are heated gently in the presence of a sulfuric acid catalyst. Esterification is a reversible reaction and has a slow rate of reaction. Figure 2 shows the structural formula chemical equation for an esterification reaction to make ethyl ethanoate.

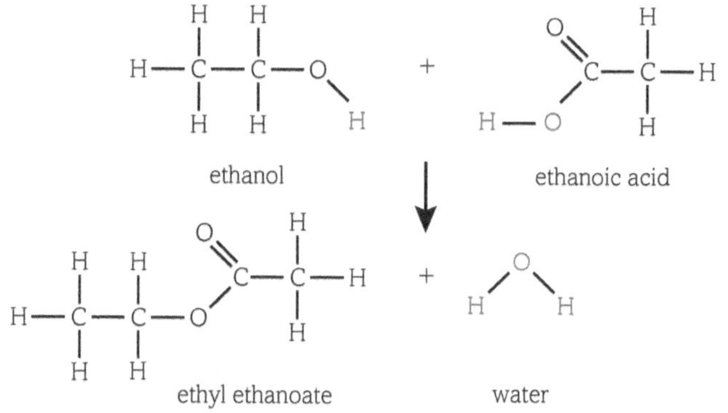

Figure 2 Reaction between ethanoic acid and ethanol to make ethyl ethanoate and water.

As the ester is volatile, with the lowest boiling point of the chemicals, it can be separated from the reaction mixture using distillation. The separation has to happen quickly to prevent the reverse reaction occurring.

To prepare larger esters, the reaction mixture will need to be heated under reflux until equilibrium has been established. The ester can be separated using fractional distillation.

This method of preparation of an ester is not suitable for phenol or its derivatives as the rate of reaction is so slow.

LEARNING TIP

When naming an ester, the alkyl part of the name comes from the alcohol. The end of the name comes from the carboxylic acid. For example when methanol reacts with propanoic acid, the resulting ester is called methyl propanoate.

Acid anhydride with alcohol

An acid anhydride is an acid derivative that is more reactive than a similar carboxylic acid. It is made by the removal of a molecule of water from two carboxylic acid molecules (Figure 3).

Acid anhydrides will react with alcohols, including phenol and its derivatives, to make an ester. This method of ester production is not reversible and therefore has a higher yield than using a carboxylic acid. The rate of reaction is still slow but can be increased by gently warming the reaction mixture. Figure 4 shows the formation of methyl ethanoate using an acid anhydride.

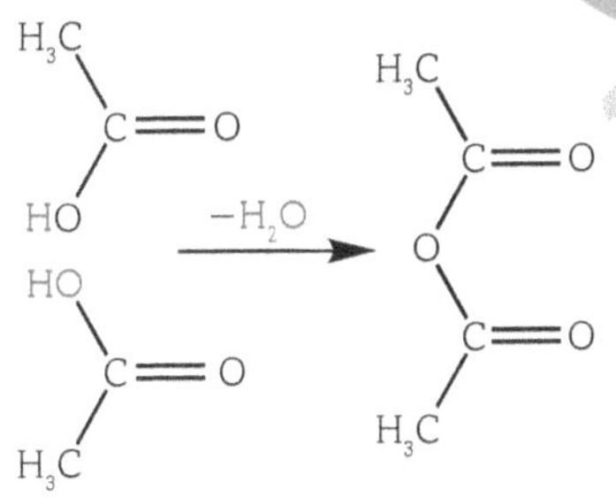

Figure 3 An acid anhydride is formed by removal of a molecule of water from two carboxylic acid molecules. The diagram shows how ethanoic anhydride, $(CH_3CO)_2O$, is related to ethanoic acid, CH_3COOH ('anhydride' means 'without water').

ethanoic anhydride   methanol   methyl ethanoate + ethanoic acid

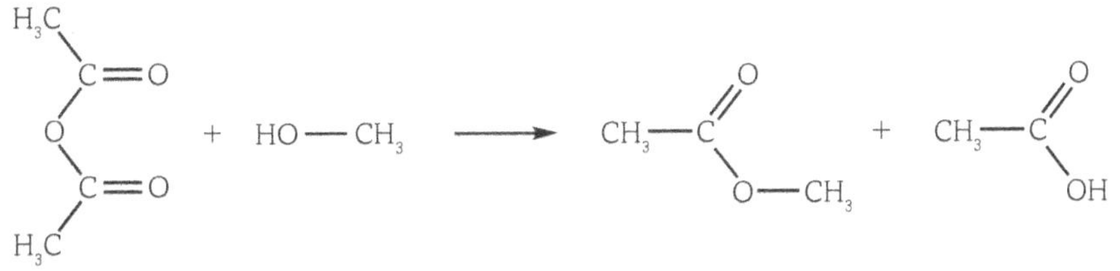

Figure 4 Ethanoic anhydride reacts with methanol to make the ester methyl ethanoate.

Hydrolysis

Hydrolysis is a chemical reaction where water breaks down another product. Esters can undergo hydrolysis to form an alcohol; it is the reverse reaction of esterification. By varying the conditions, either carboxylic acids or carboxylate salts are formed as the other product.

> **KEY DEFINITION**
>
> **Hydrolysis** is a chemical reaction where water causes the breaking of bonds, in a decomposition reaction.

Acidic conditions

When esters are refluxed with a catalyst of hot aqueous acids, such as dilute sulfuric acid or dilute hydrochloric acid, the ester will decompose reversibly into an alcohol and a carboxylic acid. The following equation illustrates the hydrolysis of propyl ethanoate.

propyl ethanoate   ethanoic acid   propan-1-ol

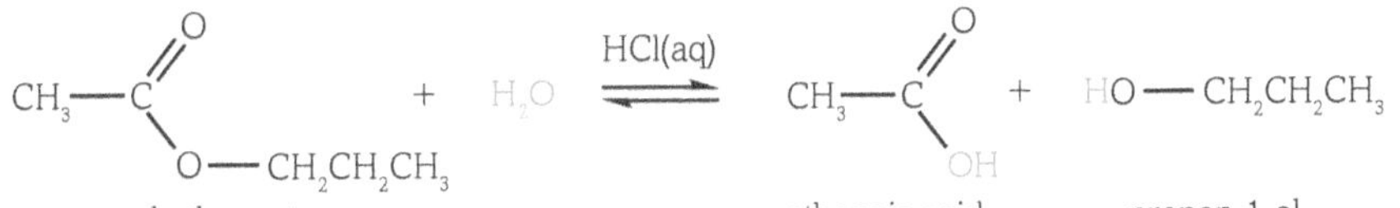

Figure 5 Acid hydrolysis of propyl ethanoate.

Alkaline conditions

Alkaline chemicals are bases (react with acids) that can dissolve in water. When an ester is refluxed with a hot aqueous alkali, such as potassium hydroxide or sodium hydroxide, it will decompose into an alcohol and a carboxylate salt. This reaction is not reversible.

Alkaline hydrolysis of esters is used to make soaps, therefore it is also called saponification (from the Latin for soap – *sapo*). The following equation illustrates the saponification of ethyl propanoate:

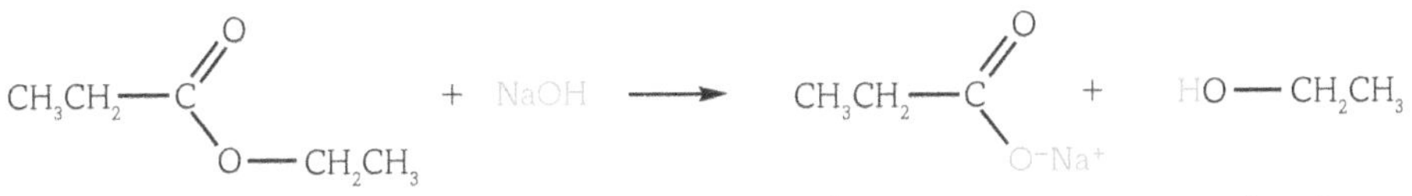

ethyl propanoate   sodium propanoate   ethanol

Figure 6 Alkaline hydrolysis of ethyl propanoate.

Questions

1 Explain why distillation must occur immediately when ethyl ethanoate is made from ethanol and ethanoic acid.

2 Describe the advantage of using an acid anhydride to make an ester rather than using a carboxylic acid.

3 Describe saponification.

Acyl chlorides

By the end of this topic, you should be able to demonstrate and apply your knowledge and understanding of:

* the formation of acyl chlorides from carboxylic acids using $SOCl_2$

* use of acyl chlorides in synthesis in formation of esters, carboxylic acids and primary and secondary amides

Acyl chlorides contain the functional group R–COCl where R is an alkyl group. This class of compounds are carboxylic acid derivatives where the –OH of the acid group has been replaced with a chlorine atom. The name of this class of compounds comes from the hydrocarbon chain, RCO, which was once known as an acyl group.

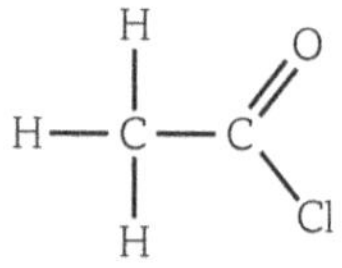

Figure 1 Ethanoyl chloride is an example of an acyl chloride.

Small acyl chlorides are fuming (a visual gas or vapour being released) colourless liquids. They are very reactive, with the chlorine atom being substituted for other groups.

Naming acyl chlorides

Look at the structure of the compound and count the number of carbon atoms to generate the root. The suffix is -oyl chloride.

Structure	Name of acyl chloride
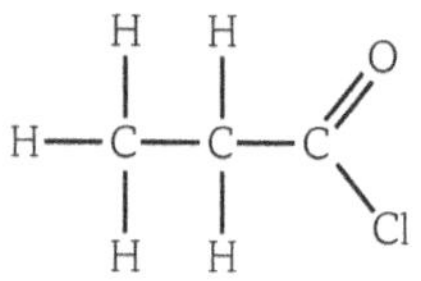	propanoyl chloride
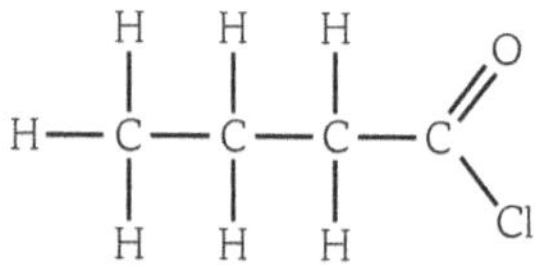	butanoyl chloride
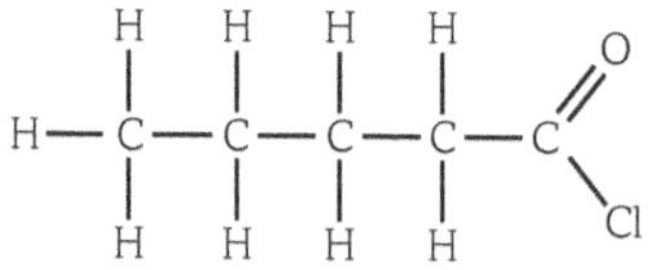	pentanoyl chloride

Preparation

To make an acyl chloride, the –OH group on a carboxylic acid must be substituted for a chlorine atom. One method is to use $SOCl_2$, which is a liquid at room temperature and readily reacts with a carboxylic acid to make the desired product. Sulfur dioxide and hydrogen chloride gases are also made. The acyl chloride is separated from the reaction mixture using distillation. The balanced chemical reaction shows the formation of ethanoyl chloride.

$$CH_3COOH + SOCl_2 \rightarrow CH_3COCl + SO_2 + HCl$$

Uses of acyl chlorides as reagents in organic synthesis

Esters

Acyl chlorides will react with alcohols to make an ester. This method of ester production is not reversible and therefore has a higher yield than using a carboxylic acid. The balanced chemical reaction shows the formation of an ester – ethyl ethanoate from an acyl chloride.

$$CH_3COCl + CH_3CH_2OH \rightarrow CH_3COOCH_2CH_3 + HCl$$

Alcohols can also react with carboxylic acids to make esters. Like alcohols, phenols have an –OH group but they do not react easily with carboxylic acids. Therefore to make an ester from phenols the acyl chloride method must be used. However, the reaction is also violent and produces corrosive fumes of HCl.

Carboxylic acids

When a small acyl chloride such as ethanoyl chloride is added to water, it quickly hydrolyses to produce a carboxylic acid. This is a very exothermic reaction and misty fumes of HCl are given off.

$$CH_3COCl + H_2O \rightarrow CH_3COOH + HCl$$

Primary amides

When an acyl chloride reacts with ammonia, a primary amide is produced. To prepare ethanamide, ethanoyl chloride is added to a concentrated ammonia solution. This quickly produces a mixture of solid ammonium chloride and ethanamide – observed as white smoke. Some of the products remain in a colourless solution.

$$CH_3COCl + 2NH_3 \rightarrow CH_3CONH_2 + NH_4Cl$$

Secondary amides

When an acyl chloride reacts with a primary amide, the product is a secondary amide, where the nitrogen has one hydrogen atom directly bonded to it. The nitrogen atom also has two organic groups attached and is often called an *N*-substituted amide.

A white solid compound of *N*-ethylethanamide can be made from ethanoyl chloride and a cold concentrated solution of ethylamine.

$$CH_3COCl + CH_3CH_2NH_2 \rightarrow CH_3CONHCH_2CH_3 + HCl$$

Questions

1. State the reagent(s) needed to make ethanoyl chloride.

2. Explain why acyl chlorides are favoured over carboxylic acids for esterification reactions.

THE IMPACT OF PESTICIDES

It would seem pretty obvious that pesticides are designed to kill pests, but many of them can also pose risks to people. In most cases the amount of pesticide that people are likely to be exposed to is extremely small and unlikely to pose a risk, but in order to determine risk both the toxicity and the likelihood of exposure must be considered. A low level of exposure to a very toxic pesticide may be no more dangerous than a high level of exposure to a relatively low-toxicity pesticide, for example, but who decides what is safe?

PESTICIDES LINKED TO VITAMIN D DEFICIENCY

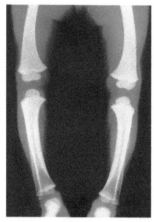

Figure 1 Vitamin D deficiency can cause chronic conditions like rickets.

Pesticides could be suppressing people's vitamin D levels, leading to deficiency and disease, say scientists. The warning follows the discovery that adults with high serum concentrations of organochlorine pesticides such as DDT have lower vitamin D levels.

Exposure to low doses of organochlorine pesticides has been previously linked to common diseases like type 2 diabetes, metabolic syndrome and cardiovascular disease. Vitamin D deficiency has similarly been associated with a rise in chronic diseases, but the two have been studied separately by researchers in different fields. 'The known associations between vitamin D deficiency and various diseases can be at least partly explained by the common exposure to organochlorine pesticides,' says senior author Duk-Hee Lee of Kyungpook National University in Korea.

The US-Korean research team studied 1275 adults in the US aged 20 years or older and checked their blood for seven organochlorine pesticides. DDT and beta-hexachlorocyclohexane levels in study volunteers showed significant associations with lower serum concentrations of a vitamin D pre-hormone, 25-hydroxyvitamin D, which is the standard way to assess vitamin D levels in the body. The study sheds no light on how pesticides might influence vitamin D levels, though.

Organochlorine pesticides were banned in the US decades ago, but are still detectable in people because they resist biodegradation in the environment, are lipophilic and accumulate in fat tissues. The World Health Organization still recommends the use of DDT to control mosquitoes in malarial regions and, while there is a global moratorium on spraying it on crops, illegal use in some countries is suspected.

Levels of these chemicals are far lower than they were in the 1960s and 1970s, but Lee believes that they may still be significant because they act as endocrine disruptors. 'One characteristic of endocrine disruptors is that they show their possible harmful effects at levels lower than those which we currently think are safe,' Lee says. 'As chemicals like organochlorine pesticides travel a long distance through a variety of ways, humans can be exposed to these kinds of chemicals even though the country where they live does not use them anymore.'

'We have known for many years that DDT causes egg shell thinning,' says David Carpenter, Director of the Institute for Health and Environment at the University of Albany, New York. 'Since egg shell thickness is regulated by vitamin D, this study shows that the same suppression of vitamin D occurs in humans.' Carpenter says he is concerned about the push to bring DDT back into use as a potent pesticide against mosquitoes and other insects. 'It is very important to communicate how harmful DDT is to humans, not just mosquitoes.'

J-H Yang *et al.*, *PLoS One*, 2012, **7**, e30093

Source
- http://www.rsc.org/chemistryworld/News/2012/January/pesticides-ddt-vitamin-d-deficiency.asp

Where else will I encounter these themes?

Book 1   5.1   5.2

Let us start by considering the nature of the writing in the article.

1. This article was originally taken from a peer-reviewed journal. Discuss the importance of this and explain why references from several different scientists are included.

Now we will look at the chemistry in, or connected to, this article. Do not worry if you are not ready to give answers to these questions yet. You may like to return to the questions once you have covered other topics later in the book. Use the timeline at the bottom of the page to help you put this work in context with what you have already learned and what is ahead in your course.

2. Do you think that the report establishes a clear causal link between raised serum levels of organochlorine molecules and lowered vitamin D serum levels? What kind of evidence might help establish a clearer causal link?

3. It has been estimated (by the National Research Data Corporation) that the half-life of DDT in human body tissue is approximately 4 years.

   a. What is meant by the term *half-life* in this context?

   b. What are the problems in determining a value for the half-life of DDT in humans?

4. A second organochlorine pesticide that has been used is pentachlorophenol. Draw the structure of this molecule.

5. Vitamin D refers to a number of related molecules, one of which, Vitamin D_3, is shown below.

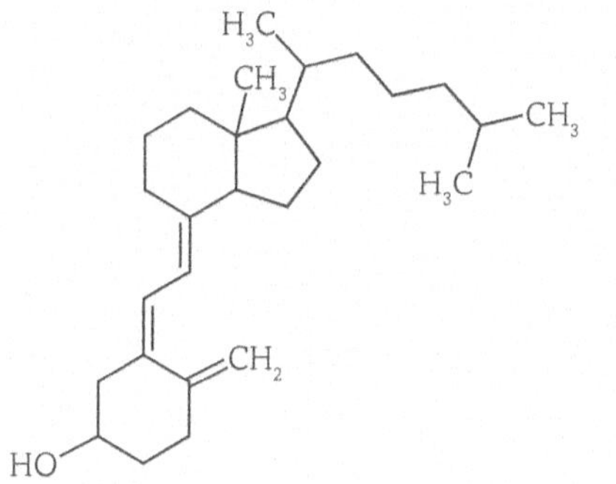

Figure 2 Vitamin D_3.

Identify all chiral carbons in this molecule.

> Remember that the skeletal formula does not show you the true 3D structure of the molecule.

> Remember, a carbon cannot be chiral if it forms a C=C bond.

Activity

A brief search of the internet for 'pesticides' will retrieve a number of concerns about their influence on conditions, including: autism, diabetes, ADHD, asthma, cancers, obesity and general developmental disorders. How are such conclusions arrived at and how valid are they? In this activity you have the chance to review online materials and present an argument on the basis of your research. You may choose a specific example or consider a more general review. Some possible suggestions are given below:

– Is there a relationship between the seasonal exposure to atrazine and birth defects?

– What is the evidence for making a causal link between autism and exposure to pesticides?

– What is meant by the term 'endocrine disruptors' and are pesticides implicated?

There are a wide range of resources including those from governmental organisations, specific pressure groups and agricultural and business corporations, so you should reflect on who is presenting the evidence and why. Aim to make your presentation between 8 and 10 slides long.

DID YOU KNOW?

Although most organochlorine species are anthropogenic in origin, the majority of chloromethane in the environment ($\approx$ 0.6 ppm) comes from natural sources including algae in the marine environment and a number of species of fungi on land.

6.1   YOU ARE HERE   6.2   6.3

Practice questions

1. Choose the correct name for the molecule shown below. [1]

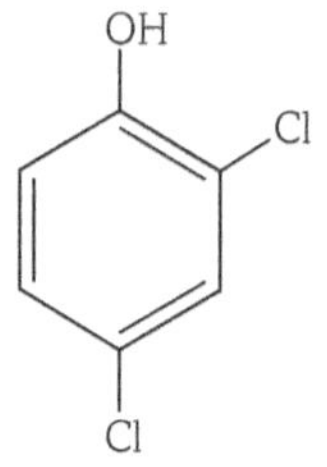

A. dichlorobenzol

B. 1,3-dichlorophenol

C. 2,4-dichlorophenol

D. 1-hydroxy-2,4-dichlorobenzene

2. What is the correct name for the organic product formed when the molecule shown below is reacted with excess acidified potassium dichromate under reflux? [1]

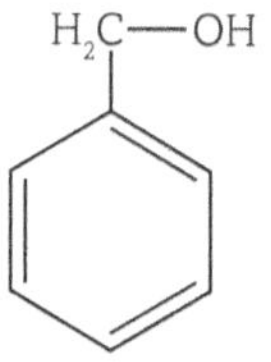

A. benzoic acid

B. benzaldehyde

C. phenol

D. methylbenzene

3. Phenol reacts with sodium hydroxide but does not react with sodium carbonate. What is the best explanation for this? [1]

A. Phenol is a weak base in aqueous solution.

B. Hydroxide ions are strong nucleophiles.

C. Sodium carbonate is thermally stable.

D. Phenol is a weak acid in aqueous solution.

4. When phenol reacts with excess nitric acid a mixture of organic products is formed. What is this mixture most likely to contain? [1]

A. Optical isomers.

B. Positional isomers only.

C. E/Z isomers only.

D. Positional and E/Z isomers.

[Total: 4]

5. A chemist was investigating the reactions of benzene, phenol and cyclohexene with bromine.

She found that they all reacted with bromine but under different conditions.

(a) The chemist found that when benzene reacts with bromine, a halogen carrier is required as a catalyst.

Write an equation for this reaction. You do **not** need to show the halogen carrier in your equation. [1]

(b) The chemist also found that when phenol or cyclohexene reacts with bromine, a halogen carrier is **not** required.

(i) The chemist observed that bromine decolourises when it reacts with phenol.

What other observation would she have made?

Draw the structure of the organic product formed. [2]

(ii) Cyclohexene also decolourises bromine.

Name the organic product formed. [1]

(iii) Explain the relative resistance to bromination of benzene compared to phenol and compared to cyclohexene.

In your answer, you should use appropriate technical terms, spelt correctly. [5]

[Total: 9]

[From Q1, F324 Jan 2010]

6. (a) Give details of the reagents and conditions that would be needed for each of the following conversions.

  (i) benzene to nitrobenzene [3]

  (ii) benzene to bromobenzene [3]

  (iii) benzene to ethylbenzene [3]

  (b) What precautions would you take to ensure that no dinitrobenzene is formed in reaction (a) (i)? [1]

  [Total: 10]

7. The standard enthalpy of reaction for the complete hydrogenation of benzene to cyclohexane is given as $-208\,kJ\,mol^{-1}$.

  (a) Write a balanced equation for this reaction. [2]

  (b) Explain in detail why the value for this reaction is less negative than might be expected. [5]

  (c) Give a chemical test that would allow you to distinguish between benzene and cyclohexene. How does the result of this test help support your answer in part (b)? [3]

  [Total: 10]

8. Look at Figure 1.

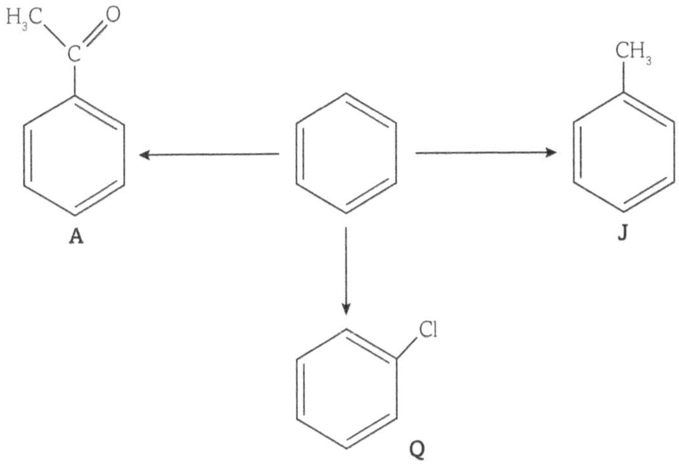

Figure 1

(a) Give the names of compounds **A, Q** and **J** shown above and in each case give details of reagents and conditions needed for each conversion. [9]

(b) Give details of a chemical test that you could use to identify compound A and the expected result. [3]

[Total: 12]

Organic chemistry and analysis

NITROGEN COMPOUNDS, POLYMERS AND SYNTHESIS

Introduction

All life on Earth contains organic chemicals with some nitrogen present. One of the ways in which scientists look for evidence of extra-terrestrial life is to use data from space probes to determine if nitrogen-containing compounds are present.

Amino acids are found in all organisms. They are the building blocks that form proteins. This essential group of chemicals is found in muscles, biological catalysts (enzymes) and even hormones. The amino acids are monomers and the proteins they make up are the polymers.

Chemists are able to make synthetic polymers that have a vast array of uses, from clothing to pharmaceuticals. Sometimes many stages are needed to take a chemical 'feedstock' and make the target molecule. This may involve changing functional groups, increasing the carbon chain or polymerisation. The reactions and processes used to make a desired chemical are known as synthetic routes.

All the maths you need

To unlock the puzzles of this chapter you need the following maths:

* Use 3D representation to show enantiomers of chiral compounds

What have I studied before?

- Applying IUPAC naming conventions
- How to identify functional groups on an organic molecule
- How to draw the monomer, repeating unit and polymer unit for addition polymers
- How to generate single-step synthesis routes

What will I study later?

- How analytical techniques can be used to differentiate between organic molecules
- How simple laboratory tests can be used to identify functional groups

What will I study in this chapter?

- To explain the basicity of related chemicals
- To describe how amines are made
- To describe how the carbon-chain length of an organic molecule can be increased
- To be able to identify chiral centres and draw enantiomers
- To be able to draw the monomer, repeat unit and polymer unit for condensation polymers
- To be able to evaluate the environmental impact of using addition and condensation polymers
- To be able to suggest multi-stage synthetic routes

1 Basicity and the preparation of amines

By the end of this topic, you should be able to demonstrate and apply your knowledge and understanding of:

* the basicity of amines in terms of proton acceptance by the nitrogen lone pair and the reactions of amines with dilute acids, e.g. HCl(aq), to form salts

* the preparation of:
  (i) aliphatic amines by substitution of haloalkanes with excess ethanolic ammonia and amines
  (ii) aromatic amines by reduction of nitroarenes using tin and concentrated hydrochloric acid

Amines are a class of compounds related to ammonia, NH_3. They are organic chemicals where one or more of the hydrogen atoms on ammonia have been replaced by alkyl chains.

There are three types of amine:

* Primary, 1° – one hydrogen atom has been substituted. The structural formula can be summarised as RNH_2 where R is the alkyl chain.
* Secondary, 2° – two hydrogen atoms have been substituted. The structural formula can be summarised as RNHR′ where R and R′ are alkyl chains, which may be the same or different.
* Tertiary, 3° – all three hydrogen atoms have been substituted. The structural formula can be summarised as RNR′R″ where R, R′ and R″ can be the same or different hydrocarbon groups.

3 H atoms:
ammonia, NH_3

1 H atom replaced:
primary amine, RNH_2

2 H atoms replaced:
secondary amine, RNHR′

3 H atoms replaced:
tertiary amine, RNR′R″

Figure 1 The relationship of ammonia with primary, secondary and tertiary amines.

Naming amines

When naming amines, the suffix is always amine. Then the alkyl chains must be determined and used to generate the root and prefix of the name.

* For primary amines – determine the root by the longest hydrocarbon chain. Add any prefixes for other groups. Finally add the suffix amine.

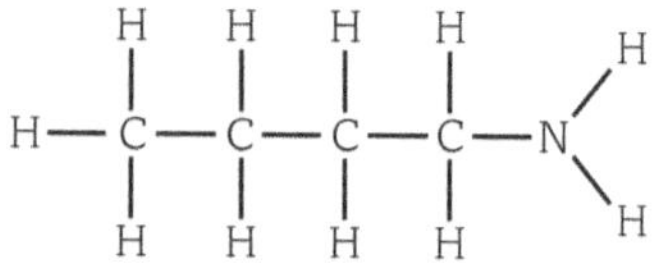

Figure 2 Butylamine is a primary amine.

* For secondary amines – determine the root by naming the two alkyl chains. Add any prefixes for additional groups and write them in alphabetical order. Finally, add the suffix 'amine'. As the alkyl groups are attached to the nitrogen atom, secondary amines are often called 'N-substituted' and this is given as a prefix to the name.

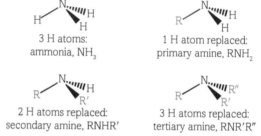

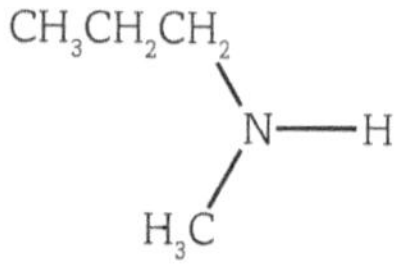

Figure 3 N-methyl propylamine is a secondary amine.

For tertiary amines – determine the root by naming the three alkyl chains. If there is more than one of the same group add the appropriate prefix; di- or tri-.

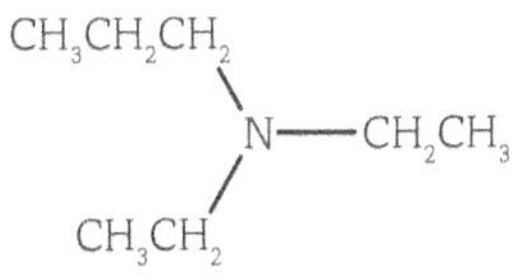

$$CH_3CH_2CH_2 \diagdown N - CH_2CH_3$$
$$CH_3CH_2 \diagup$$

Figure 4 *N,N*-diethylpropylamine is a tertiary amine.

Amines as bases

The definition of a base is different depending upon which model you use.

- A Lewis base has a lone pair of electrons for donation – ammonia and amines have a lone pair of electrons on the nitrogen atom, making them Lewis bases.

- A Brønsted–Lowry base is a proton acceptor – ammonia and amines can accept protons on the nitrogen atom and are therefore also Brønsted–Lowry bases.

Amines and ammonia are both weak bases. Using the Brønsted–Lowry model, when an amine reacts with an acid, it accepts a proton. A covalent bond is formed by the nitrogen atom donating its lone pair of electrons to the proton to form a dative covalent (or coordinate) bond.

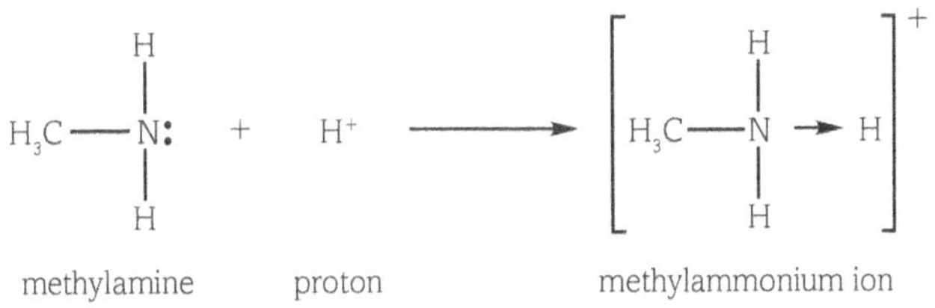

        methylamine   proton   methylammonium ion

Figure 5 Reaction of methylamine as a base.

Reaction of amines with dilute inorganic acids

HCl is a strong inorganic acid and will fully ionise in solution. When HCl(aq) reacts with a base, a chloride salt and water are produced. When primary amines react with this acid, an **alkylammonium salt** is made. This is formed by the proton in the acid being replaced by an alkylammonium ion.

$$CH_3CH_2NH_3 + HCl \rightarrow CH_3CH_2NH_3 + Cl^-$$

KEY DEFINITION

An **alkylammonium salt** is a compound where the hydrogen(s) on an ammonium ion have been substituted by alkyl chains.

DID YOU KNOW?

Secondary and tertiary amines are also bases as they have a lone pair of electrons on the nitrogen atom in their functional group. This also allows them to react with acids and form salts.

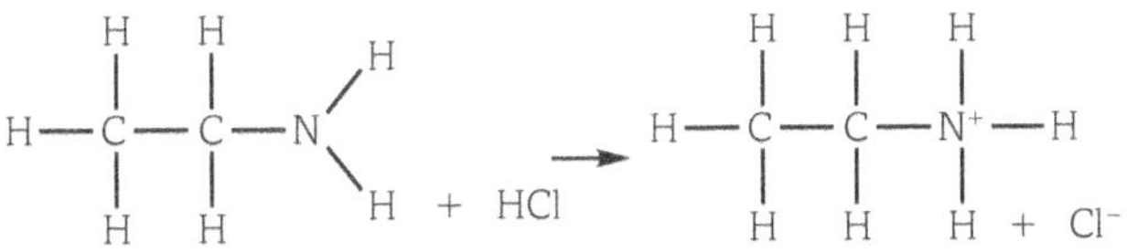

Figure 6 Reaction of ethylamine with dilute hydrochloric acid.

Other strong inorganic acids such as nitric and sulfuric acid will react in a similar way to make a salt.

When nitric acid is used instead of hydrochloric acid the nitrate ion NO_3^- is simply substituted for the chloride ion Cl^- in the balanced chemical equation.

$$CH_3CH_2NH_2 + HNO_3 \rightarrow CH_3CH_2NH_3^+NO_3^-$$

However when sulfuric acid is used instead of hydrochloric acid the balanced chemical equation is more complex.

$$2CH_3CH_2NH_2 + H_2SO_4 \rightarrow [CH_3CH_2NH_3^+]_2SO_4^{2-}$$

The table below details the naming of these salts.

Acid	Name of salt
Hydrochloric acid	Alkylammonium chloride
Nitric acid	Alkylammonium nitrate
Sulfuric acid	Alkylammonium sulfate

Table 1 The salts created when acids react with amines.

> **LEARNING TIP**
>
> When an amine reacts with sulfuric acid two molecules of amine are required for every molecule of sulfuric acid. This is because the alkylammonium sulfate produced contains two alkylammonium ions for every one sulfate ion. Alkylammonium ions have a single positive charge whilst a sulfate ion has a double negative charge.

Preparation of amines

Preparation of aliphatic amines

Using a sealed tube, a haloalkane, ammonia and ethanol are heated together to make an amine. Reflux cannot be used as the ammonia is so volatile it would escape out of the condenser rather than react.

A haloalkane, such as 1-chloropropane, will undergo nucleophilic substitution in a two-stage process to form a primary amine.

- Stage 1: The ammonia reacts with the haloalkane to make an ammonium salt.

   $$CH_3CH_2CH_2Cl + NH_3 \rightarrow CH_3CH_2CH_2NH_3Cl$$

- Stage 2: An additional ammonia molecule reacts to form the propylamine product and ammonium chloride salt.

   $$CH_3CH_2CH_2NH_3Cl + NH_3 \rightleftharpoons CH_3CH_2CH_2NH_2 + NH_4Cl$$

This is a reversible reaction, so excess ammonia will drive the reaction to the right and increase the yield of the desired primary amine product.

Additional substitution of the hydrogen atoms on the nitrogen atom can occur. In this reaction, initially N-dipropylamine, a secondary amine, is made.

$$CH_3CH_2CH_2Cl + CH_3CH_2CH_2NH_2 \rightleftharpoons (CH_3CH_2CH_2)_2NH + HCl$$

Further substitution would produce N-tripropylamine, a tertiary amine.

$$CH_3CH_2CH_2Cl + (CH_3CH_2CH_2)_2NH \rightleftharpoons (CH_3CH_2CH_2)_3N + HCl$$

The final stage is a quaternary ammonium salt, where each hydrogen on the ammonium ion has been replaced with an alkyl chain.

$$CH_3CH_2CH_2Cl + (CH_3CH_2CH_2)_3N \rightleftharpoons (CH_3CH_2CH_2)_4N^+Cl^-$$

These further substitution reactions can occur because amines have lone pairs of electrons on the nitrogen atom that can act as a nucleophile.

> **LEARNING TIP**
>
> As in some of the stages for an acid, this product can also react with ammonia or an amine to generate the ammonium salt.

When preparing amines by this method there is always a mixture of the products produced. As the second stage of the mechanism is an equilibrium reaction, using excess ammonia favours the primary amine and excess haloalkane favours the quaternary ammonium salt.

$$CH_3CH_2CH_2 - \overset{\overset{\displaystyle CH_2CH_2CH_3}{|}}{\underset{\underset{\displaystyle CH_2CH_2CH_3}{|}}{N^+}} - CH_2CH_2CH_3 \quad + \quad Cl^-$$

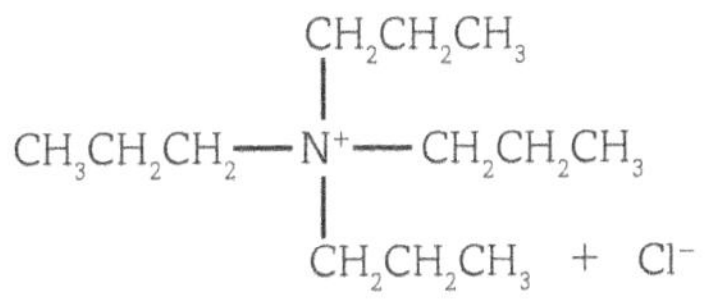

Figure 7 Tetrapropylammonium chloride is an example of a quaternary ammonium salt.

Preparation of aromatic amines

Nitroarenes, such as nitrobenzene, can be reduced to produce an amine. The reducing agent is made in situ by using a mixture of tin and concentrated hydrochloric acid. The reaction occurs under reflux at 100 °C.

After about half an hour, a strong alkali, such as sodium hydroxide, is added. This undergoes a neutralisation reaction to remove the excess hydrochloric acid and produce the amine. Separating the aromatic amine is a multi-stage process that includes steam distillation, solvent extraction and further distillation.

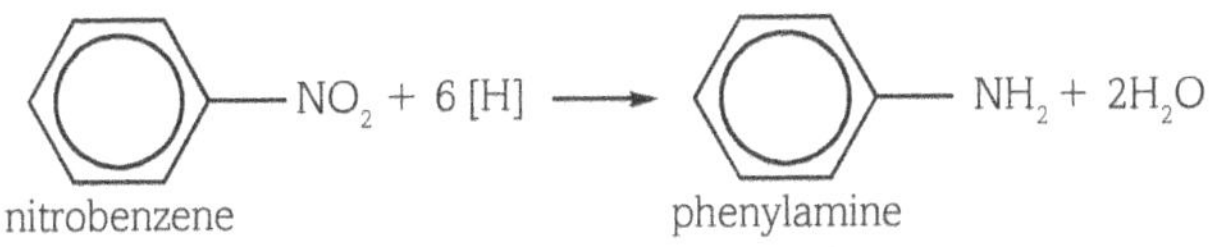

nitrobenzene   phenylamine

Figure 8 Reduction of nitrobenzene to make phenylamine.

Questions

1 Write a balanced symbol equation for the formation of ethylamine from bromoethane in ethanolic ammonia.

2 Predict the main product formed when bromoethane is reacted with ammonia in the presence of excess ammonia.

3 Predict the main product formed when bromoethane is reacted with ammonia in the present of excess bromoethane.

4 Explain why aliphatic amines are made in sealed containers when ammonia is used as the reactant.

(2) Reactions of amino acids

By the end of this topic, you should be able to demonstrate and apply your knowledge and understanding of:

* the general formula for an α-amino acid as RCH(NH$_2$)COOH and the following reactions of amino acids:
  (i) reaction of the carboxylic acid group with alkalis and in the formation of esters
  (ii) reaction of the amine group with acids

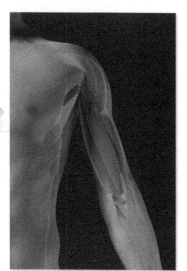

Figure 1 All proteins, like those used to make your muscles, hormones and enzymes, are made from around 20 α-amino acids.

Amino acids are a class of compounds with two functional groups: a carboxylic acid, –COOH, and an amine group, –NH$_2$. When both functional groups are attached to the same carbon atom, the compound is called an α-amino acid. This group of organic chemicals has the general formula RCH(NH$_2$)COOH.

Proteins are an essential nutrient for all animals as they are used to make, among other things, hormones, muscles and enzymes. All proteins are polymer chains made up of amino acid monomers bonded together. In human biochemistry there are around 20 α-amino acids.

Structure of an α-amino acid

The carboxylic acid functional group is a weak acid that will partially ionise in water. The nitrogen atom on the amine group has a lone pair of electrons and can act as a base. This means that amino acids are amphoteric (they can act as both an acid and a base), as their carboxylic acid group can react with bases and their amine group can react with acids.

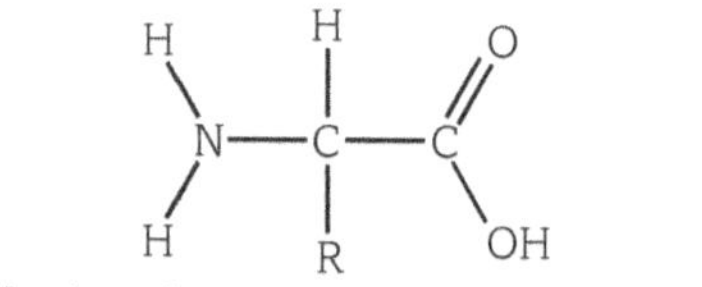

Figure 2 General formula of an α-amino acid.

As you have learned, α-amino acids have the two functional groups, –COOH and –NH$_2$, attached to the same carbon atom. α-amino acids can form zwitterions, where their two functional groups exchange a proton and make an internal salt. The carboxylic acid donates a proton to the amino group. As the two charges cancel each other out, the resultant molecule has no overall charge.

Figure 3 Formation of a zwitterion from glycine.

The isoelectric point is when there is no net electrical charge due to each zwitterion having an internal balance of charge. By changing the pH you alter the amino acid so that only one of its functional groups is charged. This means at low pH, where there is a lot of H$^+$, the carboxylic acid functional group becomes –COOH and only the amine group is charged. However, at high pH, the amine group becomes –NH$_2$ and only the carboxylic acid group is charged.

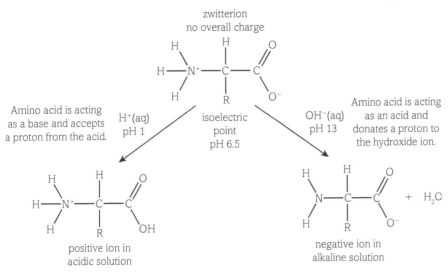

Figure 4 pH affects the structure of an amino acid.

Reactions of amino acids

Carboxylic acid functional group

The carboxylic acid functional group is a weak acid and partially ionises in solution. This functional group will undergo all the common reactions of an organic acid with:

- metal oxides – metal oxides are bases, so a neutralisation reaction occurs. The hydrogen atom on the carboxylic acid group is exchanged for a metal ion to produce a metal salt. Water is also produced in the reaction.
- alkalis – alkalis are soluble bases, so a neutralisation reaction occurs. The reaction produces a salt and water.
- carbonates – this neutralisation reaction produces a salt, water and carbon dioxide gas. As a gas is evolved, effervescence is observed.
- alcohol – this reaction produces an ester and releases a molecule of water.

Amine functional group

The amine functional group can act as a base due to the lone pair of electrons on the nitrogen atom. When an acid is added to an amino acid, the amine group accepts a proton. The result is an ammonium salt.

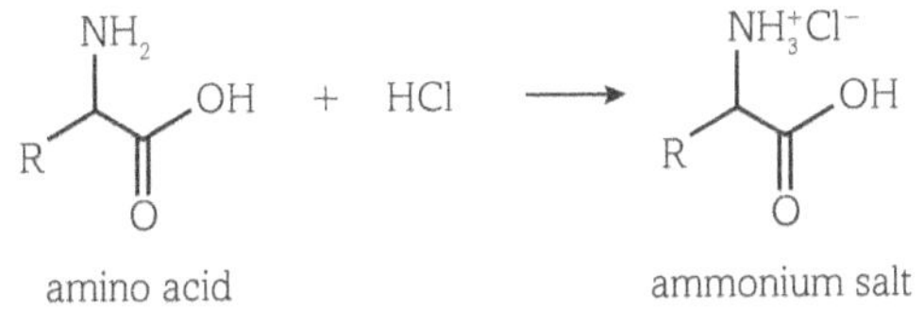

Figure 5 General equation to show the reaction between an amino acid and HCl.

Questions

1 Describe what an α-amino acid is.

2 Explain how you can change the conditions to cause an amino acid to form a positive ion.

3 Explain how you can change the conditions to cause an amino acid to form a negative ion.

4 Describe what an ammonium salt is.

5 Explain why amino acids are considered to be amphoteric.

(3) Amides

By the end of this topic, you should be able to demonstrate and apply your knowledge and understanding of:

* structures of primary and secondary amides

Amides are a class of compounds with an acyl group (RC(O)–) attached to an amine group (–NH$_2$). Organic primary amides have the general formula RC(O)NH$_2$ where R is a hydrocarbon chain or a hydrogen atom. Primary and secondary amides were introduced in topic 6.1.11.

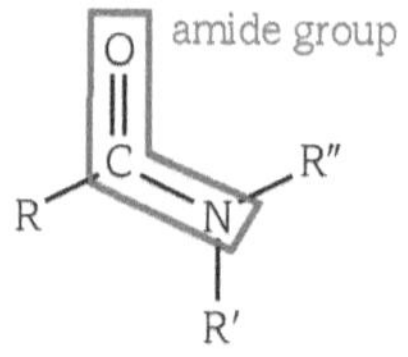

Figure 1 The amide functional group.

These compounds are related to carboxylic acids, where the hydroxyl group has been substituted for an amine group.

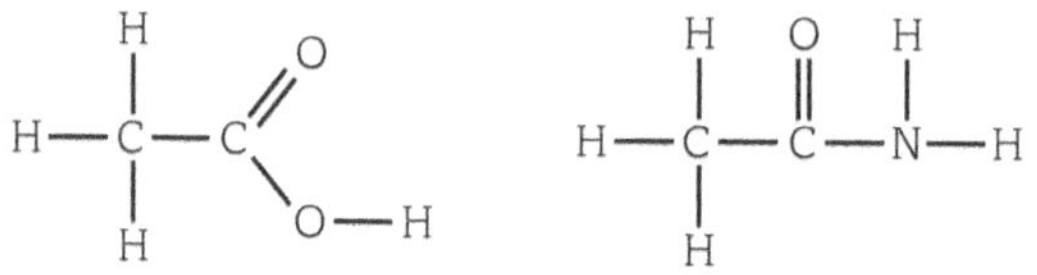

Figure 2 Ethanoic acid and ethanamide.

They are also related to ammonia, where one or more of the hydrogen atoms have been substituted for an acyl group.

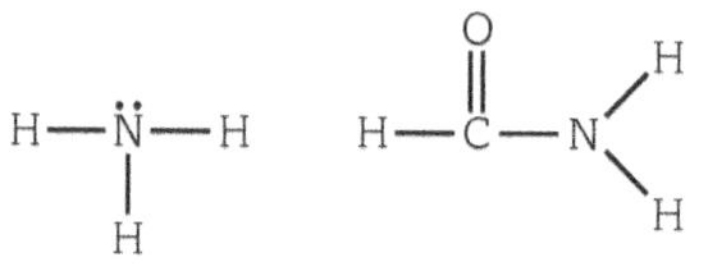

Figure 3 Ammonia and methanamide.

Structure of amides

Amides can be classified based on the groups attached to the nitrogen atom.

* Primary amides – the nitrogen atom has two hydrogen atoms and one acyl group attached.

* Secondary amides – the nitrogen atom has one hydrogen atom, one acyl group and one alkyl group attached.

* Tertiary amides – the nitrogen atom has no hydrogen atoms attached.

* Polyamides – this is a type of condensation polymer; a very long-chain molecule, with a repeating pattern of atoms. They are made from a reaction between a carboxylic acid and an amine and contain a number of secondary amide groups. You can find out more about polyamides in topic 6.2.5.

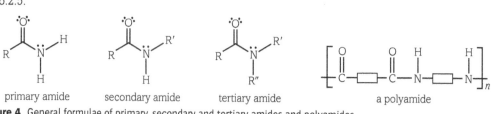

primary amide   secondary amide   tertiary amide   a polyamide

Figure 4 General formulae of primary, secondary and tertiary amides and polyamides.

Naming primary amides

When naming primary amides, the suffix is 'amide' and the root is generated by the number of carbon atoms in the parent hydrocarbon chain which has the functional group. To number prefixes, the functional group is considered to be on carbon atom 1.

Displayed	Structural formula	Name
	$HCONH_2$	methanamide
	CH_3CONH_2	ethanamide
	$CH_3CH_2CONH_2$	propanamide

Naming secondary amides

When naming secondary amides, the suffix is 'amide' and the root is generated by the number of carbon atoms in the parent hydrocarbon chain which has the functional group. The alkyl chain is attached to the nitrogen, so it is called an *N*-substituted compound.

Displayed formula	Structural formula	Name
	$CH_3CONHCH_3$	*N*-methylethanamide
	$CH_3CONHCH_2CH_3$	*N*-ethylethanamide

Questions

1. State the formula of an acyl group.

2. Describe the similarities and differences between primary and secondary amides.

3. Draw the displayed formula of butanamide and classify it as either a primary or a secondary amide.

4. Define the term *polyamide*.

5. What does the 'N' signify in the name *N*-ethylethanamide?

By the end of this topic, you should be able to demonstrate and apply your knowledge and understanding of:

* optical isomerism (an example of stereoisomerism, in terms of non-superimposable mirror images about a chiral centre)

* identification of chiral centres in a molecule of any organic compound

---

KEY DEFINITIONS

Optical isomers are molecules which are non-superimposable mirror images of each other. They have the same chemical properties but interact with polarised light differently.
A **chiral** carbon has four different groups attached to it.

Stereoisomers are a class of isomers that have a different arrangement of atoms in space. You studied *E/Z* isomerism, a type of geometric isomerism, in Book 1, topic 4.1.4; these isomers are generated because of a lack of rotation of the C=C bond.

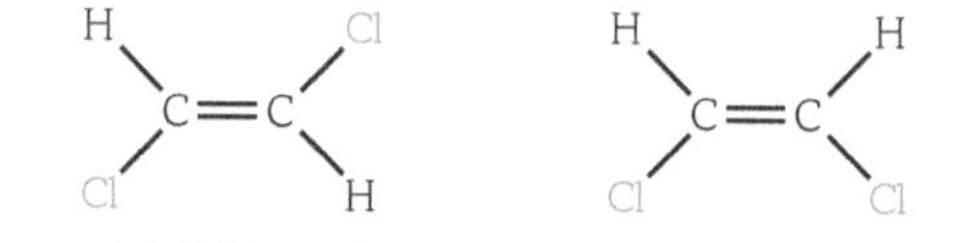

trans-1,2-dichloroethene   *cis*-1,2-dichloroethene

Figure 1 *Trans*-1,2-dichloroethene and *cis*-1,2-dichloroethene are geometric isomers of each other.

Optical isomers

Optical isomerism is another type of stereoisomerism – you learned about optical isomerism in topic 5.3.4. The isomers are non-superimposable mirror images of each other and are known as 'enantiomers'. This means that each of the two enantiomers look like mirror images of each other, and no matter how hard you try, you cannot lay them exactly on top of each other.

This type of isomerism is called *optical isomerism* as each enantiomer interacts with light in a different way. Although both enantiomers have the same chemical properties and similar physical properties, they may have different *biological* properties. One example of this is limonene (IUPAC name: 1-methyl-4-(1-methylethenyl)-cyclohexene). It is a chiral hydrocarbon that is naturally found in citrus fruit. One enantiomer smells strongly of oranges, while the other enantiomer smells of pine.

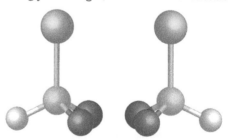

Figure 2 Simplified three-dimensional representation of two optical isomers. Note that they are mirror images of each other.

LEARNING TIP

Remember that enantiomers are chemically identical, they just interact with biological systems and light differently.

Each enantiomer can be given a prefix to denote its effect on a plane of polarised light. If the enantiomer rotates the plane of polarised light in the clockwise direction it is given the prefix '+'. The prefix '–' is given if the rotation is in the anticlockwise direction. This can only be determined by experiment and cannot be deduced from the structure.

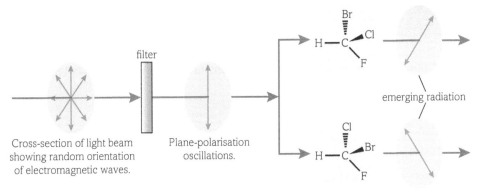

Figure 3 The effect of different enantiomers on polarised light.

A 50:50 mixture of enantiomers is called a racemic mixture or a racemate. A racemic mixture will have no effect on polarised light as the rotations of the two isomers cancel each other out.

Chirality

For optical isomers to be formed in an organic compound, there must be a carbon atom with four different groups attached. This carbon is known as the **chiral** centre and is often shown with a '*' on diagrams of the molecule. It is also possible for inorganic atoms such as nitrogen to be chiral centres.

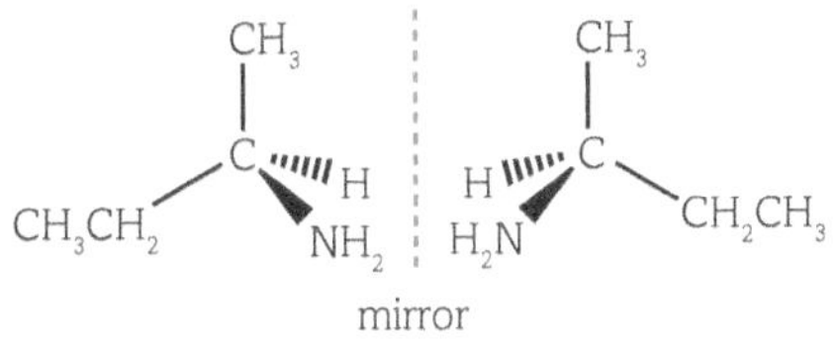

Figure 4 Optical isomers of $CH_3CH_2CH(NH_2)CH_3$.

Chiral molecules do not have a plane of symmetry due to the asymmetric, or chiral, carbon. Some molecules have more than one chiral centre. Each will be a carbon atom with four different groups attached.

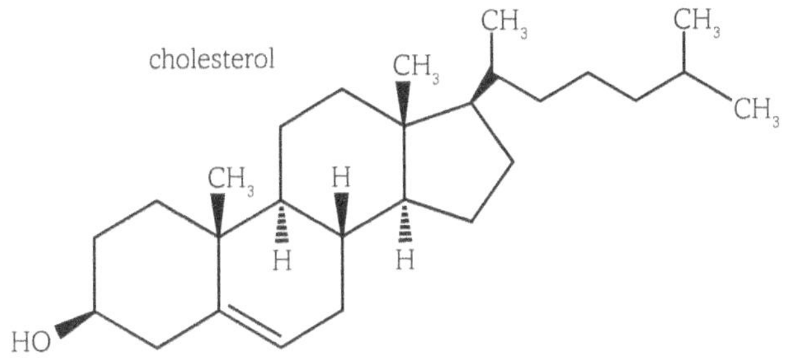

Figure 5 Cholesterol has more than one chiral carbon.

DID YOU KNOW?

The word 'chiral' is derived from the Greek for 'hand'. Our hands are non-superimposable, they are in fact mirror images of each other and therefore are chiral. There are many chiral everyday items including screws and the compounds ibuprofen and DNA.

Questions

1. (a) Identify the chiral carbon in butan-2-ol.
   (b) Draw the optical isomers of butan-2-ol.
2. Explain why butan-1-ol does not have optical activity.
3. Look at Figure 5 – how many chiral carbons does it have?

⑤ Condensation polymers

By the end of this topic, you should be able to demonstrate and apply your knowledge and understanding of:

* condensation polymerisation to form:
  (i) polyesters
  (ii) polyamides

* prediction from addition and condensation polymerisation of:
  (i) the repeat unit from a given monomer(s)
  (ii) the monomer(s) required for a given section of a polymer molecule
  (iii) the type of polymerisation

KEY DEFINITIONS

A **repeat unit** is the arrangement of atoms that occurs many times in a polymer.

Condensation polymerisation is the chemical reaction to form a long-chain molecule by elimination of a small molecule, such as water.

Polymerisation is the chemical reaction that results in the production of a very long-chain molecule with **repeating units** – a polymer. There are two types of polymerisation:

- Addition polymerisation – where monomers have at least one C=C, which breaks and joins the other monomers together. There is only one product from this type of polymerisation. Common addition polymers include polyethene, made from many ethene monomers. You learned about addition polymerisation in Book 1, topic 4.1.10.

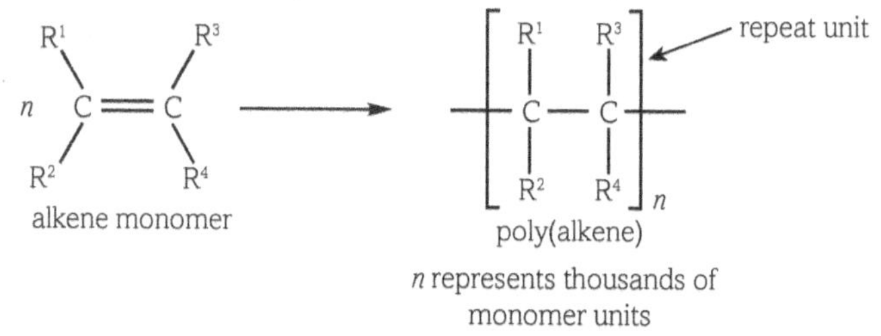

$$n \quad \underset{R^2}{\overset{R^1}{\underset{\diagdown}{\diagup}}} C = C \underset{R^4}{\overset{R^3}{\underset{\diagdown}{\diagup}}} \longrightarrow \left[\underset{\underset{R^2}{|}}{\overset{\overset{R^1}{|}}{C}} - \underset{\underset{R^4}{|}}{\overset{\overset{R^3}{|}}{C}} \right]_n \quad \leftarrow \text{repeat unit}$$

alkene monomer   poly(alkene)

n represents thousands of monomer units

Figure 1 A general equation to show addition polymerisation.

- **Condensation polymerisation** – where two different monomers, with different functional groups, react to form a polymer and release another small molecule, often water, hence the name. Common condensation polymers include nylon and polyester.

Classifying polymers

If you are given the formula of a monomer and it contains C=C, then it must form an addition polymer. If the monomer contains two different functional groups or more than one monomer is present, each with different functional groups, then they must make a condensation polymer. You do not need to recall the structures of synthetic polyesters and polyamides or their monomers.

WORKED EXAMPLE 1

Classifying a polymer and drawing its repeat units

Classify the polymerisation method and draw the repeat unit for the following monomer:

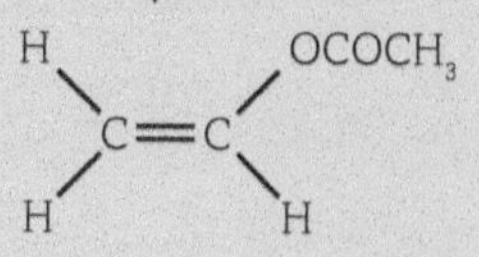

$$\underset{H}{\overset{H}{\underset{\diagdown}{\diagup}}} C = C \underset{H}{\overset{OCOCH_3}{\underset{\diagdown}{\diagup}}}$$

The poly(alkene) is formed by addition of monomers with C=C bonds to form a carbon chain. This is an example of addition polymerisation. The repeat unit is:

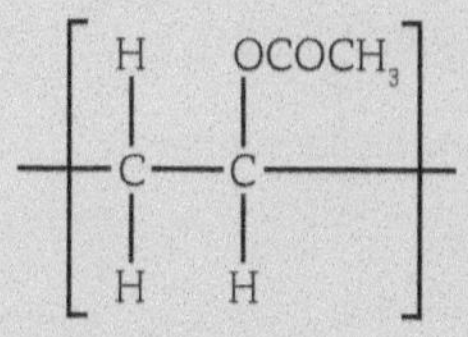

$$\left[\underset{\underset{H}{|}}{\overset{\overset{H}{|}}{C}} - \underset{\underset{H}{|}}{\overset{\overset{OCOCH_3}{|}}{C}} \right]$$

WORKED EXAMPLE 2

Classifying a polymer and drawing its monomer

Classify the polymerisation method and draw the monomer for the following polymer:

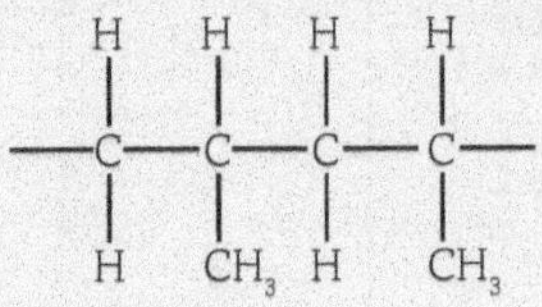

can also be
written as

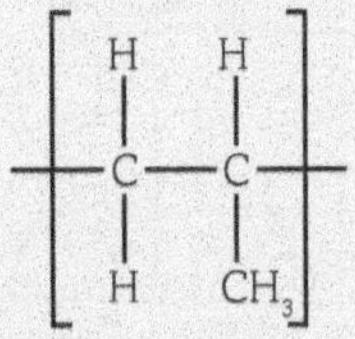

The poly(alkene) is an addition polymer. The chemical that is used to make the addition polymer must be unsaturated and contain at least one C=C. To determine the monomer, replace C–C with C=C.

The monomer is:

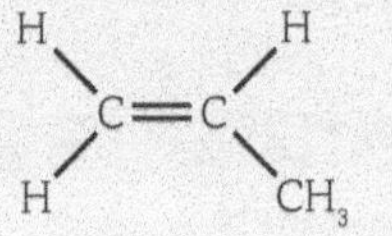

Polyesters

Polyesters are a class of condensation polymer made by the chemical reaction of a dicarboxylic acid (–COOH) and a diol (–OH). When the carboxylic acid group on one monomer reacts with the alcohol group on the other monomer, an ester link (R–COO–R′) is formed between the two molecules. This happens many times and produces a long-chain molecule, which is called a polyester.

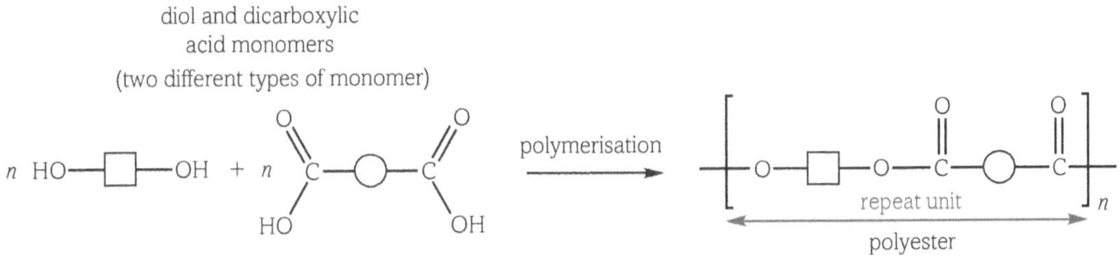

Figure 2 General equation to show diol and dicarboxylic acid monomers reacting to make a polyester.

LEARNING TIP

You need to be able to give the structure of the polymer when the monomer is given, and *vice versa*.

It is possible to have a molecule that contains both an alcohol group and a carboxylic acid group. This allows just one monomer to be used to make the polyester.

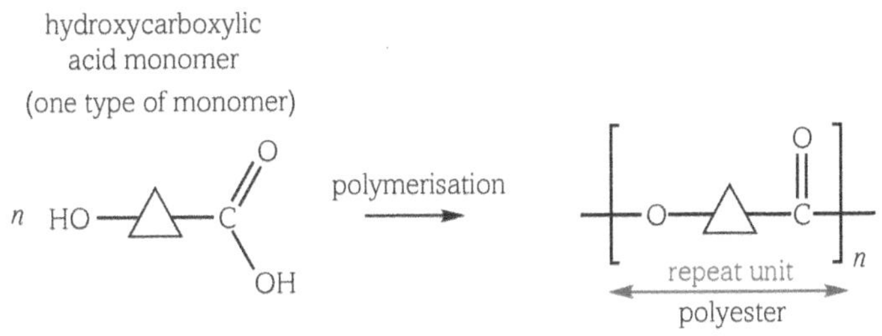

Figure 3 General equation to show a hydroxycarboxylic acid reacting to make a polyester.

Table 1 shows some examples of polyesters.

Polymer name	Polyethyleneterephthalate (PET)		Poly(lactic acid)
Polymer structure		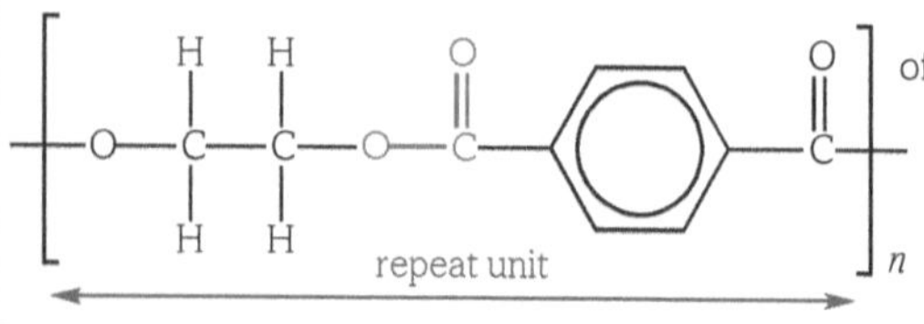 One molecule of water is formed for each ester group formed.	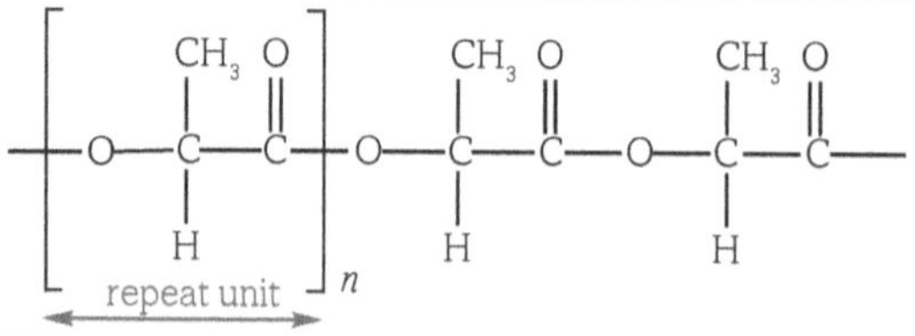
Monomer	ethane-1,2-diol benzene-1,4-dicarboxylic acid		lactic acid
Monomer structure			

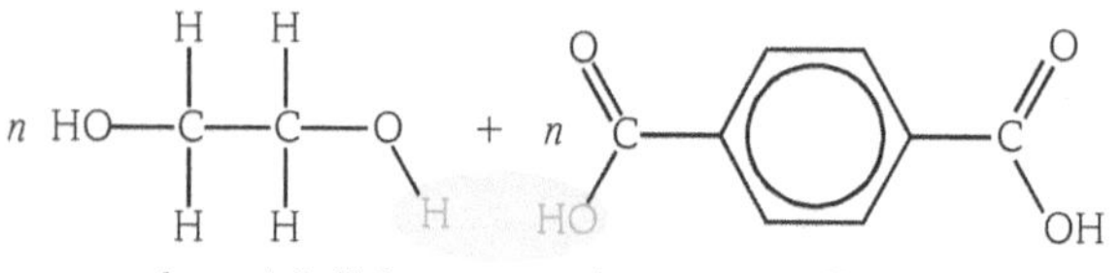

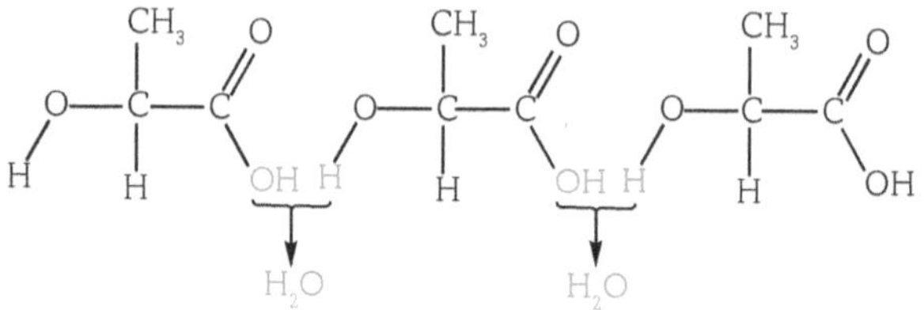

Table 1 Some examples of polyesters.

WORKED EXAMPLE 3

Classifying a polymer and drawing its repeat units

Classify the polymerisation method and draw the repeat unit for the following two monomers:

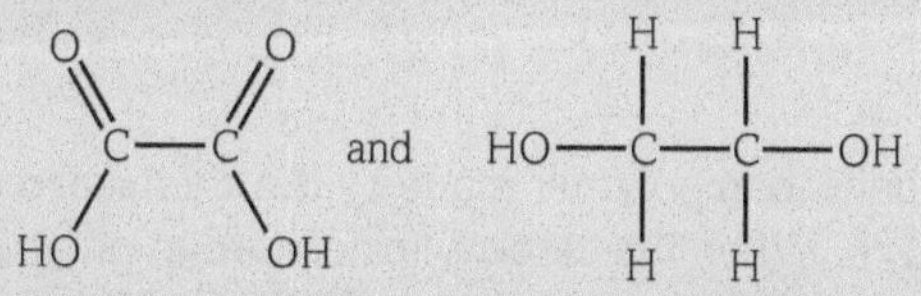

The polymer is formed from an ester link and is made by condensation polymerisation, releasing a water molecule. The repeating unit is:

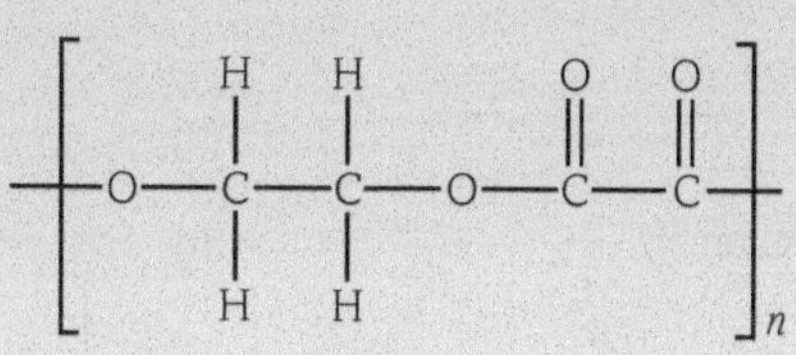

Polyamides

This class of condensation polymer is made by the chemical reaction of a dicarboxylic acid (–COOH) and a diamine (–NH$_2$). When the carboxylic acid group on one monomer reacts with the amine group on the other monomer, an amide link (R–CONH–R') is formed between the two molecules. This happens between many molecules and produces a long-chain molecule called a polyamide.

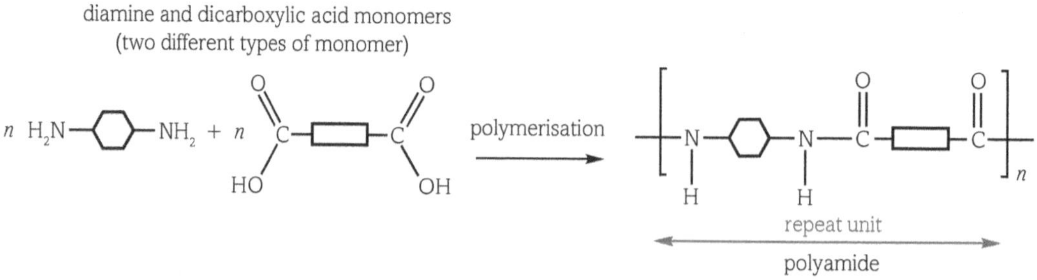

diamine and dicarboxylic acid monomers
(two different types of monomer)

Figure 4 General equation to show diamine and dicarboxylic acid monomers reacting to make a polyamide.

It is possible to have a molecule that contains both an amine group and a carboxylic acid group, as we saw when we looked at amino acids in topic 6.2.2. This allows just one monomer to be used to make the polyamide.

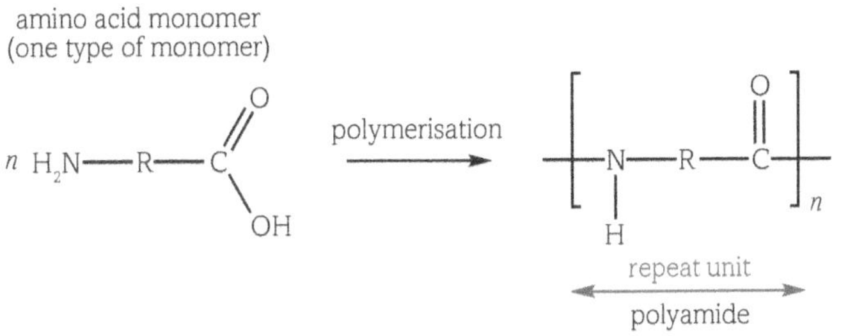

Figure 5 General equation to show an amino acid reacting to make a polyamide.

Table 2 shows some examples of polyamides.

Polymer name	Nylon-6,6	Kevlar
Polymer structure	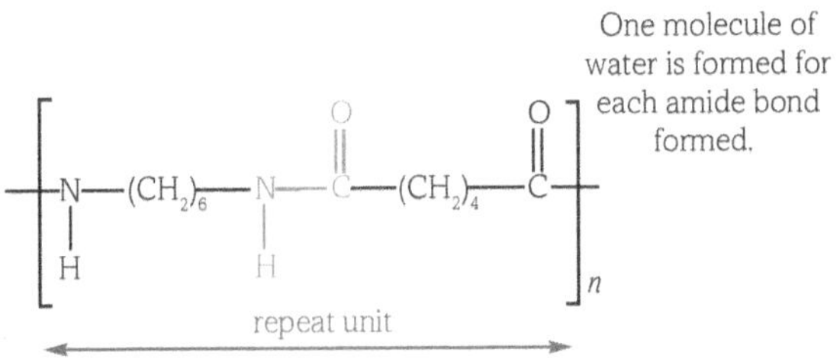 One molecule of water is formed for each amide bond formed.	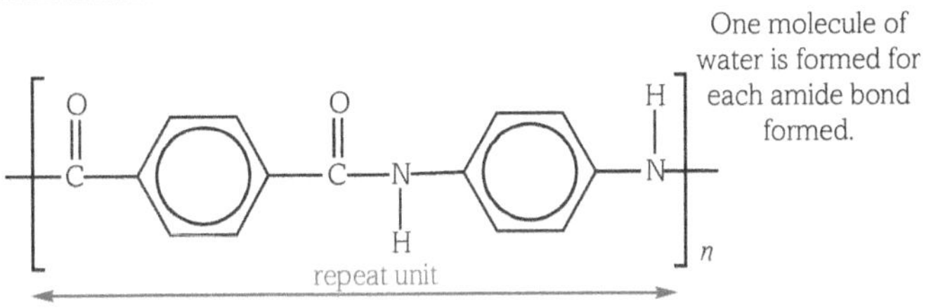 One molecule of water is formed for each amide bond formed.
Monomer	1,6-diaminohexane hexane-1,6-dioic acid	benzene-1,4-dioic acid benzene-1,4-diamine
Monomer structure	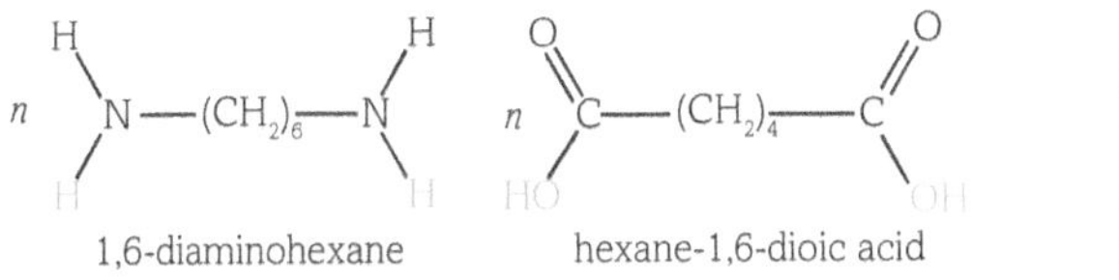	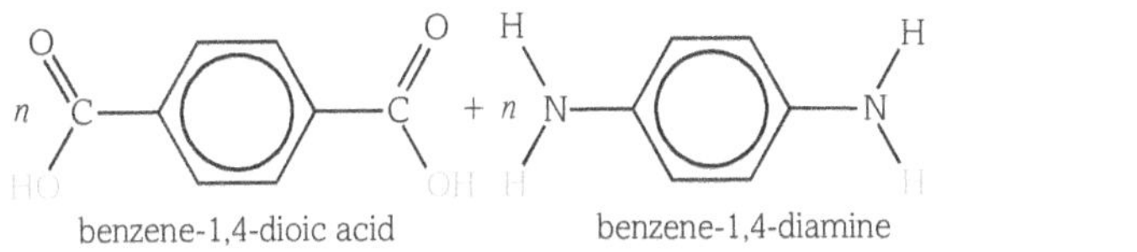

Table 2 Some examples of polyamides.

WORKED EXAMPLE 4

Classifying a polymer and drawing its monomer

Classify the polymerisation method and draw the monomer for the following polymer:

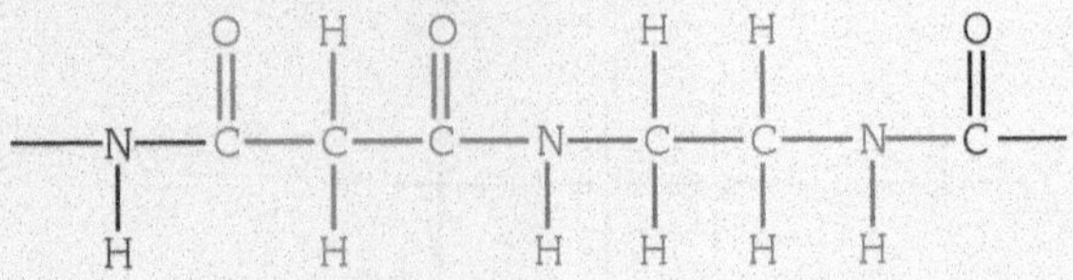

The polymer is formed from an amide link and is made by condensation polymerisation, releasing a water molecule. The two monomers are:

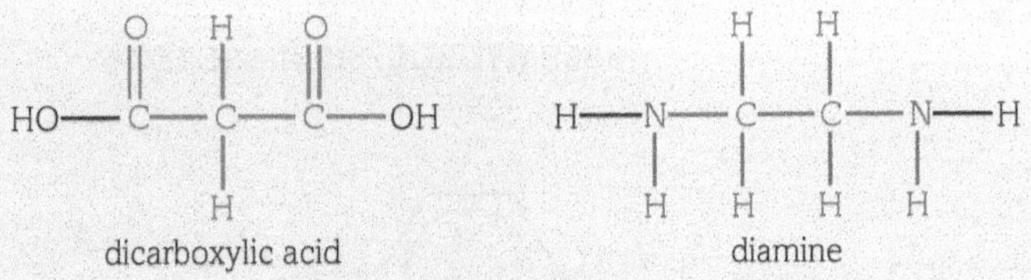

Questions

1. What are the two functional groups needed to produce:
   (a) a polyester   (b) a polyamide?

2. (a) What type of polymerisation is used to form a polymer from 3-hydroxypropanoic acid?
   (b) Draw the repeat unit of this polymer.

3. (a) Describe the differences between condensation and addition polymerisation.
   (b) Describe the similarities between condensation and addition polymerisation.

⑥ Hydrolysis of polymers

By the end of this topic, you should be able to demonstrate and apply your knowledge and understanding of:

* the acid and base hydrolysis of:
  (i) the ester groups in polyesters
  (ii) the amide groups in polyamides

Hydrolysis is a chemical reaction in which water breaks bonds. For condensation polymers, hydrolysis is the reverse reaction of polymerisation – causing the polymer chains to break down into their constituent monomers.

Hydrolysis of polyesters and polyamides with water, however, has a very slow rate of reaction. This is why condensation polymers, such as nylon, do not degrade when it rains! However, the reaction can be achieved in acidic or basic conditions and the rate can be increased further by heating the reaction mixture.

Hydrolysis of polyesters

Polyesters will undergo hydrolysis in acidic and basic conditions. The rate of reaction with basic hydrolysis is so fast that if you dropped a small amount of sodium hydroxide solution on a polyester jumper a hole would quickly be formed.

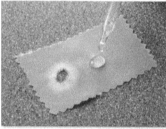

Figure 1 You must be careful working in a lab with bases as they can easily hydrolyse polyesters and cause holes in clothing.

Acid hydrolysis

When polyester is reacted with a strong aqueous acid, the reaction produces a diol and a dicarboxylic acid. The acid hydrolysis of polyester has a slow rate of reaction.

Alkali hydrolysis

When polyester is reacted with hot sodium hydroxide solution, the reaction produces the diol and the salt of the dicarboxylic acid.

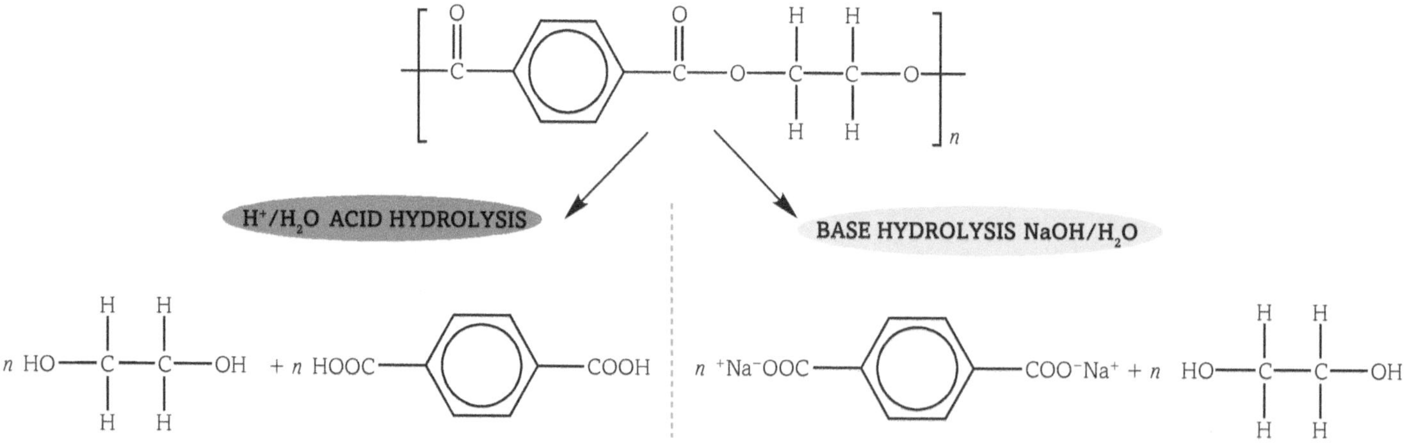

Figure 2 Acid and base hydrolysis of the polyester, Terylene.

Hydrolysis of polyamides

Polyamides will undergo hydrolysis in acidic and basic conditions. The rate of reaction of acid hydrolysis is much faster than basic hydrolysis for polyamides. Acids can easily hydrolyse nylon and cause holes in clothing, so you must be careful when working with them.

Acid hydrolysis

When a polyamide is reacted with a strong aqueous acid, the reaction produces a diammonium salt and a dicarboxylic acid.

Alkali hydrolysis

When a polyamide is reacted with hot sodium hydroxide solution, the reaction produces the diamine and the salt of the dicarboxylic acid.

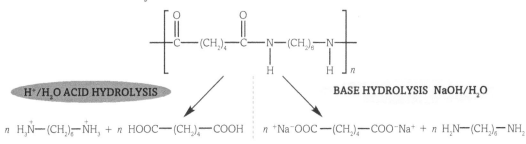

Figure 3 Acid and base hydrolysis of the polyamide, nylon-6,6.

DID YOU KNOW?

Controlling the rate of hydrolysis of condensation polymers has allowed the development of degradable polymers. Many of these polymers have ester links that can undergo hydrolysis and cause the breakdown of the polymer chains.

To make these polymers more environmentally friendly, there has been a lot of research into finding monomers from renewable sources rather than monomers derived from the non-renewable raw material, crude oil. One example is poly(lactic acid) a polyester made from lactic acid extracted from the renewable resource, corn starch.

Questions

1 What are the two products of acid hydrolysis of:
  (a) polyester
  (b) polyamide?

2 What are the two products of basic hydrolysis of:
  (a) polyester
  (b) polyamide?

3 Polyester and nylon can be used to make textiles.
  (a) Explain why nylon clothes do not hydrolyse when they get rained on.
  (b) Many washing powders are basic and have hazard labels on their packaging. Explain why soaking polyester in concentrated washing powder will damage the fabric but this will not occur if nylon is used.

Figure 4 Poly(lactic acid) is a bioplastic and used to make compostable plastic bags for packaging.

7 Extending carbon chain length

By the end of this topic, you should be able to demonstrate and apply your knowledge and understanding of:

* the use of C–C bond formation in synthesis to increase the length of a carbon chain.

* formation of C–C≡N by reaction of:
  (i) haloalkanes with CN⁻ and ethanol, including nucleophilic substitution mechanism
  (ii) carbonyl compounds with HCN, including nucleophilic addition mechanism

Organic chemists will often try to design new molecules using a variety of reactions. This is called a *synthesis*. It is possible to change functional groups in a variety of ways but it is quite difficult to change the length of the carbon chain. One way to do this is to use the cyanide ion (CN⁻) as the nucleophile.

> **LEARNING TIP**
> Remember that cyanide is the attacking species, but the reagent or chemical that you use in the reaction is HCN or KCN.

Cyanide or nitrile?

Cyanide is a nucleophile and attracted to areas of positive charge in a molecule. Other common nucleophiles include hydroxide (OH⁻) and water (H_2O).

The CN⁻ ion is known as the cyanide ion. It retains the name cyanide in inorganic chemicals like sodium cyanide (NaCN). However, when the –CN group is found in an organic chemical it becomes known as a *nitrile functional group*.

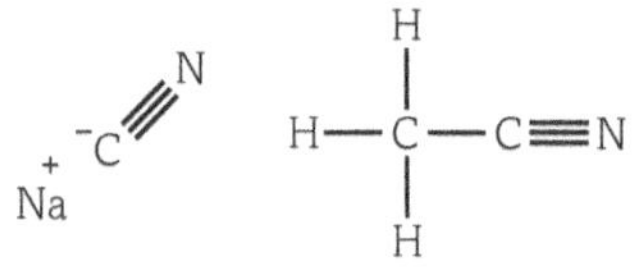

Figure 1 Structure of sodium cyanide and ethanenitrile.

Nucleophilic substitution

Haloalkanes have a dipole in the C–X bond due to the significant difference in electronegativities between carbon and the halogen. This means that the carbon is susceptible to nucleophilic attack.

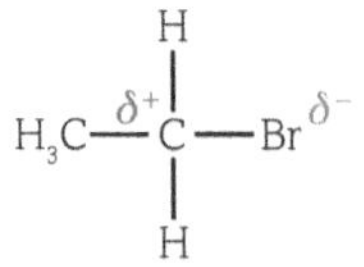

Figure 2 The carbon in the C–Br bond is susceptible to nucleophilic attack.

In the laboratory the haloalkane is mixed with potassium cyanide and heated under reflux. The solvent is ethanol and this ensures that the nucleophile is the cyanide ion. If water is used then the hydroxide ion tends to be the nucleophile and the carbon-chain length would not increase.

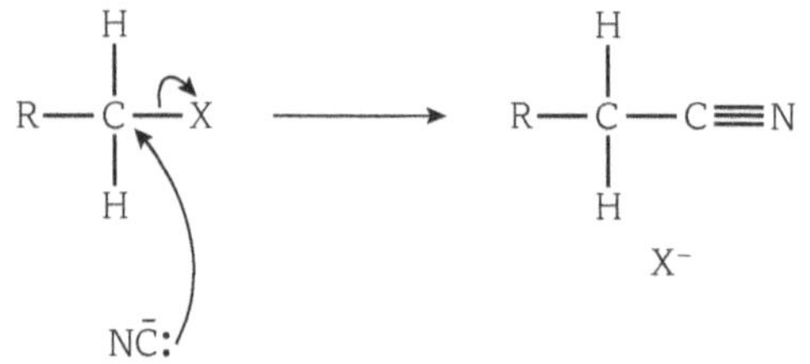

Figure 3 General mechanism for nucleophilic attack of a cyanide ion on a haloalkane.

For example, when 1-bromopropane is heated under reflux with potassium cyanide in a solution containing ethanol, the reaction will occur. The mechanism is a nucleophilic substitution where the cyanide takes the place of the halogen. The balanced chemical equation for the reaction is:

$$CH_3CH_2CH_2Br + KCN \rightarrow CH_3CH_2CH_2CN + KBr$$

The mechanism for this reaction is:

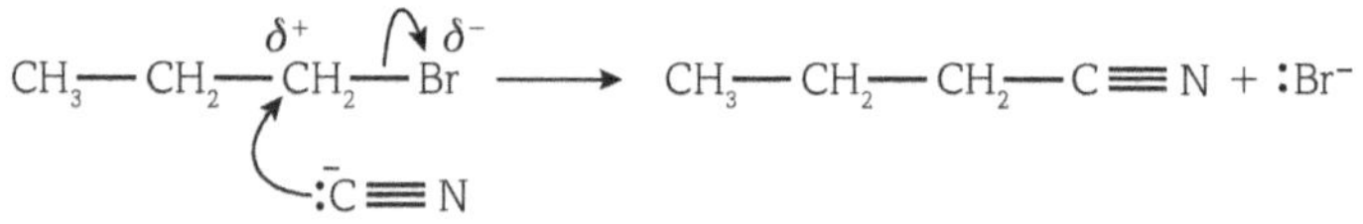

Figure 4 Mechanism for nucleophilic attack of a cyanide ion on 1-bromopropane.

This is an important reaction as the carbon chain has gone from three carbons to four carbons in length. A new C–C bond is formed between the original organic molecule and cyanide, which increases the length of the carbon chain by one carbon atom. The new nitrile functional group can undergo more reactions and make a variety of different chemicals.

Nucleophilic addition

The carbon in a carbonyl group has a slight positive charge and this makes it susceptible to nucleophilic attack. Hydrogen cyanide can be used to generate the cyanide nucleophile. The cyanide ion will be attracted to the carbon of the carbonyl group and forms a new covalent bond which extends the length of the carbon chain.

The π-bond of the C=O then opens and the oxygen accepts the extra pair of electrons. This allows the now negatively charged oxygen atom to accept a proton and become a hydroxyl group (–OH). The result is a hydroxynitrile organic compound.

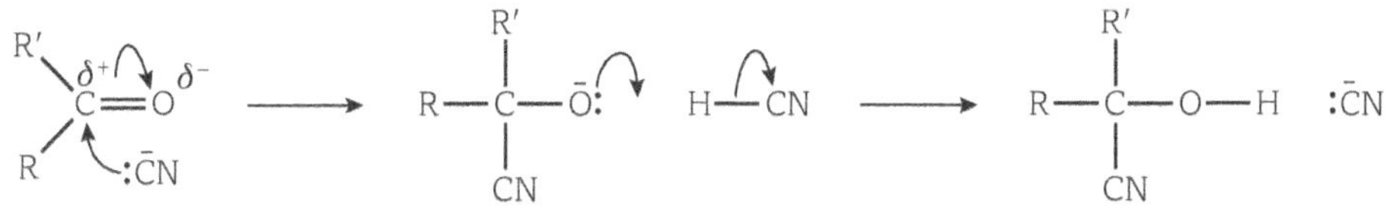

Figure 5 General mechanism for nucleophilic attack of a cyanide ion on an organic chemical having a carbonyl group.

As the C=O is a planar bond, there is an equally likely chance that the cyanide ion will attack from either side. So, a racemic mixture is produced when an asymmetric ketone or any aldehyde other than methanal is used in this reaction. You learned about optical isomerism and racemic mixtures in topic 5.3.4.

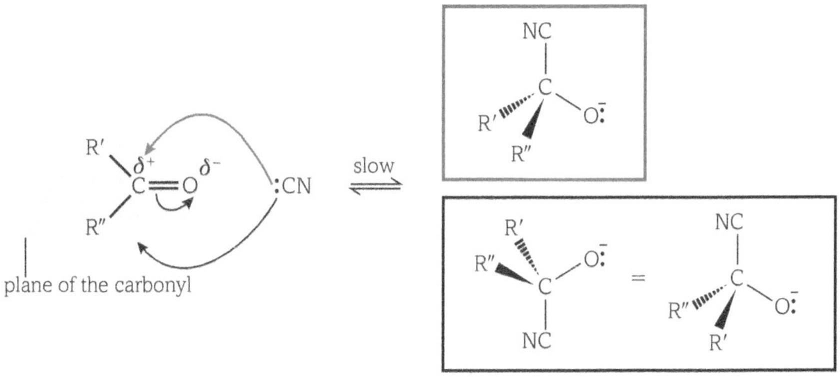

Figure 6 Enantiomers can be formed from the electrophilic addition of a carbonyl group. The red (upper) arrow gives the product in the red (upper) box and the black (lower) arrow gives the product in the black (lower) box. Since both enantiomers are formed, a racemic mixture is produced.

LEARNING TIP

HCN is a highly toxic gas in normal laboratory conditions. It stops the mitochondrial enzymes working in living cells. A safer alternative in the laboratory is to use a cyanide salt like potassium cyanide (KCN). This is a solid at room temperature and is easier to handle. The salt must be used in an acidified solution so that both CN^- and H^+ ions are present.

Questions

1. What is the name given to the 'CN' group of atoms in:
   (a) inorganic compounds
   (b) organic compounds?

2. Explain why ethanol and KCN are used, rather than HCN and water, in the laboratory reaction of a haloalkane to increase the chain length.

3. The cyanide ion is a nucleophile and will react with carbonyl compounds.
   (a) Explain how a racemic mixture can be formed when KCN is reacted with butan-2-one.
   (b) Explain why only one product is formed when methanal reacts with KCN.

8 Reactions of nitriles

By the end of this topic, you should be able to demonstrate and apply your knowledge and understanding of:

* reaction of nitriles:
  (i) by reduction (e.g. with H_2/Ni) to form amines
  (ii) by acid hydrolysis to form carboxylic acids

KEY DEFINITION

A **nitrile** is an organic chemical with a –CN functional group.

Nitriles are organic chemicals that have a –CN functional group. The –CN group can undergo other chemical reactions to form a variety of functional groups including amines and carboxylic acids.

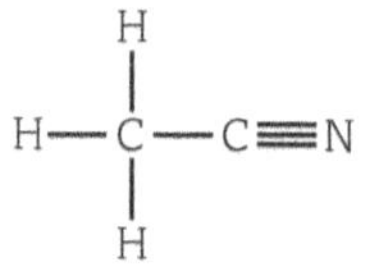

Figure 1 Ethanenitrile has a –CN functional group.

Reduction

Nitriles form amines when they undergo **reduction**. One definition of reduction is 'the addition of hydrogen'. Hydrogen gas can react directly with a nitrile to form a primary aliphatic amine. Two hydrogen atoms are added to the nitrogen atom of the nitrile group.

A transition metal catalyst, such as nickel, is needed to lower the activation energy to enable this reaction to occur. To increase the rate of reaction, the reaction mixture is heated to about 150 °C with a nickel catalyst, at raised pressure.

The general equation for this reaction is:

$$RCN + 2H_2 \rightarrow RCH_2NH_2$$

where R is a hydrocarbon or hydrogen.

DID YOU KNOW?

In industry, the reduction of a nitrile directly with hydrogen uses a Raney nickel catalyst. This is an alloy made mainly from nickel and a very small amount of aluminium. Murray Raney, an American engineer, developed this catalyst in the early twentieth century for the hydrogenation of vegetable oils to make margarine.

Reduction of a nitrile is also possible using a reducing agent such as lithium tetrahydridoaluminate(III), $LiAlH_4$.

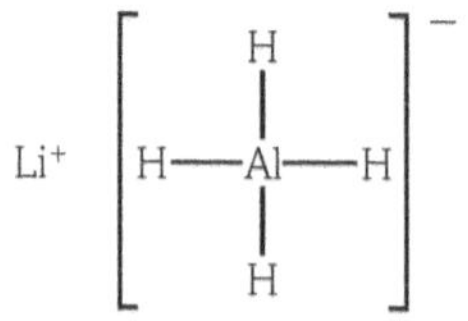

Figure 2 Lithium tetrahydridoaluminate(III) is a reducing agent.

In a chemical equation, the reducing agent is often abbreviated to [H]. The general equation for this reaction is:

$$RCN + 4[H] \rightarrow RCH_2NH_2$$

where R is a hydrocarbon or hydrogen.

Hydrolysis

Nitriles will undergo hydrolysis to initially form an amide and then an ammonium salt. However, this second reaction is so slow that it is not significant. The rate of reaction can be increased by using a strong acid as a catalyst and heating the reaction mixture under reflux. Under acid hydrolysis, nitriles react to form carboxylic acids.

When ethanenitrile undergoes acid hydrolysis, ethanoic acid and an inorganic ammonium salt are formed. The balanced chemical equation is:

$$CH_3CN + 2H_2O + HCl \rightarrow CH_3COOH + NH_4Cl$$

Organic acids (such as CH_3COOH) are formed in this reaction rather than an organic ammonium salt (CH_3COONH_4). This is because the strong acid fully ionises in solution, forming hydrogen ions and chloride ions. Once the amide is formed, the bond between the nitrogen and the carbon is broken, making the ethanoate ion. This then reacts with the proton from the strong acid to form the carboxylic acid. All organic acids are weak acids and only partially ionise in solution. In these reactions the $C\equiv N$ is completely broken through a series of steps via the amide. This gives the carboxylic acid and the ammonium salt as the major products.

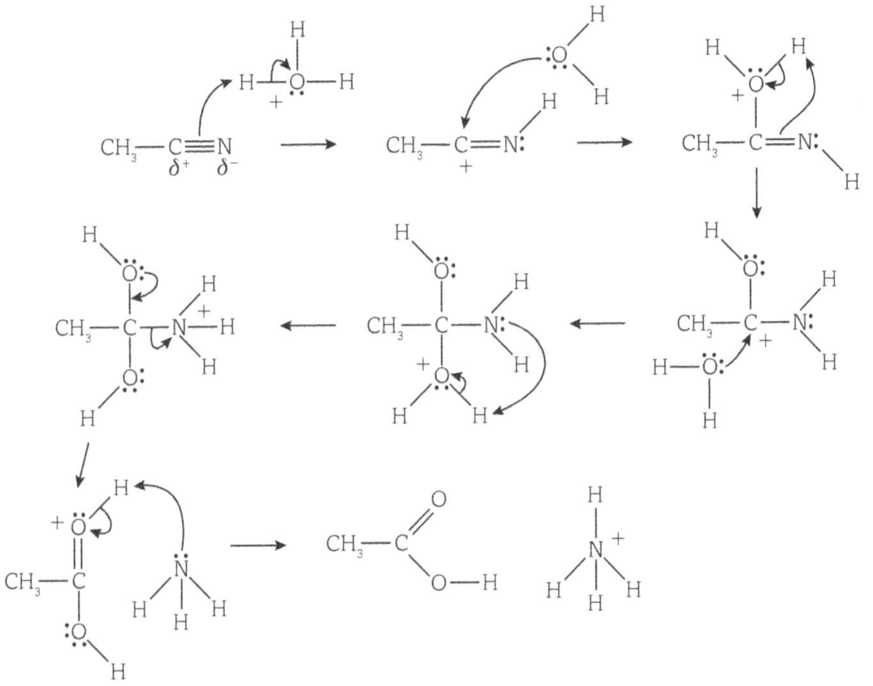

Figure 3 Mechanism for the acid hydrolysis of ethanenitrile. You do *not* need to memorise this mechanism.

LEARNING TIP

Remember that reduction has multiple definitions:

- loss of oxygen
- gain of electrons
- lowering of oxidation number
- gain of hydrogen.

Questions

1. (a) Write a balanced chemical equation for the reduction of ethanenitrile with hydrogen gas.
   (b) Write a balanced chemical equation for the reduction of ethanenitrile with lithium tetrahydridoaluminate(III), which we represent as [H] in equations.

2. (a) State the reagents needed to make propanoic acid from propanenitrile.
   (b) State the conditions needed to make propanoic acid from propanenitrile.

9 Substitution reactions in aromatic compounds

By the end of this topic, you should be able to demonstrate and apply your knowledge and understanding of:

* formation of a substituted aromatic C–C by alkylation (using a haloalkane) and acylation (using an acyl chloride) in the presence of a halogen carrier (Friedel–Crafts reaction)

KEY DEFINITION

Friedel–Crafts reactions allow electrophilic substitution to occur on an aromatic ring.

All aromatic compounds contain a six-membered carbon ring with delocalised electrons. As we learned in topic 6.1.1, this delocalisation of electrons offers additional stability to the molecule. So it will not undergo addition reactions like other unsaturated compounds.

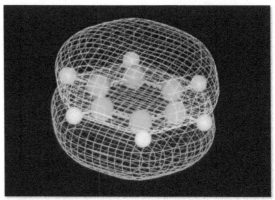

Figure 1 The delocalised electrons in an aromatic ring make it more stable and less reactive than other unsaturated compounds. This computer graphic shows the π-electrons as yellow and blue cages above and below the flat plane of the molecules.

Substitution reactions occur when one or more hydrogen atoms on the aromatic ring are exchanged for another atom or group of atoms. For arenes, these reactions occur by electrophilic substitution using a catalyst such as aluminium(III) chloride, $AlCl_3$, or iron(III) chloride, $FeCl_3$. A substituted aromatic ring is produced by forming a C–C bond.

DID YOU KNOW?

The French–American team of Charles Friedel and James Crafts developed the first substitution reactions involving benzene in the late 1900s. Although this was a breakthrough, the reactions were not very predictable or controllable. Their catalysts are also known as halogen carriers.

Alkylation

Alkylation reactions occur when hydrocarbon chains are added to an organic compound. The reagents are an aromatic compound, haloalkane and strong Lewis acid catalyst.

Benzene can react with chloromethane to make methylbenzene in the presence of $FeCl_3$ catalyst. Iron(III) chloride can be used as a **Friedel–Crafts** catalyst as it is an electron pair acceptor. The balanced chemical equation for this reaction is:

$$C_6H_6 + CH_3Cl \rightarrow C_6H_5CH_3 + HCl$$

The electrophilic substitution reaction can be explained using a general mechanism, where R is a hydrocarbon chain. It is possible for further substitution reactions to occur and so a mixture of products is produced. But this mechanism focuses on just one substitution occurring.

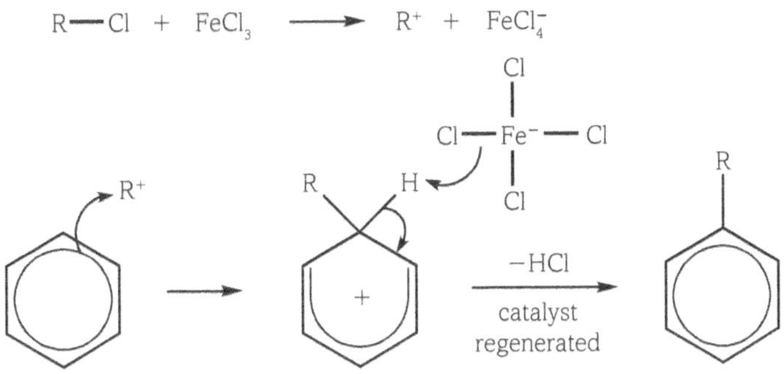

$$R-Cl + FeCl_3 \longrightarrow R^+ + FeCl_4^-$$

Figure 2 Friedel–Crafts alkylation.

Acylation

Acylation reactions occur when RCO– is added to an organic compound, where R is a hydrocarbon chain. The reagents are an acyl chloride and a strong Lewis acid catalyst.

Benzene can be gently heated under reflux with ethanoyl chloride in the presence of anhydrous aluminium(III) chloride to form phenylethanone. The balanced chemical equation for this reaction is:

$$C_6H_6 + CH_3COCl \rightarrow C_6H_5COCH_3 + HCl$$

The electrophilic substitution reaction can be explained using a general mechanism, where R is a hydrocarbon chain.

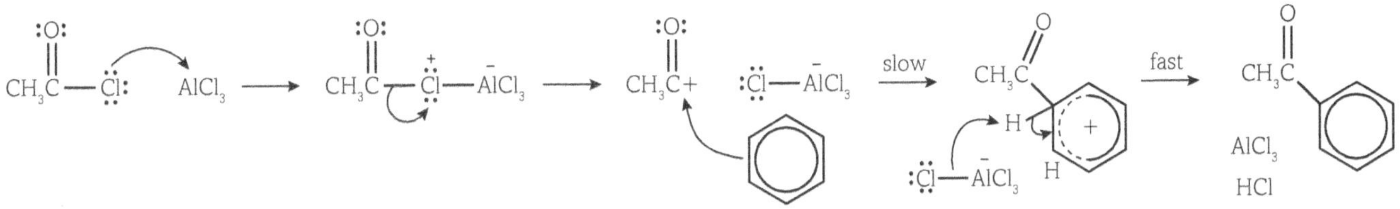

Figure 3 Friedel–Crafts acylation.

LEARNING TIP
You learned about electrophilic substitution reactions in topic 6.1.3.

Questions

1 (a) Describe the features of a Friedel–Crafts reaction.
   (b) Give **two** examples of Friedel–Crafts catalysts.

2 Draw a mechanism for the Friedel–Crafts reaction to form 1,1-dimethylethylbenzene from benzene and 2-chloro-2-methylpropane.

3 State the:
   (a) similarities
   (b) differences
   between alkylation and acylation of aromatic compounds.

(10) Practical skills for organic synthesis

By the end of this topic, you should be able to demonstrate and apply your knowledge and understanding of:

* the techniques and procedures used for the preparation and purification of organic solids involving use of a range of techniques including:

  (i) organic preparation (use of Quickfit apparatus; distillation and heating under reflux)

  (ii) purification of an organic solid (filtration under reduced pressure; recrystallisation; measurement of melting points)

KEY DEFINITIONS

Distillation is a technique used to separate miscible liquids or solutions.
Heating under **reflux** is a technique used to ensure that volatile compounds are not lost from the reaction mixture.
Recrystallisation is a method for purifying organic compounds.

Quickfit

As you learned in Book 1, topic 4.2.6, synthesis of organic molecules is often carried out using Quickfit apparatus. It is a selection of heat-resistant glassware with connectors that can be easily fixed in a variety of arrangements.

Distillation

Distillation is often used to separate an organic product from the reaction mixture. Figure 2 shows a distillation set-up using Quickfit equipment. Each connector is carefully greased or held with a plastic clip to ensure it fits well.

The bottom of the condenser is connected to the cold water supply and the top connector is connected to a hose that drains water into the sink. A few anti-bumping granules are added to the distillation flask to ensure smooth boiling.

To complete the distillation, the mixture is heated. As the product is collected, the temperature on the thermometer can be used to identify the chemical.

Heating under reflux

Heating under **reflux** is a technique used in organic synthesis to ensure that volatile compounds are not lost from the reaction mixture. Volatile reagents are returned to the reaction mixture, rather than escaping before they can react. Figure 3 shows a reflux set-up using Quickfit apparatus.

Purification of an organic solid

Filtration under reduced pressure

Organic solids will often crystallise out of solution after an organic synthesis. Filtration under reduced pressure is one method of collecting the product. This is done as follows:

1. Connect thick-walled rubber tubing to the vacuum pump and check that there is suction.
2. Put a Büchner funnel into the top of a filter flask. For small quantities, a Hirsch funnel can be used.
3. Connect the tubing from the vacuum pump to the side arm of the clamped filter flask and start the suction.
4. Place a piece of filter paper, just big enough to cover the holes, into the top of the funnel.
5. Using distilled water or the solvent from your organic synthesis if this is different, dampen the filter paper so that it sticks over the holes of the funnel.
6. Slowly pour your reaction mixture into the centre of the funnel.

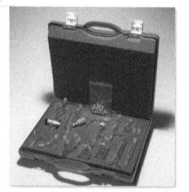

Figure 1 Quickfit apparatus often comes in a kit with a variety of commonly used equipment.

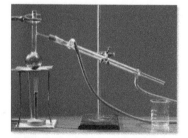

Figure 2 Quickfit apparatus for distillation.

Figure 3 Quickfit apparatus for reflux.

7. Wash out the reaction vessel and the funnel with the solvent and add this to the funnel.

8. Rinse the collected solid with more solvent and maintain suction for a minute after all the washings have been added.

9. Turn off the suction by removing the rubber tube from the side arm of the filtration flask.

10. Invert the funnel onto a watch glass to collect the organic solid.

Recrystallisation

In organic synthesis, often a mixture of different organic chemicals is produced. **Recrystallisation** is a technique that allows an organic product to be purified from unreacted starting materials, catalysts and unwanted side products. It has three stages:

1. The crude product is dissolved into the minimum volume of hot solvent. Hot solvent is used as the solubility of solids is higher. If the product is coloured, activated charcoal can be heated with the mixture to remove the coloured impurities.

2. The hot solution should then quickly undergo gravity filtration using fluted filter paper. The residue of the used activated charcoal and insoluble impurities can be disposed of.

   The hot filtrate should be allowed to cool. This reduces solubility and crystals of the product will form. If no crystals appear to be forming, scratching the sides of the conical flask with a clean glass rod can form nucleation sites and aid crystallisation. Crash cooling the mixture in an ice bath can also be used to speed up this stage. Any soluble impurities should remain dissolved.

3. The purified organic product can be collected by vacuum filtration. The crystals should be washed with a small amount of cold solvent to remove any of the filtrate which contains dissolved impurities.

Checking purity

Measuring melting points

All pure chemicals have distinct melting and boiling points. By measuring these values and comparing them to information in a data book or database, the chemical can be identified.

By measuring the melting point of a known chemical, the purity can be determined. The pure chemical will have the exact melting point as listed in a trusted source. However, the more impure the product is, the greater the range of the melting point will be.

There are two methods for measuring melting point:

1. *Melting point apparatus:* In a sealed capillary tube, put a few grains of the organic solid. Gently insert the tube into the melting point machine and add an accurate thermometer with a suitable range.

   Switch on the machine and turn the heating dial to about 4. For lower melting point compounds, do not be tempted to use a higher setting or the boost function as you are likely to miss the melting point.

   Look through the lens and carefully watch the crystals. When they start to fall away from the sides of the capillary tube, melting has begun. Record this temperature. Continue to watch until the sample has liquefied and note the new temperature. These two values give your melting point range.

2. *Thiele tube:* In a sealed capillary tube, put a few grains of the organic solid. Using a small rubber band, attach the tube to a thermometer. Submerge the thermometer into the oil of the Thiele tube, ensuring that the rubber band is above the oil line.

   Using a microburner with a small gentle flame, heat the side arm of the Thiele tube. Note the temperature range of when the organic solid starts and finishes melting.

Figure 4 Filtration under reduced pressure.

LEARNING TIP

To dry the solid, put the sample in a drying oven. Ensure that the oven temperature is set to below the melting point of your compound.

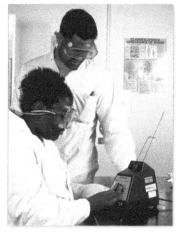

Figure 5 Melting point apparatus.

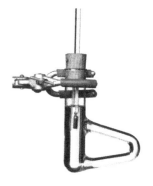

Figure 6 A Thiele tube is a special shape to allow convection currents to flow easily around the oil and ensure the same temperature throughout.

Questions

1. State how you would decide whether to use a Büchner or a Hirsch funnel.

2. (a) Explain why a hot solvent should be used for the first stage of recrystallisation.
   (b) Explain why a cold solvent should be used for washing the crystals in the last stage of recrystallisation.

3. Explain how melting point data can indicate purity.

11 Synthetic routes in organic synthesis

By the end of this topic, you should be able to demonstrate and apply your knowledge and understanding of:

* for an organic molecule containing several functional groups:
  (i) identification of individual functional groups
  (ii) prediction of properties and reactions
* multi-stage synthetic routes for preparing organic compounds

Identifying functional groups

> **KEY DEFINITION**
>
> A **synthetic route** (or synthetic pathway) is a series of reactions that can be used to change a starting chemical into a target molecule.

Organic molecules often have more than one functional group. It is important that you can identify them and suggest the properties that the group will have.

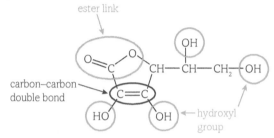

Figure 1 Vitamin C contains multiple functional groups including unsaturation, ester link and hydroxyl groups.

The table below summarises the reactions of the common functional groups.

Functional group	Reactions
Unsaturated containing at least one C=C	Can undergo addition reactions with: • bromine • hydrogen halides • themselves, to form an addition polymer.
Haloalkanes	• React with ammonia in ethanol to form amines. • Can be reduced by alkaline hydrolysis to form alcohols.
Carbonyl	• Aldehydes can be reduced to primary alcohols or oxidised to carboxylic acids. • Ketones can be reduced to secondary alcohols.
Carboxylic acids	Reactive with: • reactive metals to make a salt and hydrogen • metal carbonate to make a salt, water and carbon dioxide • alcohols to form esters.
Alcohol	• Primary alcohols can be oxidised to aldehydes. • Secondary alcohols can be oxidised to ketones. • Tertiary alcohols cannot be oxidised.
Esters	• Undergo alkaline hydrolysis to form a carboxylate and an alcohol. • Undergo acid hydrolysis to form a carboxylic acid and an alcohol.

Table 1 Reactions of common functional groups.

Synthetic routes

The flow charts in Figures 2 and 3 show how the different functional groups can be interconverted. This outline can help you to design a series of reactions to form a target molecule. This is known as a **synthetic route** (or synthetic pathway).

Aliphatic functional groups

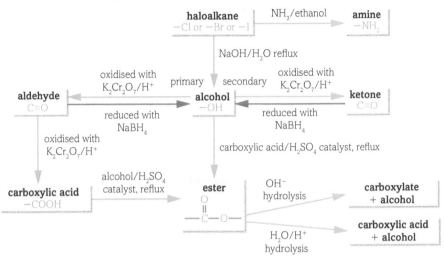

Figure 2 Summary of the reactions of aliphatic functional groups.

Aromatic functional groups

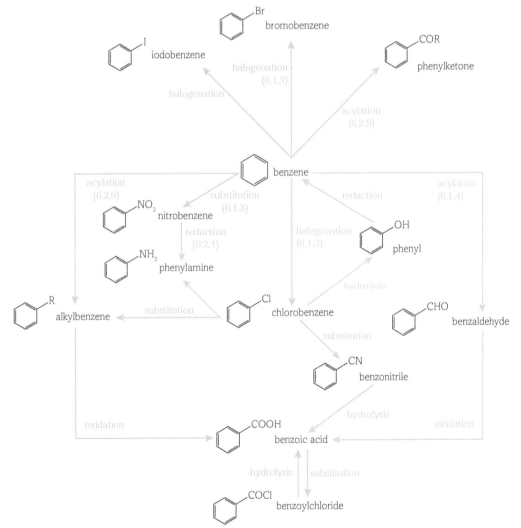

Figure 3 Summary of the reactions of aromatic functional groups.

Designing synthetic routes

You need to follow a number of steps to design and check a synthetic route.

1. Compare the starting material and the target product. Focus on the changes that have happened to the molecule.

    (a) An increase in carbon chain length by one atom means the reaction involved cyanide.

    (b) An increase in carbon chain length by more than one atom means a reaction involving two carbon-containing compounds.

2. Think about the reactions in steps, to change the starting chemical to the target molecule. Sometimes it helps to focus on the product and think backwards; this is known as retro-synthesis.

3. Write down all the stages and check that they are all balanced, with every atom accounted for.

WORKED EXAMPLE 1

Outline a suitable synthetic route to make 3-aminopropan-1-ol from 3-chloropropanal. Your answers should include all reagents and conditions.

Step 1: Draw the structures of the starting material and the target product. The chain length is the same but the chlorine atom is substituted by an amine group and aldehyde has become an alcohol.

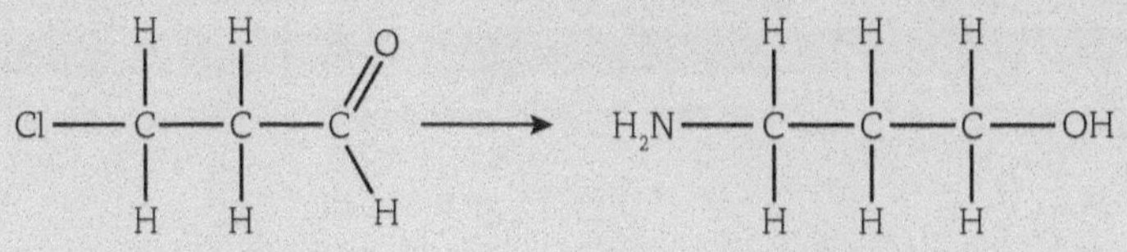

3-chloropropanal   3-aminopropan-1-ol

Figure 4 Synthesis of 3-aminopropan-1-ol from 3-chloropropanal.

Step 2: The aldehyde can be reduced to form an alcohol using $NaBH_4$/ethanol. The sodium borohydride is very reactive in water and forms boric acid so the reactions are carried out in ethanol. Haloalkanes can be converted to amines using excess ammonia in an ethanol solvent heated under reflux.

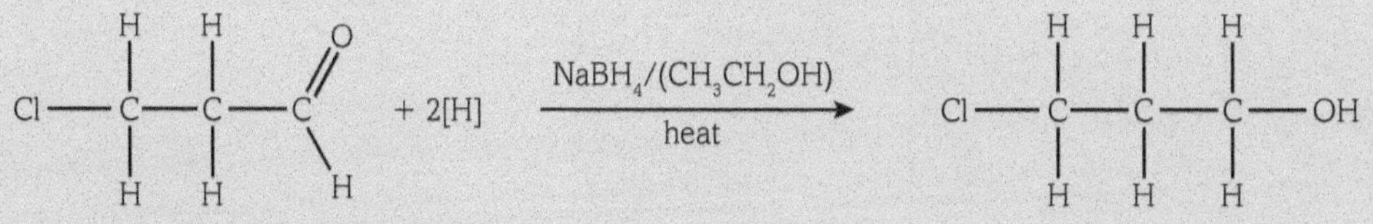

3-chloropropanal   3-chloropropan-1-ol

Figure 5 Converting an aldehyde to an alcohol.

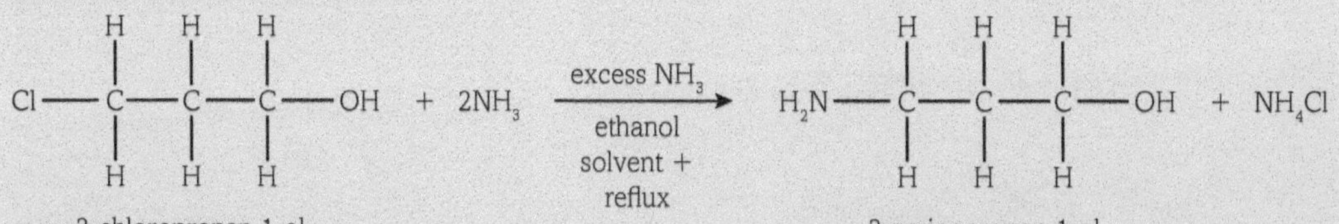

3-chloropropan-1-ol   3-aminopropan-1-ol

Figure 6 Converting halo group to amine.

WORKED EXAMPLE 2

Design a synthetic route to prepare compound B from compound A, as shown in Figure 7. Your answer should include reagents.

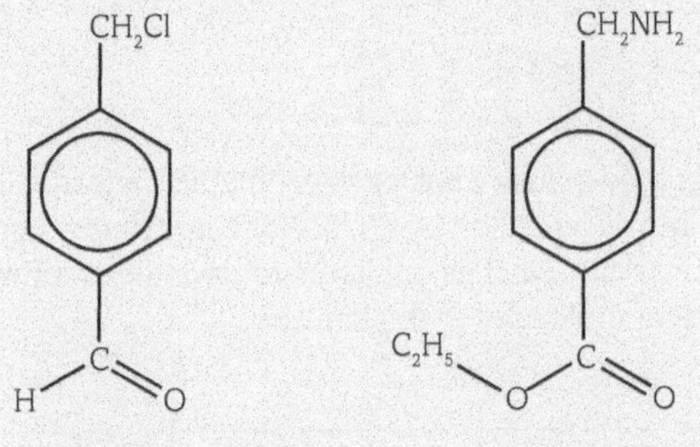

Figure 7 Compound A, the starting material, and compound B, the target molecule.

Step 1: The structures are given in the question. More than one carbon atom has been added to the structure. Therefore, a reaction between two carbon-containing compounds has occurred.

Step 2: The chlorine atom has been substituted for an amine group, which can be achieved with excess ammonia in ethanol solvent.

The aldehyde functional group has become an ester. This must be a two-stage reaction. (1) The aldehyde is oxidised using acidified potassium dichromate under reflux. (2) Distillation is used to separate the resulting carboxylic acid, which is heated with ethanol and a sulfuric acid catalyst to form the ester group with the appropriate carbon chain length.

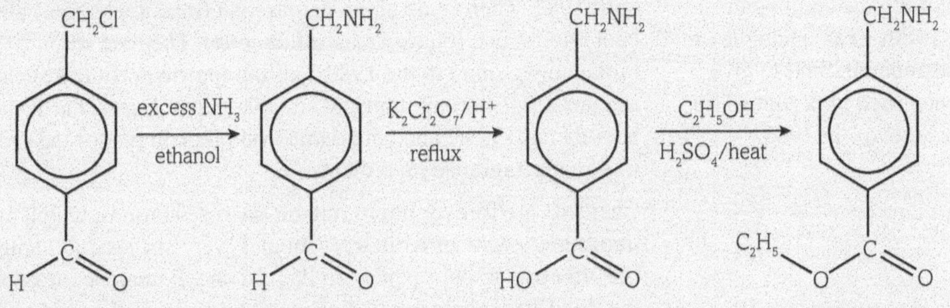

Figure 8 Multi-stage aromatic synthetic route.

Questions

1. State what is meant by a *synthetic route*.

2. Explain why a reduction using sodium borohydride is completed in ethanol rather than in water solution.

3. Design a synthetic route for the preparation of 2-aminoethanoic acid from 2-chloroethanol.

4. Design a synthetic route for the preparation of 3-chloronitrobenzene from benzene. Your answer should include reagents and conditions.

THINKING BIGGER

CHART TOPPERS

The global value of the fragrance industry has been estimated at nearly 38 billion USD in 2014 and is predicted to grow to nearly 43 billion by 2020. Not surprisingly then, a new fragrance can be something of a 'money spinner' if it catches on! In the following extract we will look at the chemistry of some of the most important molecules in the industry.

MOLECULES OF FRAGRANCE AND TASTE

To make a fine fragrance, new molecules are mixed with old favourites. But which fragrance compounds are most valued by industry – a kind of fragrance top 10? It's a difficult question to answer. Gautier has no doubt about the value of Hedione (methyl dihydrojasmonate), used in almost all fine fragrances and Firmenich's top seller in terms of volume. The compound was discovered at Firmenich in the early 1960s as an analogue of methyl jasmonate, a key component of jasmine oil. Said to give a warm floral-jasmine note, Hedione has been used in perfumes for over 40 years, first starring in Dior's *Eau Sauvage* in 1966.

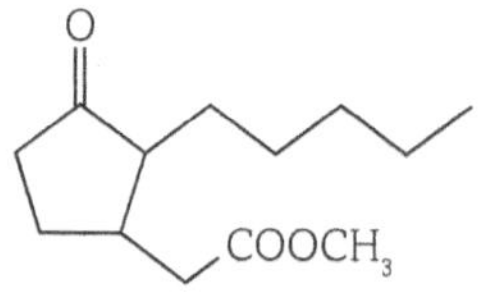

Figure 1 Hedione, from jasmine oil, is used in almost all fine fragrances.

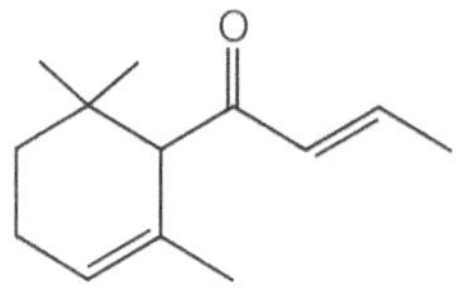

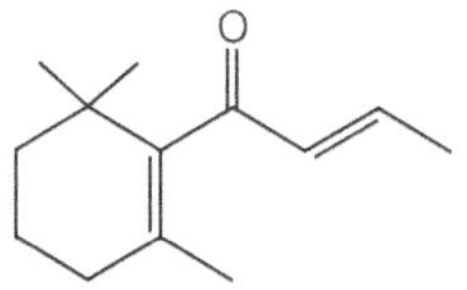

Figure 2 Structure of alpha damascone (top) and beta damascone (bottom).

'The synthesis of Hedione gives four stereoisomers,' explains Gautier. 'Only one of them truly contributes to the odour – (+)-cis-methyl dihydrojasmonate – while the others are thought to modulate other fragrances. Over the years we've been able to synthesise new versions of Hedione which have a much higher quantity of the right isomer.'

Gautier also has a soft spot for the rose ketones: damascenone, and alpha and beta damascone. Firmenich is famed for their discovery and synthesis in the 1960s during a quest to identify the characteristic smell of Bulgarian rose oil. But batch purity problems meant that damascenone was not released commercially until 1982. Then came alpha-damascone (rose-apple note) and beta-damascone (blackcurrant-plum note). The rose ketones broke new ground in the 1980s, giving female perfumes such as Dior's *Poison* their unusual and distinctive fragrance. Today, beta-damascenone and beta-damascone remain two of the most important fragrance ingredients.

Chemists are forever improving on the rose ketones, which are still trendsetters. One of Kraft's recent and 'very successful' captives is Pomarose, a 'cut-open' seco-damascone. Pomarose introduced the dried-fruit character of *Poison* to the male market and had its debut in DKNY's male fragrance *Be delicious Men*. 'It has a very specific note of cooked apple, rose, and dried plums. It's very diffusive so it gives a lot of bloom to a fragrance,' enthuses Kraft. Givaudan's process development team had to try 19 different synthesis routes before it managed to produce Pomarose. 'Initially we thought we couldn't produce it. It was a crazy idea that originated from a 1 per cent impurity for which different structures were proposed based on the NMR spectrum. After we had discovered the correct structure we synthesised the 'wrong proposal' for fun but it turned out to be so powerful,' recalls Kraft. Interestingly, it wasn't the impurity itself that became Pomarose, but one of the alternative structures.

Source
- http://www.rsc.org/chemistryworld/Issues/2009/February/TheSweetScentOf-Success.asp

Where else will I encounter these themes?

Book 1   5.1   5.2

Let us start by considering the nature of the writing in the article.

1. This article is written for members of an international chemistry association and thus relies on the audience having a high degree of scientific literacy. Having read the article a few times, attempt to rewrite the article for a less scientifically literate audience. Can you get the main ideas across without using chemical structures and terminology to the same degree?

Now we will look at the chemistry in, or connected to, this article. Use the timeline at the bottom of the page to help you put this work in context with what you have already learned and what is ahead in your course.

2. a. Work out the molecular formula of hedione from the skeletal structure shown in Figure 1.

   b. Identify the **two** chiral centres in the molecule hedione and explain how **four** isomers are possible.

   c. Write an equation for the hydrolysis reaction of hedione and 2M sodium hydroxide solution and draw structures of the products.

3. a. Figure 2 shows the molecular isomers alpha-damascone and beta-damascone, both found in rose oil. Using the structures, identify which of the two is able to form optical isomers and explain why the other isomer is not.

> In order to work out the molecular formula, it may help to convert the skeletal formula into a full displayed formula.

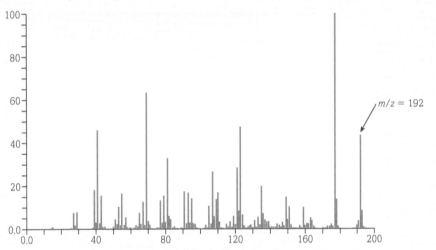

Figure 3 The mass spectrum for alpha-damascone.

> In predicting molecular fragments, it helps to look at the molecule and think about which are the most exposed parts of the molecule that might be knocked off.

   b. The mass spectrum shown in Figure 3 is for alpha-damascone. The molecular ion peak (M) $m/z = 192$ is shown. Suggest what species are responsible for peaks at $m/z = 193$ and 194. Can these peaks be used to help you determine any information about the molecule?

   c. Suggest what species is responsible for the peak at $m/z = 177$. Can you suggest why this ion fragment is so abundant in the spectrum?

DID YOU KNOW?

Ambergris, which can occasionally be found on UK beaches, is in fact sperm whale vomit. As unpleasant as this sounds, people begin to show rather more interest when they realise that it can fetch £60–£70 per gram. Why is it so valuable? Well, it has been used for centuries in the perfume industry as a fixative, which is a substance that 'holds on' to the more volatile components of a perfume thus improving the longevity of the perfume's smell.

Activity

Many different molecules are now used in adding, or augmenting, both the flavour and aroma of foods. Choose one of the following groups of flavour molecules and write a report on their use as flavouring and smell additives:

Thiazoles – Pyrazines – Ketones and aldehydes – Furanones and lactones

Your report should include molecular structures, some examples of their usage, and information about how they are extracted or synthesised.

1. The molecule shown in Figure 1 is an amino acid and is known by the name glutamic acid. What is the IUPAC name for this molecule? [1]

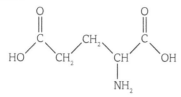

Figure 1

A. dihydroxybutylamine

B. 2-amino pentanendioic acid

C. 1,4-dicarboxypentylamine

D. 1,4-dicarboxybutylamide

2. Nylon-6,6 can be made by a condensation polymerisation reaction between hexane-1,6-diamine and which of the following molecules? [1]

A. hexane-1,6-dioic acid

B. hexane-1,6-dial

C. butane-1,4-dioic acid

D. hexane-1,6-diol

3. Look at the molecule shown in Figure 2. Which of the following statements about this molecule is true? [1]

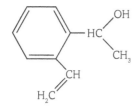

Figure 2

A. The molecule has geometric isomers and positional isomers only.

B. The molecule has optical isomers and positional isomers only.

C. The molecule has optical, positional and geometric isomers.

D. The molecule has geometric and positional isomers only.

4. Why is 4-aminophenol more soluble in an acidic aqueous solution than in a neutral aqueous solution? [1]

A. The OH group is protonated in aqueous acid.

B. The OH group is deprotonated in aqueous acid.

C. The NH_2 group is deprotonated in aqueous acid.

D. The NH_2 group is protonated in aqueous acid.

[Total: 4]

5. Aspirin and paracetamol are commonly available painkillers.

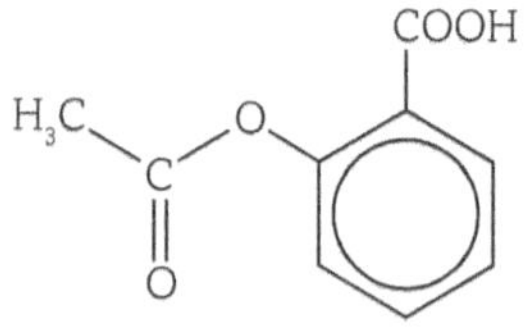

Figure 3 Aspirin

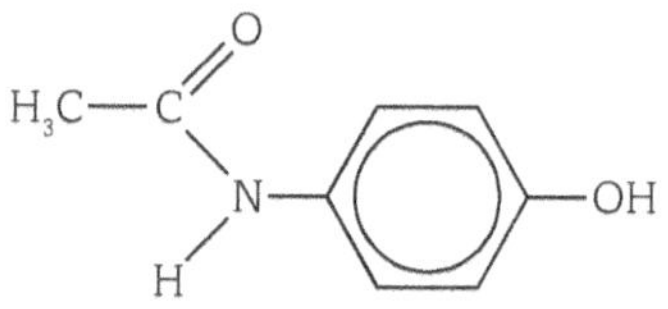

Figure 4 Paracetamol

Aspirin and paracetamol can be prepared using ethanoic anhydride, $(CH_3CO)_2O$.

Some examples of the reactions of ethanoic anhydride are shown below.

reaction 1
$(CH_3CO)_2O + CH_3OH \rightarrow CH_3COOCH_3 + CH_3COOH$

reaction 2
$(CH_3CO)_2O + CH_3NH_2 \rightarrow CH_3CONHCH_3 + CH_3COOH$

reaction 3
$(CH_3CO)_2O + C_6H_5OH \rightarrow CH_3COOC_6H_5 + CH_3COOH$

(a) Draw the structure of a compound that could react with ethanoic anhydride to form aspirin. [1]

(b) Ethanoic anhydride can react with 4-aminophenol to produce paracetamol.

(i) Write an equation, showing structural formulae, for this formation of paracetamol. [2]

(ii) An impurity with molecular formula $C_{10}H_{11}NO_3$ is also formed.

Draw the structure of this impurity. [1]

(iii) Explain why it is necessary for pharmaceutical companies to ensure that drugs and medicines are pure. [1]

(c) Name the functional groups in aspirin and in paracetamol. [2]

(d) A student carried out some reactions with samples of aspirin and paracetamol in the laboratory. Their structures are shown in Figures 3 and 4.

The student tried to react each of the reagents **A**, **B** and **C** with aspirin and paracetamol.
- Reagent **A** reacted with aspirin **and** with paracetamol.
- Reagent **B** reacted **only** with aspirin.
- Reagent **C** reacted **only** with paracetamol.

Suggest possible identities of reagents **A**, **B** and **C** and the organic products that would be formed.

   (i) Reagent **A**:
     Organic product with aspirin:
     Organic product with paracetamol: [3]

   (ii) Reagent **B**:
      Organic product with aspirin: [2]

   (iii) Reagent **C**:
       Organic product with paracetamol: [2]

[Total: 14]

[Q5, F324 Jan 2010]

6. A student was investigating the reactions and uses of organic amines.

(a) The student found that amines such as ethylamine, $C_2H_5NH_2$, and phenylamine, $C_6H_5NH_2$, both behave as bases.

   (i) Explain why amines can behave as bases. [1]

   (ii) The student reacted an excess of $C_2H_5NH_2$ with two different acids.

      Write the formulae of the salts that would be formed when an **excess** of $C_2H_5NH_2$ reacts with:
      - sulfuric acid
      - ethanoic acid. [2]

(b) The student reacted phenylamine with a mixture of $NaNO_2(aq)$ and $HCl(aq)$ whilst keeping the temperature below 10 °C. A diazonium ion was formed. The student then reacted the diazonium ion with compound **B**. After neutralisation, compound **A** was formed.

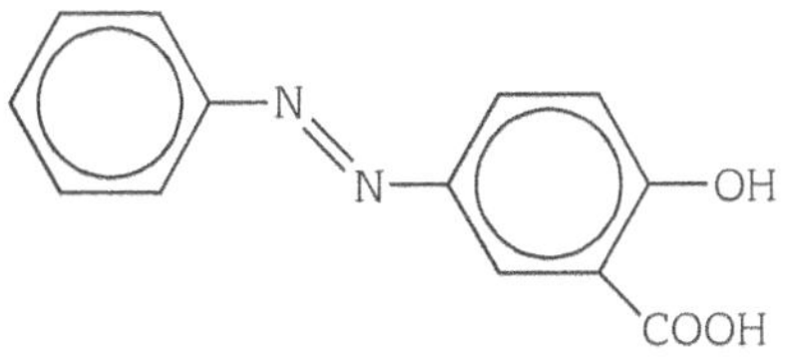

compound A

(i) Draw the structures of the diazonium **ion** and compound **B**.

Display the functional group in the diazonium ion. [2]

(ii) State the conditions required for the reaction of the diazonium ion with compound **B** and state a possible use for compound **A**. [1]

(iii) The student added Na_2CO_3 to a solution of compound **A**.

Draw the structure of the organic product and state the formulae of any other products from this reaction. [2]

(c) The student repeated the experiment in part **(b)** but allowed the temperature to rise above 10 °C.

Under these conditions, the diazonium **ion** in **(b)(i)** reacts with water to produce phenol. A gas with molar mass of $28.0 \, g \, mol^{-1}$ and one other product are also formed.

Construct an equation for this reaction. [1]

[Total: 9]

[Q2, F324 June 2011]

7. This question concerns the amino acid phenylalanine shown below:

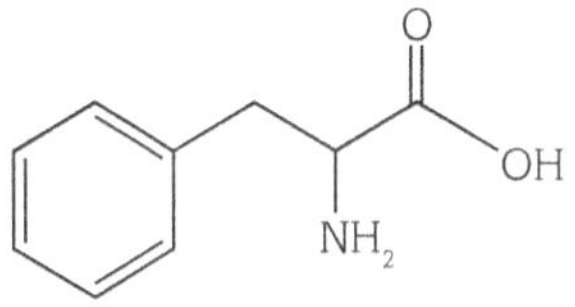

(a) Give the IUPAC name for phenylalanine. [1]

(b) Uing one or more molecules of phenylalanine explain, using diagrams where appropriate, what is meant by the following:

   (i) a zwitterion

   (ii) a chiral centre

   (iii) an amide bond. [6]

Organic chemistry and analysis

ANALYSIS

Introduction

Industry is always developing chemicals for existing, or new, applications. Computational chemists use advanced computer software to design new molecules. These engineered molecules have diverse markets, from new antibiotic medicines to super-strong adhesives. The substances are specifically designed with their end use in mind; the structure and functional groups they contain are carefully chosen as these determine their properties. Process chemists look at the target molecule and design synthetic routes in order to take a cheap, easily available chemical and change it to the desired product.

Once the synthetic route has been designed, chemical engineers will test the process. Analysis is then completed on samples from different stages, to ensure that the desired chemical reactions are occurring. Any modifications take place and the synthesis is scaled up to an industrial production. Analysis is key in quality control to ensure the yield and product quality are maintained in the plant.

Analysis techniques are also very important in crime detection. Simple laboratory tests can be used by Scene of Crime Officers (SOCO) to indicate the presence of chemicals such as gunshot residue or drugs. Then, the more expensive and sensitive instrumental analysis is completed in a laboratory to gain quantitative information.

All the maths you need

To unlock the puzzles of this chapter you need the following maths:

* Substitute numbers into formulae to calculate quantities (*e.g. calculation of R_f values in chromatography*)
* Understand and use ratios (*e.g. understanding integration values on NMR trace to determine the number of equivalent atoms*)
* Interpret data (*e.g. using the n+1 rule to determine the neighbouring equivalent protons in NMR spectroscopy*)

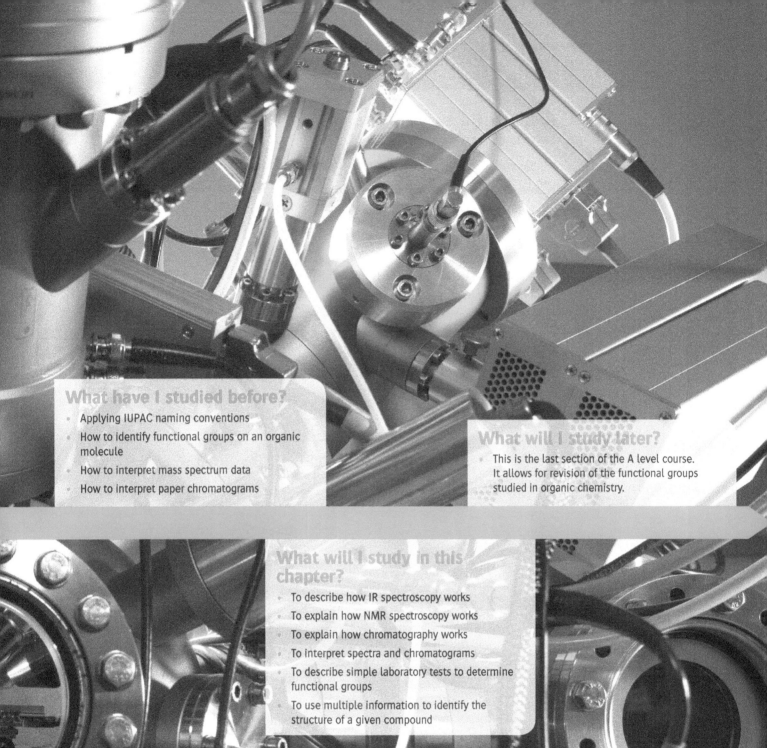

What have I studied before?

- Applying IUPAC naming conventions
- How to identify functional groups on an organic molecule
- How to interpret mass spectrum data
- How to interpret paper chromatograms

What will I study later?

- This is the last section of the A level course. It allows for revision of the functional groups studied in organic chemistry.

What will I study in this chapter?

- To describe how IR spectroscopy works
- To explain how NMR spectroscopy works
- To explain how chromatography works
- To interpret spectra and chromatograms
- To describe simple laboratory tests to determine functional groups
- To use multiple information to identify the structure of a given compound

(1) Chromatography

By the end of this topic, you should be able to demonstrate and apply your knowledge and understanding of:

* interpretation of one-way TLC chromatograms in terms of R_f values

* interpretation of gas chromatograms in terms of:
  (i) retention times
  (ii) the amounts and proportions of the components in a mixture

KEY DEFINITIONS

R_f **value** is a comparison between how far a component has moved compared to the solvent in thin layer chromatography.
Retention time is the time taken for a component to travel from the inlet to the detector in a gas chromatograph.

Chromatography

Chromatography is a separation technique that can be used to identify chemicals in a mixture. It requires two phases.

* **Stationary phase** – fixed in place. This could be a liquid, as in some gas chromatography (GC) systems, where the mixture components have different attractions to each phase. Or it could be a solid, as in thin layer chromatography (TLC), where the components of the mixture are adsorbed by different amounts.
* **Mobile phase** – moves in a definite direction. This could be an inert gas as in GC or a liquid solvent as in TLC.

The separation occurs as each chemical in a mixture has different attractions to each phase. A substance that is very attracted to the mobile phase but not to the stationary phase will travel a long way up the chromatogram but a chemical that is very attracted to the stationary phase will not travel far.

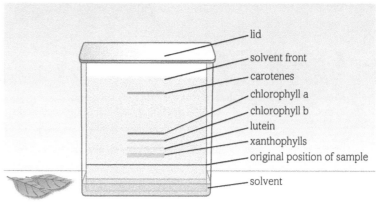

Figure 1 Separation of plant pigments by TLC as each component of the mixture is attracted to the stationary and mobile phases by different amounts.

One-way thin layer chromatography (TLC)

TLC is a quick and easy technique that can be used to check the purity of a sample or to determine the extent of a chemical reaction.

The stationary phase is a thin piece of inert material, such as glass, covered with an absorbent chemical, such as aluminium oxide. The mobile phase is usually an organic solvent, which will move in one direction, vertically, up the plate and develop the chromatogram.

Figure 2 The chromatogram develops on the TLC plate in a tank with the solvent at the bottom. The lid reduces the rate at which the solvent is absorbed and allows for a better separation.

R_f values

An R_f **value** is a quantity that shows how far a chemical has moved up a chromatogram compared to the solvent front. It can be expressed as:

$$R_f = \frac{\text{distance moved by component}}{\text{distance moved by solvent}}$$

Data tables list R_f values for a wide variety of chemicals and this allows the identification of substances in a sample. The values are different if the phases are changed, so it is important to use the correct table for the mobile and stationary phases in your experiment. Alternatively, control spots can be run on the same TLC plate and a direct comparison can be made.

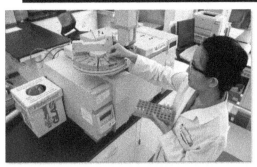

Figure 4 GC machine.

WORKED EXAMPLE

Look at the chromatogram shown in Figure 3. The solvent has moved 4.85 cm. Use this data to calculate the R_f values of each component of the mixture.

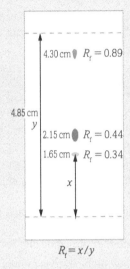

Figure 3

The green spot has travelled 1.65 cm from the base line, so its R_f value $= \dfrac{1.65}{4.85} = 0.34$

The pink spot has travelled 2.15 cm from the base line, so its R_f value $= \dfrac{2.15}{4.85} = 0.44$

The blue spot has travelled 4.30 cm from the base line, so its R_f value $= \dfrac{4.30}{4.85} = 0.89$

LEARNING TIP

Similar compounds will have similar R_f values and also chemicals that are very attracted to the mobile phase may not be separated using this method.

Gas chromatography (GC)

The GC technique is used to separate volatile components in a mixture. The gas chromatograph is often connected to a mass spectrometer so that the separate parts of the mixture can undergo further analysis.

The stationary phase is a solid or liquid coating on a coiled tube; this coating is usually a hydrocarbon with a high boiling point. The mobile phase is an unreactive carrier gas such as helium or nitrogen.

Results

GC will produce a chart of absorption against time. The x-axis is the **retention time** of a chemical or the time it takes for a component to pass from the column inlet to the detector. This value can be compared to known values to identify the component. As with TLC, retention times for the same chemical will be different when different phases are used, so it is important to compare with the correct data table.

The area under the absorption peaks is proportional to the concentration of each component. The greater the area under the peak, the larger the amount of that chemical in the original mixture.

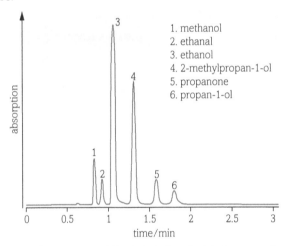

1. methanol
2. ethanal
3. ethanol
4. 2-methylpropan-1-ol
5. propanone
6. propan-1-ol

Figure 5 Gas chromatogram of blood alcohols and related compounds showing retention time. You can estimate the relative concentrations of each compound by comparing the peak areas.

Questions

1 Calculate the R_f value for a component that has moved 2.5 cm, when the solvent has moved 5.0 cm in a TLC experiment.

2 Explain what the position of, and area beneath, a peak in a gas chromatogram shows.

3 Suggest why GC is usually used as part of another analytical technique.

Tests for organic functional groups

By the end of this topic, you should be able to demonstrate and apply your knowledge and understanding of:

* qualitative analysis of organic functional groups on a test-tube scale

KEY DEFINITION

Qualitative analysis is an observable change and does not involve observations using numerical values.

Simple laboratory tests for organic functional groups

In Book 1, topic 4.2.7 you learned that simple practical tests can be used to show the presence of certain functional groups. The table below summarises these test-tube reactions and the resulting observations (**qualitative analysis**):

Functional group	Reagents	Observations
Unsaturation (alkenes)	Bromine water	Add a few drops of bromine water to the sample and shake well. If the bromine water decolourises then the compound was unsaturated.
Haloalkanes	Aqueous silver nitrate and ethanol	If a precipitate forms then the halogen can be inferred. White precipitate is silver chloride and implies a chloroalkane. Cream precipitate is silver bromide and implies a bromoalkane. Yellow precipitate is silver iodide and implies an iodoalkane.
Carbonyl	Brady's reagent (2,4-DNP)	A yellow or orange precipitate is formed when a ketone or aldehyde is present.
Aldehydes	Fehling's solution Tollen's reagent Acidified potassium dichromate	With Fehling's solution, the blue solution forms a dark red precipitate. With Tollens' reagent, a silver mirror is produced. With acidified potassium dichromate, the orange solution turns green.
Aliphatic carboxylic acids	Universal indicator or pH probe Reactive metal, e.g. magnesium Metal carbonate, e.g. calcium carbonate	Check that the pH is that for a weak acid. Carboxylic acids will cause effervescence with reactive metals or metal carbonates.
Phenols	Universal indicator or pH probe No reaction with carbonate ion (CO_3^{2-})	Check that the pH is that for a weak acid. No effervescence with the carbonate ion.
Alcohols	Acidified potassium dichromate	Colour change from orange to green as first an aldehyde is formed and then a carboxylic acid. Secondary alcohols will be oxidised to a ketone but not a carboxylic acid and will also produce this colour change. Tertiary alcohols cannot be oxidised in this way and there is not an observable change.

Table 1 Simple practical tests to identify functional groups.

LEARNING TIP

You may need to perform more than one test-tube reaction to determine all the functional groups present in an organic molecule.

Questions

1. An organic chemical produces a cream-coloured precipitate when reacted with aqueous silver nitrate and ethanol. It also has a pH of 5 but does not react with calcium carbonate. Suggest what this information tells us about the organic chemical.

2. Explain why phenol has a pH of 5 but does not react with sodium carbonate.

3. (a) Explain why refluxing with acidified potassium dichromate could not be used to distinguish between propan-1-ol and propan-2-ol.

   (b) Explain how you could distinguish between propan-1-ol and propan-2-ol using a series of chemical tests.

③ Introduction to nuclear magnetic resonance

By the end of this topic, you should be able to demonstrate and apply your knowledge and understanding of:

* the use of tetramethylsilane, TMS, as the standard for chemical shift measurements
* the need for deuterated solvents, e.g. $CDCl_3$, when running an NMR spectrum

---

KEY DEFINITIONS

TMS is an internal standard for both carbon and proton NMR.
Chemical shift is the scale that compares the frequency of NMR absorption with the frequency of the reference peak of TMS.
Deuterium is an isotope of hydrogen and does not produce a signal in the proton NMR spectrum.

Nuclear magnetic resonance

Nuclear magnetic resonance (NMR) is a non-destructive analytical technique. The spectrum can confirm the formula mass of an organic molecule and give clues to the structure of the chemical.

Some isotopes, such as ^1H and ^{13}C, have a property known as spin. In an NMR spectrometer, organic compounds containing these isotopes are put in a large magnetic field. The direction of spin of the nucleons aligns with the direction of the magnetic field. Energy in the form of radio waves is used to force the spins to flip and change direction. The energy required to change the alignment depends on the environment that the atom is in and can be compared to a standard.

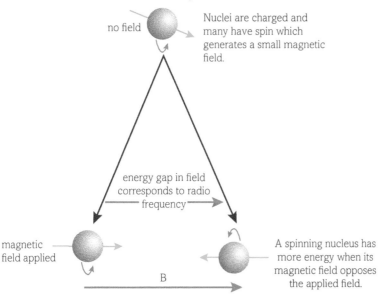

Figure 2 Atoms with spin will align themselves to a magnetic field. Energy can be used to flip the alignment. This information can be used to produce an NMR spectrum.

Figure 1 NMR spectrometer. A superconducting magnet is cooled to 4 K by liquid helium. The large cylinder is a giant thermos flask used to keep the helium liquid.

DID YOU KNOW?

^1H, ^{13}C, ^{19}F and ^{31}P have the property of spin. But in A level chemistry we only study ^1H, known as proton NMR, and ^{13}C, known as carbon NMR.

Internal standard

A reference chemical is added to the NMR spectrometer to compare the values from the sample. In both proton (^1H) and carbon (^{13}C) NMR, tetramethylsilane is used. This chemical is also known as **TMS** and has the formula $(CH_3)_4Si$. The spectrometer makes a relative scale (chemical shift) for the x-axis of the spectrum. The TMS reference peak is given the value 0 ppm and all other peaks are placed on the x-axis compared to it.

Figure 3 Tetramethylsilane (TMS), $(CH_3)_4Si$.

TMS is a good internal standard because:

- it contains both carbon and hydrogen so can be used for carbon and proton NMR
- it produces one sharp signal as hydrogen atoms are in a single environment. As there is only one carbon atom, this will also produce one signal in carbon-13 NMR
- it is non-toxic
- it has a low boiling point and is volatile so it can easily be removed from the sample, although the sample is rarely recovered
- it is inert so unlikely to react with the chemical that is being investigated.

Chemical shift

Chemical shift is the x-axis variable on an NMR spectrum. It is given the symbol delta, δ, and is measured in parts per million (ppm). TMS, which is the internal standard, is given 0 ppm on the far right of the axis. In proton NMR (^{1}H) the values are from 0 to about 15 ppm, but in carbon NMR (^{13}C) the scale is from 0 to about 220 ppm.

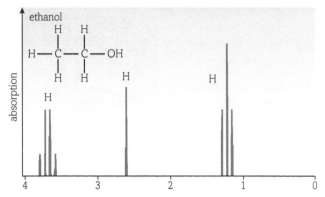

Figure 4 Proton NMR of ethanol.

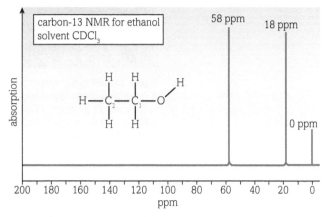

Figure 5 Carbon NMR spectrum of ethanol.

The more similar the protons or carbon atoms are to the symmetrical, non-polar TMS atoms, the lower the chemical shift values. So, a proton in ethane would have a low chemical shift, but a proton on the end of the alcohol group in ethanol would have a very high chemical shift.

Solvents

Most NMR spectroscopic analysis is achieved by first dissolving the molecule in a solvent. As this technique is mainly used to investigate organic compounds, organic solvents like benzene and trichloromethane are needed. However, these chemicals contain carbon and hydrogen and they would produce signals that would swamp the spectrum and make it unreadable.

Hydrogen has three isotopes; but only ^{1}H has the property of spin and is active in NMR. So, we use organic solvents that contain hydrogen isotopes that are not NMR active. The **deuterium** isotope (^{2}H) is used and is shown by using a 'D' in the formula. As deuterium has a greater atomic mass compared to the more abundant ^{1}H isotope, these deuterated solvents are often called 'heavy'. Therefore, trichloromethane, $CHCl_3$ (also known as chloroform), would become heavy trichloromethane, $CDCl_3$.

Figure 6 NMR samples are prepared using a deuterated solvent and a tiny amount of TMS. Both the solvent and TMS can be evaporated off after the analysis and this recovers the original sample. Sample recovery is only done if the yield of product was very small or further analysis is required.

LEARNING TIP

A ^{1}H atom is known as a proton in NMR spectroscopy.

Questions

1. State the formula of the deuterated solvents for:
   (a) water
   (b) benzene.

2. State the unit of chemical shift.

3. Explain why an internal standard is added to NMR machines.

4. Suggest why water cannot be used as a solvent for proton NMR but can for carbon-13 NMR.

(4) Carbon-13 NMR spectroscopy

By the end of this topic, you should be able to demonstrate and apply your knowledge and understanding of:

* analysis of a carbon-13 NMR spectrum of an organic molecule to make predictions about:
  - (i) the number of carbon environments in the molecule
  - (ii) the different types of carbon environment present, from chemical shift values
  - (iii) possible structures for the molecule
* prediction of a carbon-13 or proton NMR spectrum for a given molecule

Carbon-13 NMR spectroscopy

Carbon has three isotopes but only ^{13}C has the property of spin and can be detected in an NMR spectrometer. The NMR spectrum gives information about the number of carbon environments but not the ratio of atoms in each environment.

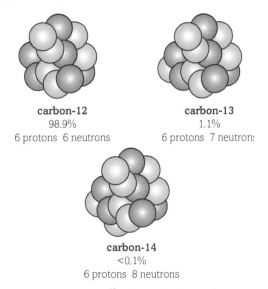

carbon-12
98.9%
6 protons 6 neutrons

carbon-13
1.1%
6 protons 7 neutrons

carbon-14
<0.1%
6 protons 8 neutrons

Figure 1 Although carbon has three isotopes it is only ^{13}C that can be detected in an NMR spectrometer.

TMS is used as the internal standard and the chemical shift has large range (0–220 ppm). There is one signal as there is only one atom found.

In a sample, carbon atoms in the same environment are known as *equivalents*. They are bonded to the same atoms and feel the same magnetic field in the NMR spectrometer.

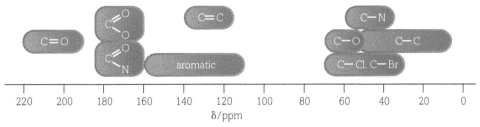

Figure 2 Carbon-13 chemical shifts.

Interpreting carbon-13 NMR spectra

On the back of your data sheet in your examination, you are given a chart that shows you the range of chemical shifts for carbon atoms in different environments. This can be used to give clues to the structure of the organic chemical.

WORKED EXAMPLE 1

The carbon-13 NMR spectrum for propan-1-ol is shown in Figure 3.

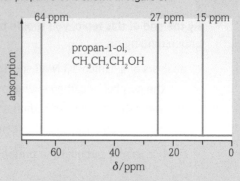

Figure 3

Propan-1-ol has the structural formula $CH_3CH_2CH_2OH$. It contains three carbon atoms, all in different environments. This accounts for the three peaks.

The carbon atom attached to the alcohol functional group will have the largest chemical shift as it is the most different from the internal standard TMS. Using the chart on the data sheet we can see that this carbon atom should have a shift of between 50 and 70 ppm. Looking at the spectrum, we can see the actual value is 64 ppm and within this range.

The other two carbon environments are more similar to the carbon atom in TMS and will have a shift of between 10 and 55 ppm. Looking at the spectrum they have a value of 27 ppm for the $-CH_2-$ carbon and 15 ppm for the $-CH_3$ carbon. This is due to the CH_3 groups giving signals at lower frequency than CH_2 groups.

WORKED EXAMPLE 2

Propan-2-ol has the structural formula $CH_3CH(OH)CH_3$. Although it contains three carbon atoms, due to the plane of symmetry, it only has two carbon environments. This means there will be two peaks, as there are two equivalent carbon atoms. These are drawn in a 2 : 1 ratio that shows how the peak height corresponds to the number of atoms.

The carbon-13 NMR spectrum for propan-2-ol is shown in Figure 4.

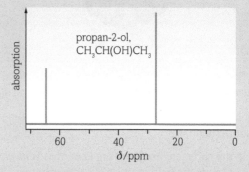

Figure 4

LEARNING TIP

You do not need to learn the chemical shift values as they are given to you as a chart on the data sheet in the examination.

Making predictions

The ^{13}C spectrum along with the molecular formula can be used to determine the structure of a compound. It is important to analyse the spectrum closely, considering the number of carbon environments and using the data sheet to suggest the functional groups that the carbon atoms may be joined to.

WORKED EXAMPLE 3

Determine the structure of a carbonyl compound with the molecular formula C_3H_6O and the ^{13}C NMR spectrum shown in Figure 5.

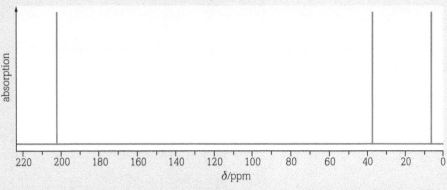

Figure 5

In the question we were told that the compound has a carbonyl group, C=O. Using the molecular formula we also know there are three carbon atoms so it could be propanal or propanone. The spectrum shows three signals, showing all the carbon atoms are in different environments. So, the chemical must be propanal.

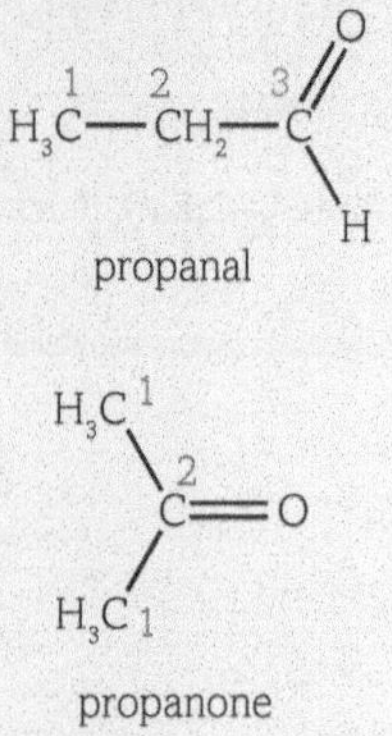

propanal

propanone

Figure 6

Using the data sheet, we can assign each peak to the relevant carbon atom, as shown in Table 1.

Chemical shift/ppm	Group	Carbon number
205	C=O	3
37	CH₂	2
6	CH₃	1

Table 1

WORKED EXAMPLE 4

An aromatic compound has the molecular formula C_8H_{10}. Use the carbon-13 NMR spectrum of the compound (shown in Figure 7) to identify the compound.

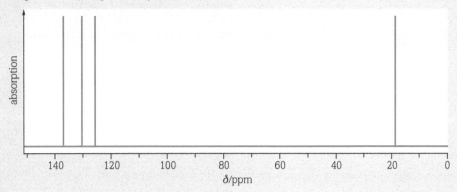

Figure 7

From the question we know that the compound has an aromatic ring which will be made of 6 carbon atoms. Using the molecular formula you can deduce that there are two further carbon atoms that must be substituted onto the aromatic ring. This could be as two methyl groups or an ethyl group.
Draw all the possible isomers and consider the number of peaks that they would form in the ^{13}C NMR spectrum.

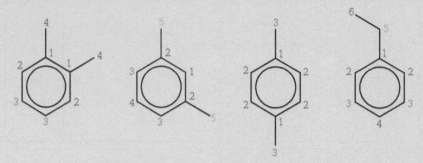

Figure 8

Looking at the ^{13}C NMR spectrum, there are 4 peaks and so the compound must be 1,2-dimethylbenzene.

Questions

1 State the internal standard that is used for carbon-13 NMR.

2 State which isotope of carbon is detected in NMR.

3 A carbonyl compound has the molecular formula C_4H_8O. Its carbon-13 NMR spectrum is shown in Figure 9.

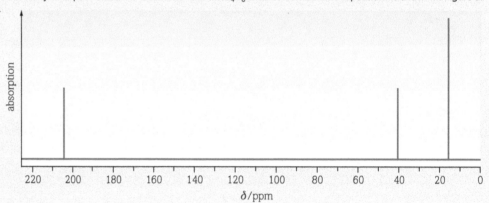

Figure 9

(a) Draw the possible structural isomers of the carbonyl compound, C_4H_8O.

(b) State how many carbon environments each isomer has.

(c) Identify the isomer of C_4H_8O that produces the carbon-13 spectrum above.

(d) Explain whether or not the other isomers of C_4H_8O that are carbonyl compounds could be identified from their carbon-13 NMR spectra.

(5) Proton NMR spectroscopy

By the end of this topic, you should be able to demonstrate and apply your knowledge and understanding of:

* analysis of a high-resolution proton NMR spectrum of an organic molecule to make predictions about:

    (i) the number of proton environments in the molecule

    (ii) the different types of proton environment present, from chemical shift values

    (iii) the relative numbers of each type of proton present from relative peak areas, using integration traces or ratio numbers, when required

    (iv) the number of non-equivalent protons adjacent to a given proton from the spin–spin splitting pattern, using the $n + 1$ rule

    (v) possible structures for the molecule

* prediction of a carbon-13 or proton NMR spectrum for a given molecule

---

KEY DEFINITION

Equivalent protons are hydrogen atoms bonded to the same atoms that therefore experience the same magnetic field in the NMR spectrometer.

Proton NMR spectroscopy

The most common hydrogen isotope, ^1H, has the property of spin and can be detected in an NMR machine. The NMR spectrum gives information about the number of proton environments and the area under each peak is the ratio of protons in each environment.

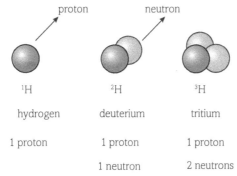

^1H	^2H	^3H
hydrogen	deuterium	tritium
1 proton	1 proton	1 proton
	1 neutron	2 neutrons

Figure 1 Although hydrogen has three isotopes, it is only ^1H that can be detected in an NMR machine.

TMS is used as the internal standard as this molecule has 12 **equivalent protons** and gives an intense single peak, which can be assigned as 0 ppm. The range of chemical shifts for proton NMR is much smaller than for carbon NMR.

For proton NMR spectroscopy a much smaller sample size can be used compared to ^{13}C NMR spectroscopy. There will be a signal for each environment that a hydrogen atom is found in. The area under each signal gives the ratio of atoms in each environment. This is often shown as an integration trace added by the NMR machine directly onto the spectrum.

Example 1

Ethanal has the structural formula CH_3CHO. Although there are four hydrogen atoms, there are actually only two proton environments. So, the proton NMR spectrum will contain two signals: one signal will be close to TMS from the $–CH_3$ protons and one signal will have a higher chemical shift from the CHO proton.

The integration trace shows that the area under the CHO signal is $\frac{1}{3}$ of the area under the $-CH_3$ signal as the ratio of these protons is $1:3$.

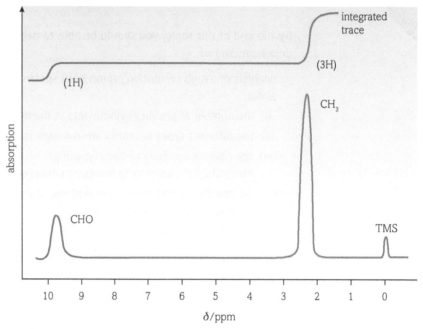

Figure 2 Low-resolution proton NMR for ethanal.

LEARNING TIP

On the back of your data sheet in your examination, you are given a chart which shows you the range of chemical shifts for hydrogen atoms in different environments. This can be used to give clues to the structure of the organic chemical.

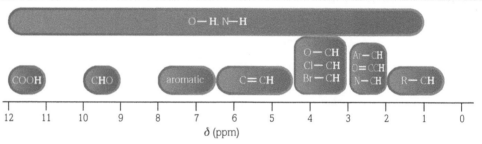

Figure 3 Proton chemical shifts.

Spin–spin coupling

In high-resolution NMR spectra it is possible to see that some signals are actually made from a cluster of peaks. This is known as a splitting pattern caused by spin–spin coupling of neighbouring protons.

When the organic chemical is put into the NMR spectrometer, it experiences a large magnetic field. However, the protons on the neighbouring carbon atoms also affect the magnetic field. This affects the alignment of the proton and causes a split in the signal. Different numbers of adjacent protons will cause different numbers of splits.

The splitting pattern can be predicted using a simple rule:

Number of peaks in the splitting pattern = number of protons on adjacent carbon atom + 1

This is often expressed as the $n + 1$ rule, where n is the number of equivalent adjacent hydrogen atoms on a neighbouring carbon.

The different splitting patterns are summarised in Table 1.

n	$n + 1$	Multiplet	Ratio of peak areas within multiplet
0	$0 + 1 = 1$	Singlet	1
1	$1 + 1 = 2$	Doublet	$1:1$
2	$2 + 1 = 3$	Triplet	$1:2:1$
3	$3 + 1 = 4$	Quartet	$1:3:3:1$

Table 1 Spin–spin coupling patterns and the $n + 1$ rule. You don't need to memorise the ratio of peaks but do you notice its Pascal triangle?

Example 2

The high-resolution proton NMR spectrum for methyl propanoate ($CH_3CH_2COOCH_3$) is shown in Figure 4.

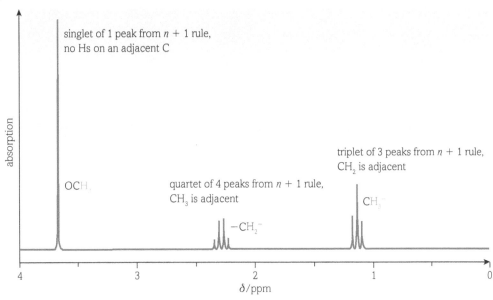

Figure showing labels: "singlet of 1 peak from $n + 1$ rule, no Hs on an adjacent C", "OCH_3", "quartet of 4 peaks from $n + 1$ rule, CH_3 is adjacent", "$-CH_2-$", "triplet of 3 peaks from $n + 1$ rule, CH_2 is adjacent", "CH_3". Axis labelled absorption (vertical) and δ/ppm (horizontal, 4 to 0).

Figure 4 Proton NMR spectrum of methyl propanoate showing splitting patterns.

The singlet is produced by the $-OCH_3$ as there are no neighbouring carbon atoms and hence no adjacent protons. So, $n = 0$ and $n + 1 = 1$, hence a singlet.

The quartet is produced by the CH_2 as $4 = n + 1$, so $n = 3$. This implies that there are 3 adjacent protons.

The triplet is produced by the CH_3 as $3 = n + 1$, so $n = 2$. This implies that there are 2 adjacent protons.

> **LEARNING TIP**
> * A singlet does not necessarily mean there is always only one hydrogen atom in that environment. There are two hydrogens in a $-NH_2$ singlet and three hydrogens in an $-OCH_3$ singlet.
> * A doublet always implies that there is one hydrogen attached to the neighbouring atom.
> * A triplet does not necessarily mean there is always a neighbouring $-CH_2$ group. There could be one hydrogen attached to a neighbouring atom and another hydrogen attached to a different neighbouring atom.
> * A quartet does not necessarily mean there is always a neighbouring $-CH_3$ group. There could be one neighbouring $-CH_2$ group and another hydrogen attached to a different neighbouring atom.
> * The triplet–quartet pairing is very common and often implies a CH_3CH_2- section in the molecule.

Questions

1 State what is meant by an *equivalent proton*.

2 Explain what a triplet-quartet pairing suggests in a proton NMR.

3 For each low-resolution proton NMR spectrum below, identify the proton(s) responsible for each labelled peak.

(a) $CH_3CHBrCH_3$

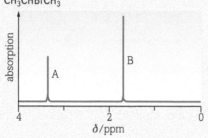

(b) $CH_3COOCH_2CH_3$

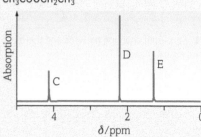

4 In the spectra below, identify the sequence in the molecule that produces the splitting patterns.

(a)

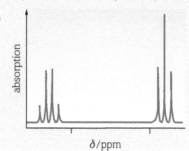

(b)

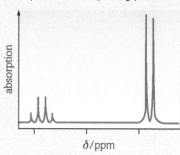

(c)

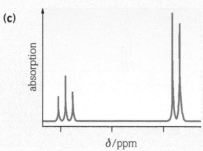

(d)

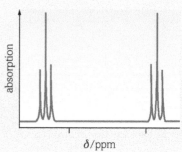

(6) NMR spectra of –OH and –NH protons

By the end of this topic, you should be able to demonstrate and apply your knowledge and understanding of:

* the identification of O–H and N–H protons by proton exchange using D_2O

NMR spectra of compounds with –OH and –NH protons

We have not yet looked at proton NMR spectra of compounds containing protons in hydroxyl groups, –OH, or amino groups, –NH.

It can be difficult to identify –OH and –NH protons because:

* peaks can appear over a wide range of different chemical shift values, depending on the solvent used and the concentration of the sample
* the signals are often broad
* there is usually no splitting pattern.

These factors not only make it difficult to assign an absorption to an –OH or –NH proton, but may also lead to confusion with other peaks. Consequently an –OH or –NH absorption may mistakenly be assigned to another type of proton on the basis of an unpredictable chemical shift value.

What is needed is a way of filtering out absorptions due to –OH and –NH protons – and this is where deuterium oxide, D_2O, comes in.

Use of D_2O

In the previous topic you met the use of $CDCl_3$ as a solvent in NMR spectroscopy. Deuterium oxide, D_2O, is the same chemical compound as water, H_2O, but with the isotope deuterium, ^2H or D, in place of ^1H, the common isotope of hydrogen. This is why the common name for deuterium oxide is 'heavy water'.

Water and heavy water react identically in chemical reactions, but they have different physical properties. D_2O has an M_r value of 20.0 and a density of $1.11\ g\ cm^{-3}$, so it is truly heavier than normal water. D_2O melts at 3.8 °C and boils at 101.4 °C. Most importantly for NMR spectroscopy, the deuterium nucleus does not produce a signal in a proton NMR spectrum (remember that an odd number of nucleons is needed for NMR).

D_2O is used by following these stages:

* first a proton NMR spectrum is run
* a small amount of D_2O is added to the sample solution and the mixture is shaken (vigorous shaking is essential, as $CDCl_3$ and D_2O are immiscible)
* then a second proton NMR spectrum is run, and any peak due to –OH or –NH protons disappears.

This works because the deuterium in D_2O exchanges with the H present in –OH and –NH.

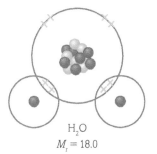
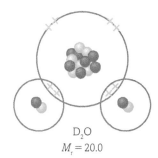

H_2O
$M_r = 18.0$

D_2O
$M_r = 20.0$

Figure 1 Water, H_2O, and heavy water, D_2O. Note that not all of the nucleons in the oxygen atom are visible in this illustration.

For example, when D_2O is added to ethanol, CH_3CH_2OH, deuterium exchanges with the –OH proton:

$$CH_3CH_2OH + D_2O \rightleftharpoons CH_3CH_2OD + HOD$$

When the second NMR spectrum is run, the material being analysed is now CH_3CH_2OD rather than CH_3CH_2OH. In the absence of the –OH proton there will obviously be no –OH signal.

These illustrations show that absorptions due to –OH and –NH protons can be easily identified by simply comparing the two spectra.

Splitting from –OH and –NH protons

As a general rule, NMR peaks for –OH or –NH protons are not split. An –OH or –NH peak usually shows as a singlet, which may be broad.

* It is difficult to get solvents that are absolutely dry – traces of water in the solvent form hydrogen bonds with –OH and –NH protons in the compound being analysed. This results in the broadening of –OH or –NH signals.

* Protons on adjacent carbon atoms are not split by the –OH or –NH; neither is the –OH or –NH proton split itself.

> **LEARNING TIP**
> As far as splitting is concerned, just ignore –OH and –NH.

In the NMR spectra of ethanol in Figure 2, the protons in adjacent CH_3 and CH_2 groups are split:

* the CH_3 signal at $\delta = 1.1$ ppm is split into a triplet by the adjacent CH_2 ($n + 1$ rule: $2 + 1 = 3$)

* the CH_2 signal at $\delta = 3.6$ ppm is split into a quartet by the adjacent CH_3 ($n + 1$ rule: $3 + 1 = 4$).

The adjacent OH group does not split this signal.

WORKED EXAMPLE

Compare the spectra of ethanol, C_2H_5OH, in Figure 2(a) with no D_2O, and Figure 2(b) with D_2O added.

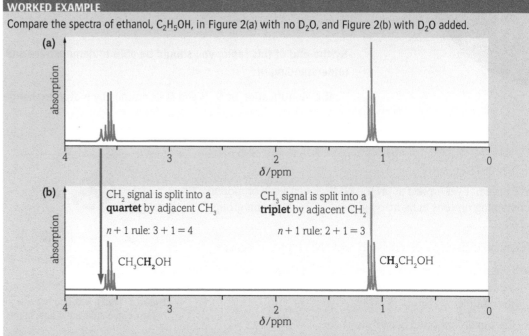

(b)

CH$_2$ signal is split into a **quartet** by adjacent CH$_3$

$n + 1$ rule: $3 + 1 = 4$

CH$_3$**CH$_2$**OH

CH$_3$ signal is split into a **triplet** by adjacent CH$_2$

$n + 1$ rule: $2 + 1 = 3$

CH$_3$CH$_2$OH

Figure 2 Proton NMR spectrum of ethanol, C_2H_5OH; (a) without D_2O and (b) with D_2O.

After D_2O is added, the —OH peak disappears.

Questions

1 State why D_2O is used in NMR spectra.

2 Explain how heavy water is:
(a) similar to water   (b) different from water.

3 How many peaks, and in what ratio, would you expect in the proton NMR spectra of the following compounds run without D_2O and with D_2O?
(a) glycolic acid, $HOCH_2COOH$
(b) glycine, H_2NCH_2COOH
(c) lactic acid, $CH_3CHOHCOOH$

4 The proton NMR spectra in Figure 3 are for a compound with molecular formula $C_3H_7NO_2$. Analyse the spectra to suggest a structure for this compound.

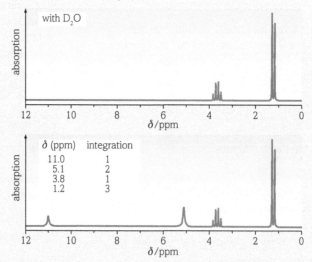

with D_2O

δ (ppm)	integration
11.0	1
5.1	2
3.8	1
1.2	3

Figure 3 Compound with molecular formula $C_3H_7NO_2$. The upper spectrum has been run after addition of D_2O.

By the end of this topic, you should be able to demonstrate and apply your knowledge and understanding of:

* deduction of the structures of organic compounds from different analytical data including:
  (i) elemental analysis
  (ii) mass spectra
  (iii) IR spectra
  (iv) NMR spectra

Mass spectrometry

Chemical analysis of an unknown compound can provide the percentage by mass of each element. The empirical formula can be determined from the results. A mass spectrum gives the relative molecular mass of a compound. Using the empirical formula, the molecular formula can then be found.

Unfortunately, the same M_r value could apply to more than one compound with the same molecular formula – for example, CH_3COOH and $HCOOCH_3$ *both* have an M_r value of 60.0.

Fragmentation patterns also give clues about the carbon skeleton in the molecules of a compound.

Infrared spectroscopy

In Book 1, topic 4.2.7, we saw how simple chemical tests can be used to identify functional groups. However, if multiple functional groups are present, these methods can be time-consuming and the chemical that is being analysed can be chemically changed in the process. However, infrared spectroscopy gives information about the bonds present in a molecule and the likely functional groups present, while being a non-destructive technique.

However, different members of a homologous series have the same functional groups – for example, different alcohols have C–O and O–H absorptions in their IR spectra.

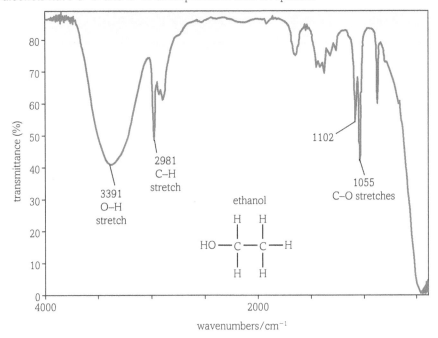

Figure 1 An IR spectrum.

NMR spectroscopy

A carbon-13 NMR spectrum gives information on the number and types of carbon environments in a molecule.

Proton NMR spectroscopy gives additional information about the number of each type of proton environment and the number of protons on adjacent carbon atoms.

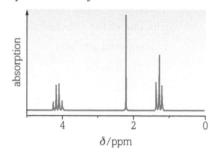

Figure 2 An NMR spectrum.

Table 1 compares the information that can be obtained from carbon-13 and proton NMR spectra.

Feature	Carbon-13 NMR	Proton NMR
number of peaks	number of types of C	number of types of H
chemical shift	type of C δ range 0–220 ppm	type of H δ range 0–12 ppm
peak area	relative number of carbon atoms of each type	relative number of protons of each type
spin–spin coupling	–	number of protons on adjacent carbon

Table 1 Comparison of information from ^{13}C and ^{1}H NMR spectra.

Often, a single technique is inconclusive and a combination of techniques is used in practice. This topic looks at how combining several different techniques can lead to the identification of an organic compound, and builds upon the techniques covered in Book 1, topic 4.2.12.

Overall, NMR spectroscopy gives more information about organic molecules than MS or IR spectroscopy.

WORKED EXAMPLE

The empirical formula of an unknown compound is C_2H_4O. Use the three spectra in Figure 3 to suggest a possible structure for the compound.
To solve the problem, evidence has been annotated on each spectrum.

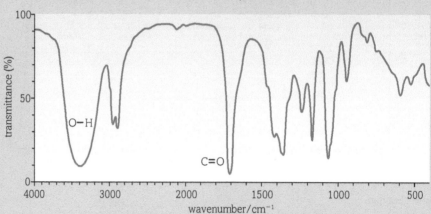

LEARNING TIP

Use your data sheet to find key absorptions that you would expect on an IR spectrum. There are three common ones to look out for:

- C=O around 1700 cm^{-1}
- O–H around 3200–3500 cm^{-1}
- O–H (carboxylic acid) around 2500–3300 cm^{-1}.

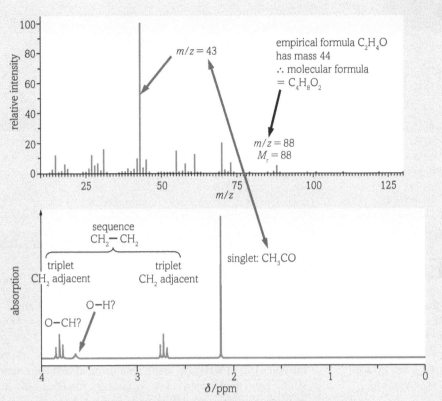

Figure 3 IR spectrum, mass spectrum and proton NMR spectrum for an unknown compound.

Tying together the information is like doing a jigsaw puzzle.

- Look at the different pieces of information and get a structure that fits. In this example, when contrasting the IR spectrum, there are –OH and C=O, which suggests that the molecular formula is at least double the empirical formula.
- The proton NMR gives the most information, but D_2O would have helped to confirm the O–H.
- CH_3CO from the NMR is backed up by the fragment ion in the mass spectrum at $m/z = 43$, suggesting CH_3CO^+.

Linking everything together suggests the structure in Figure 4.

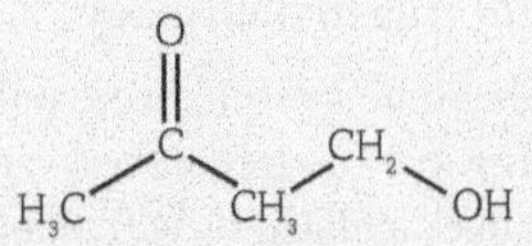

Figure 4 Suggested structure for the unknown compound.

LEARNING TIP

In a mass spectrum, first find the peak with the highest m/z value to get the mass.
Then look for fragment ions to identify parts of the skeleton of the molecule.

LEARNING TIP

In an NMR spectrum, look for δ values for types of proton or carbon; look for splitting for protons on adjacent carbon atoms; use the integration trace for number of each type of proton.

Questions

1. Discuss the benefits of using IR spectra rather than simple laboratory experiments to identify a molecule.
2. Explain why more than one analytical technique is often used, rather than just one technique in isolation.
3. Suggest values for the relative peak areas for the protons in the NMR spectrum in Figure 3.

THINKING BIGGER

MAKING PAIN HISTORY

In the last few years, newspaper articles have regularly appeared urging 'Daily aspirin dose for "everyone over 45"' and 'If aspirin is a miracle drug, shouldn't we all be taking it?' and even 'daily aspirin "prevents and possibly treats cancer"'. Aspirin has a long history of usage in medicine, which the following extract considers.

THE DEVELOPMENT OF ASPIRIN

Around 250 years ago, a riverside walk was the first step in the story of aspirin. The story takes in the rise of the global pharmaceutical industry, two world wars, one Nobel prize and the drug is still the most widely taken in the world, with new therapeutic uses still emerging. But it all started when Edward Stone announced that powdered willow bark was a helpful treatment for malaria.

This pharmaceutical revolution had a long prehistory. Ancient Sumerian and Egyptian texts recommended willow bark for various complaints, and Greek, Roman and Islamic medical authors noted its power to relieve pain and reduce fevers. However, its tendency to cause inflammation – and occasionally bleeding – of the stomach lining considerably diminished its utility. Possibly for this reason, medieval and early modern European physicians usually ignored it (though in some areas it survived as a folk remedy).

Once the botanical resources of the Americas were accessible, Peruvian quinine became the preferred treatment for fevers. But quinine was expensive, so Stone was pleased to find a locally available remedy for the malarial fever – or 'ague', as it was then called – which afflicted him. His success encouraged other sufferers to try it, even though he was merely an amateur scientist without medical qualifications.

Ague, Stone noted, was generally associated with marshy environments and their bad air (malaria) – this was long before mosquitoes were identified as the vectors of malaria. He was also aware of the ancient herbalists' doctrine of signatures: 'many natural maladies carry their cures along with them, or that their remedies lie not far from their causes'. And so, during a riverside walk in 1757, he tried chewing some willow bark. Since its bitter taste resembled that of quinine, he hoped it might relieve his discomfort.

Stone then decided to pursue a more systematic approach. After drying the bark over a baker's oven, he pounded and sieved it thoroughly. Then, by taking gradually increasing doses of it, he showed that two scruples (about 2.5 g) every four hours dispelled his fever. Over the next five years he treated about fifty other sufferers with his preparation, and although it could not eradicate their malaria entirely, it gave welcome relief from the symptoms.

From tree to pharmacy

After Stone's results were published, some apothecaries began using willow bark to treat fevers, and in 1828 the German chemist Joseph Buchner isolated its active ingredient. He called the yellowish, bitter-tasting substance salicin (salix being the Latin for willow).

Source

- http://www.rsc.org/chemistryworld/2012/12/aspirin-history

Where else will I encounter these themes?

Book 1   5.1   5.2

Let us start by considering the nature of the writing in the article.

1. Who do you think is the intended audience for this article?

Now we will look at the chemistry in, or connected to, this article.

2. Give the IUPAC name for salicylic acid and calculate its relative molecular mass.

3. How many different aromatic proton environments are there in salicylic acid? Explain your answer.

4. Draw a structural isomer of salicylic acid that has only two aromatic proton environments and explain why this is the case.

5. A sample of acetyl salicylate dissolved in a deuterated solvent analysed using ^{1}H-NMR spectroscopy shows a peak at about 12 ppm, which gradually disappears over time.

   a. Explain what is meant by a deuterated solvent and explain why such a solvent is used in ^{1}H-NMR spectroscopic analysis.

   b. Suggest which proton gives rise to the peak at 12 ppm and why it gradually disappears over time.

6. Acetyl salicylate (2-ethanoyloxobenzene carboxylic acid) is a monoprotic acid with a K_a of $2.4 \times 10^{-4}\,\mathrm{mol\,dm^{-3}}$.

   a. Explain what is meant by the term monoprotic acid.

   b. Calculate the pH of a 0.002 M solution of acetyl salicylate.

It will help to draw out the full displayed formula of the molecule showing all of the H atoms.

You can assume acetyl salicylate is a **weak** acid.

Activity

In recent times many claims have been made about aspirin as a 'wonder drug' including as a cure for cancers, such as lung and skin, and a way of preventing stroke and heart disease. Equally, some concerns have been expressed about the damaging effects of taking aspirin – such as pancreatitis.

Carry out a web survey about the various claims and counterclaims made for aspirin. How do you go about evaluating the reliability of the claims and counterclaims? Are there any vested interests in promoting aspirin as a 'miracle cure'? You could choose to present either a balanced argument OR select your information to promote one side of the argument or the other.

Practice questions

1. Compound X is a colourless liquid. It forms an orange precipitate with 2,4-DNP but does not react with Tollens' reagent. Proton NMR analysis shows a single peak. What could X be? [1]

   A. propanone

   B. methanal

   C. phenol

   D. trichloromethanol

2. A mixture of three compounds P, Q and R are put on to a GC column for analysis. The compounds emerge in the order R, P, and finally Q. What does this suggest? [1]

   A. Q has the largest molar mass.

   B. P is more volatile than R.

   C. R is the least soluble in the stationary phase.

   D. R is the least soluble in the mobile phase.

3. Compound G contains carbon, hydrogen and oxygen only. 4.8 g of compound G undergo combustion analysis to give 11.73 g of CO_2 and 4.8 g of H_2O. Mass spectrum analysis of compound G gives a molecular ion peak at $m/z = 72$. What could compound G be? [1]

   A. butan-2-ol

   B. cyclobutanone

   C. ethyl ethanoate

   D. butanone

4. Which types of molecular isomer are least likely to be distinguishable using proton NMR spectroscopy alone? [1]

   A. geometric isomers

   B. optical isomers

   C. functional group isomers

   D. positional isomers

   [Total: 4]

5. α-Amino acids are found in human sweat. A student had read that chromatography could be used to separate and identify the amino acids present in human sweat.

   (a) The student used thin-layer chromatography (TLC) to separate the α-amino acids in a sample of human sweat and discovered that three different α-amino acids were present.

   (i) Name the process by which TLC separates α-amino acids. [1]

   (ii) The chromatogram was treated to show the positions of the separated α-amino acids. Explain how the student could analyse the chromatogram to identify the three α-amino acids that were present. [2]

   (iii) Several α-amino acids have structures that are very similar. Suggest why this could cause problems when using TLC to analyse mixtures of α-amino acids. [1]

   (b) Some of the α-amino acids found in human sweat are shown in **Table 1**.

α-amino acid	R group
glycine	H
leucine	$CH_2CH(CH_3)_2$
isoleucine	$CH(CH_3)CH_2CH_3$
alanine	CH_3
valine	$CH(CH_3)_2$
lysine	$(CH_2)_4NH_2$
glutamic acid	$(CH_2)_2COOH$

   Table 1

   (i) State the general formula of an α-amino acid. [1]

   (ii) There are four stereoisomers of isoleucine. One of the stereoisomers is shown below.

   Draw 3D diagrams for the other **three** stereoisomers of isoleucine. [3]

   (c) α-Amino acids form different ions at different pH values. Zwitterions are formed when the pH is equal to the isoelectric point of the α-amino acid.

   The isoelectric points of three α-amino acids are given below:

alanine, **pH = 6.0**	**glutamic acid,** **pH = 3.2**	**lysine,** **pH = 9.7**

   Draw the structures of the ions formed by these α-amino acids at the pH values below. Refer to **Table 1** above.

alanine at **pH = 6.0**	**glutamic acid** **at pH = 10**	**lysine at** **pH = 2.0**

   [3]

   (d) α-Amino acids can react to form polypeptides. A short section of a polypeptide is shown below.

Name the α-amino acid sequence in this section of the polypeptide. Refer to **Table 1**. [1]

(e) Synthetic polyamides, such as nylon, contain the same link as polypeptides. Nylon is the general name for a family of polyamides. A short section of a nylon polymer is shown below.

$$\cdots\overset{O}{\overset{\|}{C}}-(CH_2)_8-\overset{H}{\overset{|}{C}}-\overset{H}{\overset{|}{N}}-(CH_2)_6-\overset{H}{\overset{|}{N}}-\overset{O}{\overset{\|}{C}}-(CH_2)_8-\overset{H}{\overset{|}{C}}-\overset{H}{\overset{|}{N}}-(CH_2)_6-\overset{H}{\overset{|}{N}}\cdots$$

Draw the structures of **two** monomers that could be used to make this nylon. [2]

[Total: 14]

[Q3, F324 Jan 2010]

6. 'Methylglyoxal', CH_3COCHO, is formed in the body during metabolism.

Describe **one** reduction reaction and **one** oxidation reaction of methylglyoxal that could be carried out in the laboratory.

Your answer should include reagents, equations and observations, if any. [5]

[Total: 5]

[Q4, F324 June 2011]

7. Pyruvic acid, shown below, is an organic compound that has a smell similar to ethanoic acid. It is extremely soluble in water.

$$H_3C-\overset{O}{\overset{\|}{C}}-\overset{\overset{O}{\|}}{\underset{}{C}}-OH$$

pyruvic acid

(a) Explain why pyruvic acid is soluble in water. Use a labelled diagram to support your answer. [2]

(b) Pyruvic acid can be prepared in the laboratory by reacting propane-1,2-diol with excess acidified potassium dichromate(VI). The reaction mixture is heated under reflux.

Write an equation for this oxidation. Use **[O]** to represent the oxidising agent and show structural formulae for organic compounds. [2]

(c) Pyruvic acid can also be reduced by $NaBH_4$ to form $CH_3CH(OH)COOH$.

Outline the mechanism for this reduction. Use curly arrows and show relevant dipoles. [4]

(d) Compound **A**, shown below, is a structural isomer of pyruvic acid.

$$\overset{O}{\overset{\|}{C}}\!\!\begin{matrix}H\\|\end{matrix}\!\!-\overset{H}{\underset{H}{\overset{|}{C}}}-\overset{O}{\overset{\|}{C}}\begin{matrix}\\OH\end{matrix}$$

compound A

Describe a chemical test that could be carried out in a laboratory to distinguish between samples of pyruvic acid and compound **A**.

Your answer should include reagents, observations, the type of reaction and the organic product formed.

In your answer, you should use appropriate technical terms, spelled correctly. [3]

(e) Compound **B** is an organic compound used to make some cosmetics. Compound **B** contains C, H and O only.

Elemental analysis shows that **B** has the percentage composition by mass: C, 55.81%; H, 7.02%; O, 37.17%.

The mass spectrum of compound **B** is shown below.

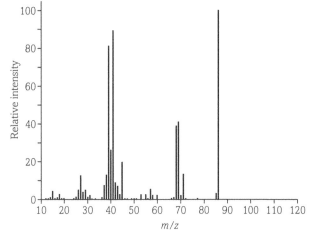

(i) Determine the **molecular** formula of compound **B**. Show all of your working. [2]

(ii) Compound **B** is an *E*-stereoisomer.

Compound **B** effervesces with aqueous Na_2CO_3 to form organic compound **C**.

Compound **B** decolourises Br_2 to form compound **D**.

Compound **B** polymerises to form polymer **E**.

- Draw structures of compounds **B**, **C** and **D**.
- Draw **one** repeat unit for polymer **E**. [4]

[Total: 17]

[Q1, F324 Jan 2013]

Maths skills

In order to be able to develop your skills, knowledge and understanding in chemistry, you will need to have developed your mathematical skills in a number of key areas. This section gives more explanation and examples of some key mathematical concepts you need to understand. Further examples relevant to your A Level Chemistry studies are given throughout the book and in Book 1.

Using logarithms

Calculating logarithms

Many formulae in science and mathematics involve powers. Consider the equation:

$$10^x = 62$$

The value of x obviously lies between 1 and 2, but how can you find a precise answer? The term logarithm means 'index' or 'power'. Logarithms enable you to solve such equations. You can take the logarithm to base 10 of each side of the equation using the **log** button of a calculator.

> **EXAMPLE**
> $$10^x = 62$$
> $$\log_{10}(10^x) = \log_{10}(62)$$
> $$x = 1.792392...$$

You can calculate logarithms using any number as the base by using the **log$_x$(y)** button.

> **EXAMPLE**
> $$2^x = 7$$
> $$\log_2(2^x) = \log_2(7)$$
> $$x = 2.807355...$$

Many equations relating to the natural world involve powers of e. These are called exponentials. The logarithm to base e is referred to as the natural logarithm and written **ln**.

Using logarithmic plots

An earthquake measuring 8.0 on the Richter scale is much more than twice as powerful as an earthquake measuring 4.0 on the Richter scale, because the units used for measuring earthquakes are logarithmic. Logarithmic scales in charts and graphs can accommodate enormous increases or decreases in one variable as another variable changes.

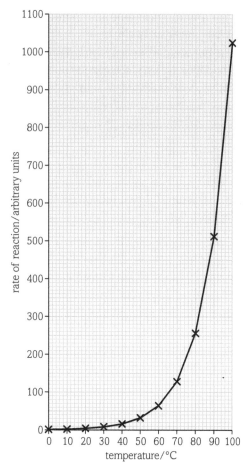

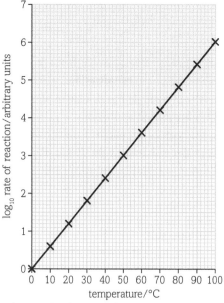

Figure 1 Logarithmic scales (lower graph) are useful when representing a very large range of values, such as in the case of rates of reaction.

Graphs

Understanding that $y = mx + c$ represents a linear relationship

Two variables have a linear relationship if they increase at a constant rate in relation to one another. If you plot a graph with one such variable on the x-axis and the other on the y-axis, you get a straight line.

Any linear relationship can be represented by the equation $y = mx + c$, where the gradient of the line is m and the value at which the line crosses the y-axis is c. An example of a linear relationship is the relationship between degrees Celsius and degrees Fahrenheit, which can be represented by the equation $F = 9/5C + 32$, where C is temperature in degrees Celsius and F is temperature in degrees Fahrenheit.

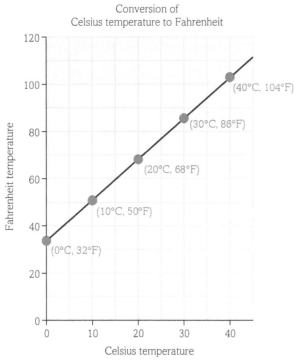

Figure 2 Linear relationship between Fahrenheit and Celsius temperature measurements.

Using the gradient as a measure of a rate of change

Sir Isaac Newton drew tangents to curves on graphs to find the rates of change of the variables as part of his journey towards discovering calculus – a fascinating branch of mathematics. He stated that the gradient of a curve at a given point is exactly equal to the gradient of the tangent to the curve at that point.

To find the gradient at a point on a curve:

1. Use a ruler to draw a tangent to the curve at that point.

2. Calculate the gradient of the tangent using the equation for a linear relationship. This gradient is equal to the gradient of the curve at the point of the tangent.

3. Include the unit with your answer.

Applying your skills

You will often find that you need to use more than one maths technique to answer a question. In this section, you will look at two example questions and consider which maths skills are required and how to apply them.

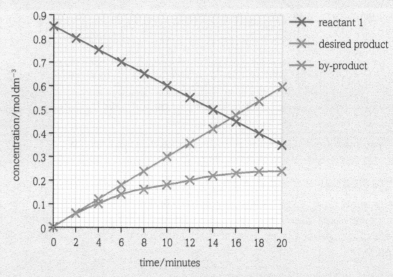

Take care to use the correct units. Concentration is in $mol\,dm^{-3}$ but time is in minutes, so you must convert time to seconds to get the correct answer.

concentration after 8 minutes $= 0.24\,mol\,dm^{-3}$

concentration after 12 minutes $= 0.36\,mol\,dm^{-3}$

change of concentration $= 0.12\,mol\,dm^{-3}$

time taken for change $= 4 \times 60 = 240\,s$

$$\text{rate of change} = \frac{\text{change in concentration of by-product}}{\text{time taken for change}}$$

$$= \frac{0.12\,mol\,dm^{-3}}{240\,s} = 0.0005\,mol\,dm^{-3}\,s^{-1}$$

(c) To find the rate of change of desired product is a little trickier because the line is curved. To do this, you need to draw a tangent to the curve then calculate the gradient of the tangent. A simple way to do this is to draw a triangle with the tangent as its hypotenuse, centred on the 10-minute point as asked (Figure 4).

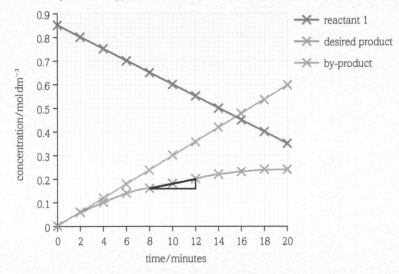

Figure 4 Finding the rate of change of concentration of the desired product by drawing a tangent.

change in concentration between 8 and 12 minutes $= 0.04\,mol\,dm^{-3}$

time taken for change $= 4 \times 60 = 240\,s$

$$\text{rate of change} = \frac{\text{change in concentration of desired product}}{\text{time taken for change}}$$

$$= \frac{0.04\,mol\,dm^{-3}}{240\,s} = 0.0001666... = 0.0002\,mol\,dm^{-3}\,s^{-1}$$

Preparing for your exams

Introduction

- A Level students will sit three exam papers, each covering content from both years of A Level learning. All three papers will include synoptic questions that may draw on two or more different topics.

- A Level students will also have their competency in key practical skills assessed by their teacher in order to gain the Practical Endorsement in Chemistry. The endorsement will not contribute to the overall grade but the result (pass or not classified) will be recorded on the certificate.

The table below gives details of the exam papers for the A Level qualification.

A Level exam papers

A Level Chemistry A (H432)					
ASSESSMENT OVERVIEW					
Paper			Marks	Duration	Weighting
Paper 1	Periodic table, elements and physical chemistry (content from Modules 1, 2, 3 and 5, including experimental methods)		100	2 hr 15 min	37%
	Section A	Multiple choice	15		
	Section B	Structured response and extended writing including two 'level of response' questions	85		
Paper 2	Synthesis and analytical techniques (content from Modules 1, 2, 4 and 6, including experimental methods)		100	2 hr 15 min	37%
	Section A	Multiple choice	15		
	Section B	Structured response and extended writing including two 'level of response' questions	85		
Paper 3	Unified chemistry (content from Modules 1–6, including experimental methods)		70	1 hr 30 min	26%
	Structured response and extended writing including two 'level of response' questions		70		
Non-exam assessment	Practical Endorsement in Chemistry		Pass/Not classified	N/A	Reported separately
	Assessed by teacher throughout course. Does not count towards A Level grade but result (pass or not classified) will be reported on A Level certificate. You will need to maintain a separate record of practical activities carried out during the course.				

Exam strategy

Arrive equipped

Make sure that you have all of the correct equipment needed for your exam. In a transparent bag or pencil case, you should have at least:

- pen (black, ink or ball-point)
- pencil (HB)
- 30 cm rule
- eraser which will not smudge
- scientific calculator.

Ensure your answers can be read

Your handwriting needs to be clear so that the examiner can read it easily. Even when you are in a hurry, make sure that the key words are easy to read.

Plan your time

Think about how many minutes you have to complete the paper and how many marks are available in total. This will give you an idea of how long to spend on each question. Use a rough guide of a minute per mark but remember that short answer and multiple choice questions may be quicker, while longer answers may require more time. Be sure to leave some time at the end of the exam for checking your answers.

Understand the question

Always read the question carefully and spend a few moments working out what you are being asked to do. The command word used will give you an indication of what is required in your answer.

Be scientific and accurate and use the technical terms you've been taught. Always show your workings for any calculations:

marks may be available for individual steps, correct units, or recall of relevant formulae, as well as for the final answer. Remember also that you may receive marks for the correct technique even if there is a numerical error in your calculations.

Plan your answer

For the extended response questions, marks will be awarded for your ability to structure your answer logically. You should show how the points are related to or follow on from each other. Read the question carefully and plan what you want to say before beginning the full answer.

Make the most of graphs and diagrams

Diagrams and sketch graphs can be used to gain marks, sometimes more quickly and clearly than a prose explanation. However, they must be drawn carefully.

If you are asked to read a graph, pay attention to the labels and numbers on the x- and y-axes. Remember that each axis is a number line.

If you are asked to draw or sketch a graph, make sure you use a sensible scale and label both axes with quantities and units. If plotting a graph, use a pencil and draw small crosses for the points.

Diagrams must always be neat, clear and fully labelled.

Check your answers

For extended response questions, check that you have made a distinct point for each mark available. For calculations, read through each stage of your working. You may be able to check your final answer by substituting it into the original question. Another simple strategy is to consider whether the answer seems sensible. Always check the units and the degree of accuracy that you have reported your answer to.

1. Copper(II) ions form an aqueous complex ion, $CuCl_4^{2-}$, with chloride ions. Read the following statements.

    (i) $CuCl_4^{2-}$ has a yellow colour

    (ii) $CuCl_4^{2-}$ has optical isomers

    (iii) $CuCl_4^{2-}$ has a square planar shape

    Which statement(s) is/are true?   [1]

    A. (i), (ii) and (iii)

    B. Only (i) and (ii)

    C. Only (ii) and (iii)

    D. Only (i)

> Consider each of the three statements separately and then make your choice from answer options A–D.
>
> $CuCl_4^{2-}$ does have a yellow colour, so statement (i) is correct.
>
> $CuCl_4^{2-}$ does not have any optical isomers because there is no asymmetric centre to the complex, so statement (ii) is incorrect.
>
> The chloride ion is bulky and so chloride complexes are tetrahedral. Statement (iii) is incorrect.

Question analysis

- Multiple choice questions always have one mark and the answer is given. For this reason, students often make the mistake of thinking that they are the easiest questions on the paper. Unfortunately, this is not the case: these questions often require you to work out several answers, and an error in one answer may lead you to choose the wrong answer to the question as a whole. In addition, the incorrect answers supplied (distractors) will include the answers you will arrive at if you make typical or common errors. The trick with multiple choice questions is to answer the question before you look at any of the answers.

- This type of multiple choice question is particularly tricky, since you need to consider all three statements before you can select the correct answer. Do not be afraid to mark the exam paper and write 'correct' or 'incorrect' next to each statement. Make sure that you have plenty of practice with this type of question.

- If you have any time left at the end of the paper, multiple choice questions should be high on your list of priority for checking.

Sample student answer

B

Verdict

This is a weak answer because although the student has recalled that this copper complex is yellow, they also think that it has optical isomers. This cannot be the case because there is no asymmetric centre.

Sample question – structured response

2. Chlorine and bromine are elements in Group 7 of the Periodic Table. Both elements exist in a number of different oxidation states and are therefore involved in many redox reactions. Chlorine dioxide reacts with cold, dilute aqueous sodium hydroxide. The equation for the reaction is:

$$2ClO_2(aq) + 2OH^-(aq) \rightarrow ClO_2^-(aq) + ClO_3^-(aq) + H_2O(l)$$

Using oxidation numbers, explain why the chlorine in ClO_2 has undergone disproportionation. [3]

Disproportionation is the simultaneous oxidation and reduction of an element in the same reaction. In order to show that disproportionation has occurred, you will need to calculate the oxidation state of chlorine in the reacting ClO_2 and in the products, ClO_2^- and ClO_3^-. If disproportionation has occurred, the oxidation state will have increased for one of the products and decreased for the other.

Question analysis

The command word in this question is 'explain'. This requires a justification or exemplification of a point – in this case, that chlorine has undergone disproportionation. In some such questions, a mathematical explanation could also be used.

Sample student answer

In ClO_2 chlorine has an oxidation state of +4. When it reacts, its oxidation state changes to +3 in ClO_2^- and +5 in ClO_3^-. This means that its oxidation state has both increased and decreased in the same reaction: it has been oxidised and reduced at the same time.

Verdict

This is an excellent answer because the student has calculated all three oxidation states correctly and shown that they understand the term disproportionation.

Sample question – calculation

3. A buffer consists of 0.200 mol dm^{-3} ethanoic acid and 0.100 mol dm^{-3} sodium ethanoate. What is the pH of the buffer? (K_a for ethanoic acid is 1.7×10^{-5}) [2]

The important thing with calculations is that you must show your working clearly and fully. The correct answer will gain all the available marks; however, an incorrect answer can gain all but one of the available marks if your working is shown and is correct. Show the calculation that you are performing at each stage and, when you have finished, look at your result and see if it seems sensible.

Question analysis

The command word in this question is 'calculate'. This means that you need to obtain a numerical answer to the question, showing relevant working. If the answer has a unit, this must be included.

To calculate the pH of a buffer, you need to write the expression for the equilibrium constant and then substitute in the known values to work out the hydrogen ion concentration. Remember, it is assumed that all of the ethanoate ions come from the sodium ethanoate and none from the ethanoic acid (hence the concentration in this case is 0.100 mol dm^{-3}). It is also assumed that the ethanoic acid is completely undissociated so that the concentration of HA is 0.200 mol dm^{-3}. The expression pH $= -\log[H^+]$ can then be used to calculate the pH.

Sample student answer

$K_a = [H^+(aq)][A^-(a)]/[HA(aq)] = 1.7 \times 10^{-5}$

$[H^+] = 0.100 \text{ mol dm}^{-3}$

$[A^-(aq)] = 0.200 \text{ mol dm}^{-3}$

$1.7 \times 10^{-5} = [H^+] \times 0.1/0.2$

$[H^+] = 3.4 \times 10^{-5}$

$pH = -\log_{10} 3.4 \times 10^{-5} = 4.47$

Verdict

This is a good answer. The student has correctly used the equilibrium expression for the reaction to calculate the H^+ concentration. For this, they obtained the first of the two available marks. They then went on to use the expression $pH = -\log[H^+]$ to calculate the pH and gain the second mark. The student completed all steps correctly and showed their working.

Sample question – practical

4. Describe the practical steps you would take to determine the standard electrode potential of the zinc half-cell. [5]

Questions asking you to outline a practical procedure usually attract a lot of marks. In the exam, it is easy to miss an obvious step, so it is a good idea to plan your answer before you write it out in full. Run through the process in your head and consider the steps you would take if you were actually carrying out the procedure in the laboratory.

Question analysis

- The command word in this question is 'describe'. This means that you need to give a detailed account of how you would carry out this procedure.

Sample student answer

Prepare the zinc/zinc ion half-cell by placing a strip of zinc into a solution of a zinc salt with a concentration of 1 mol dm^{-3}. Connect the zinc half-cell to the standard hydrogen half-cell. The hydrogen must be at 100 kPa pressure, in contact with a solution with a hydrogen ion concentration of 1 mol dm^{-3}. The temperature must be at 298 K for standard conditions. Connect the metal electrodes of the half-cells to a voltmeter using wires and record the standard cell potential from the voltmeter.

Verdict

This is a strong answer. The student has clearly learned the definitions of standard half-cells because all conditions are quoted correctly. Their answer gains four out of the available five marks because:

* they have correctly described how to prepare a zinc half-cell (1 mark)
* they have correctly identified that this needs to be connected to the standard hydrogen electrode and they have described that electrode correctly (1 mark)
* they have identified that everything needs to be at standard conditions (1 mark)
* they have described how the components need to be incorporated into a circuit, and stated that the reading on the voltmeter will give the standard electrode potential of the zinc half-cell (1 mark).

However, the student loses one mark because they have not mentioned that a salt bridge is needed in order to complete the circuit.

Sample question – extended response

5. Read the following information about organic acid, A, and then determine and draw a possible structure for this molecule. Explain your reasoning using the evidence provided. [5]

   * The molecular formula of A is $C_xH_yO_2$
   * The mass spectrum of A has a molecular ion peak at $m/z = 148$
   * A reacts by both electrophilic substitution and electrophilic addition
   * The 13C NMR spectrum of A contains seven peaks

Think about each piece of information in turn and note down what it means. This alone will gain you a number of marks, even if you are unable to go on to deduce the formula.

For example, the fact that the mass spectrum of A has a molecular ion peak at $m/z = 148$ tells us that the relative mass of the compound is 148.

Question analysis

* This is an extended writing question, so there are a lot of marks available. It is a good idea with this type of question to think about the number of marks available and consider how they might be distributed.
* It is very important that you plan your answer before you write it down. There is always space in the exam paper for you to do this, so jot down the points that you want to make before you begin writing your answer in full. This will help you to structure your answer in a coherent and logical way and ensure you don't end up contradicting yourself. Once you have written your answer, go back and cross out your notes so it is clear that they are not part of your answer.
* The first command word in this question is 'determine'. This means that you need to work something out.
* The second command word in this question is 'explain'. This requires a justification or exemplification of a point.

The mass spectrum of A has a molecular ion peak at $m/z = 148$. This means that the relative mass of A is 148.

A reacts by both electrophilic substitution and electrophilic addition. This tells me that it must contain a benzene ring, because benzene rings undergo electrophilic substitution.

The 13C NMR spectrum of A contains seven peaks. This tells me that there are 7 different types of carbon atom in the molecule.

The molecular formula of A is $C_xH_yO_2$??

Verdict

This is a weak answer because although the student has managed to interpret some of the information given, they have not made much progress towards determining the structure. They have, in fact, gained only one mark for identifying that there is a benzene ring present.

To obtain the other four marks, the student should have included the following information in their answer.

- The fact that the structure undergoes electrophilic addition means that an alkene group is present (1 mark).
- From the relative mass and the relative formula, the student should have been able to deduce that the molecular formula is $C_9H_8O_2$ (1 mark). They should also have drawn a correct structure for A (1 mark).
- For the final mark, they should have shown that there are seven different carbon environments in the structure.

Sample question – extended response

6.*The equilibrium reaction between carbon dioxide and water is important for maintaining the pH of the body by buffering.
The equation for the reaction is:

$$CO_2 + H_2O \rightleftharpoons HCO_3^- + H^+$$

Explain what is meant by the term buffer solution and explain how a buffer solution works by referring to this equation. [6]

This question has two main parts. First, you should explain what a buffer solution is. Then you should use the equation in the question to explain how a buffer solution works. It is vital that you use this specific reaction to explain how a buffer solution works; otherwise, you will not gain full credit for your answer.

Question analysis

- The * beside the question number tells you that this is a 'level of response' question. This type of question is marked holistically. You are not awarded marks for specific points within your answer; instead, your answer is assigned a level (1, 2 or 3) depending on your understanding and explanation of the relevant science content. You are then awarded a higher or lower mark within that level depending on the quality of your answer – whether it is structured logically, whether relevant information has been used and linked together, whether scientific terms are used correctly, and so on.

In a level of response question, the quality of your answer's structure will be assessed explicitly, so it is essential that you plan your answer before you write it down. There is always space in the exam paper for you to do this, so jot down the points that you want to make before you begin writing your answer in full. This will help you to structure your answer in a coherent and logical way and ensure you don't end up contradicting yourself. Once you have written your answer, go back and cross out your notes so it is clear that they are not part of your answer.

The command word in this question is 'explain'. This requires a justification or exemplification of a point.

Sample student answer

A buffer solution is a solution that resists changes in pH when a small amount of acid or alkali is added. So, if an acid is added then the pH won't change very much and if an alkali is added then the pH won't change very much. Buffers can do this because they contain something that can react with any acid added and they contain another chemical that can react with any alkali added so effectively these are removed and this means that the pH doesn't change very much.

Verdict

This is a poor answer: although the student gives a good explanation of what a buffer solution is, they do not use the equation given in the question to explain how a buffer solution works in practice. Instead, they simply repeat the information they have already stated. This answer would be classified as Level 1 (the lowest level) and would gain 1 or 2 marks.

In order to answer the question fully, the student should have made the following points.

This reaction works as a buffer because the equilibrium moves towards the left-hand side when acid is added.

This is because the acid reacts with the HCO_3^+ ion and is removed.

A reasonable amount of acid can be added before the buffer solution fails because there is a large concentration of HCO_3^+ to 'mop up' any hydrogen ions added.

In order to gain full marks, this information would need to be presented logically, making good use of relevant technical language.

Glossary

acid dissociation constant, K_a: the acid dissociation constant of an acid HA is defined as $K_a = [H^+(aq)][A^-(aq)]/[HA(aq)]$.
$K_a = 10^{-pKa}$
$pK_a = -\log_{10} K_a$

activation energy: the minimum energy required to start a reaction by breaking bonds in the reactants.

addition polymer: a very long molecular chain formed by repeated addition reactions of many unsaturated alkene molecules (monomers).

addition reaction: a reaction in which a reactant is added to an unsaturated molecule to make a saturated molecule.

aliphatic hydrocarbon: a hydrocarbon where carbon atoms are joined together in straight or branched chains.

alkali: a base that dissolves in water forming $OH^-(aq)$ ions.

alkanes: the homologous series with the general formula C_nH_{2n+2}.

alkylammonium salt: a compound where the hydrogen(s) on an ammonium ion have been substituted by alkyl chains.

alkylation: addition of hydrocarbon chains to an organic compound.

alkyl group: an alkane with a hydrogen atom removed, e.g. CH_3, C_2H_5; often shown as 'R'.

amide: a class of compound with a functional group made of an acyl group, which is directly attached to an amine.

amount of substance: the quantity whose unit is the mole. Chemists use 'amount of substance' as a means of counting atoms.

amphoteric chemicals: chemicals that can react with both acids and bases.

anion: a negatively charged ion.

anhydrous: a substance that contains no water molecules.

Arrhenius plot: a graph of $\ln k = \ln A - \dfrac{E_a}{R} \times \dfrac{1}{T}$ where $\ln k$ is plotted against $\dfrac{1}{T}$.

atomic orbital: a region of space where it is likely that you will find electrons. Each orbital can hold up to two electrons, with opposite spins.

atomic (proton) number: the number of protons in the nucleus of an atom.

average bond enthalpy: the mean energy needed for 1 mole of a given type of gaseous bonds to undergo homolytic fission.

Avogadro constant: the number of atoms per mole of the carbon-12 isotope (6.02×10^{23} mol^{-1}).

base: a chemical that will react with an acid.

benzene: a naturally occurring aromatic compound, which is a very stable planar ring structure with delocalised electrons.

benzene derivative: a benzene ring that has undergone a substitution reaction.

Boltzmann distribution: the distribution of energies of molecules at a particular temperature, often shown as a graph.

Brønsted–Lowry acid: a proton, H^+, donor.

Brønsted–Lowry base: a proton, H^+, acceptor.

buffer solution: a mixture that minimises pH changes on addition of small amounts of acid or base. The word 'minimises' is essential to this definition.

carbocation: an organic ion in which a carbon atom has a positive charge.

catalyst: a substance that increases the rate of a reaction without being used up in the process.

cation: a positively charged ion.

chemical shift: the scale that compares the frequency of NMR absorption with the frequency of the reference peak of TMS.

chiral carbon: a chiral carbon has four different groups attached to it.

chromatogram: a visible record showing the result of separation of the components of a mixture by chromatography.

***cis–trans* isomerism:** a type of E/Z isomerism where each carbon of the C=C double bond carries the same atom or group.

complex ion: a transition metal ion bonded to one or more ligands by coordinate bonds (dative covalent bonds).

condensation polymerisation: the chemical reaction to form a long-chain molecule by elimination of a small molecule, such as water.

conjugate acid–base pair: two species that transform into each other by gain or loss of a proton.

coordinate bond: *see* **dative covalent bond**.

coordination number: the total number of coordinate bonds formed between a central metal ion and its ligands.

covalent bond: a bond formed by a shared pair of electrons between nuclei.

curly arrow: a symbol used in reaction mechanisms to show the movement of an electron pair.

dative covalent bond: a bond formed by a shared pair of electrons that has been provided by one of the bonding atoms only. Also known as a coordinate bond.

degradable polymer: a polymer that breaks down into smaller fragments when exposed to light, heat or moisture.

delocalised electrons: electrons that are shared between more than two atoms.

deuterium: an isotope of hydrogen and does not produce a signal in the proton NMR spectrum.

directing effect: how a functional group attached directly to an aromatic ring affects which carbon atoms are more likely to undergo substitution.

displayed formula: a formula which shows the relative positions of atoms and the bonds between them.

disproportionation: the oxidation and reduction of the same element in a redox reaction.

distillation: a technique used to separate miscible liquids or solutions.

dynamic equilibrium: the equilibrium that exists in a closed system when the rate of the forward reaction is equal to the rate of the reverse reaction and all the chemicals have their concentrations maintained.

electron structure or configuration: the arrangement of electrons in an atom or ion.

electrophile: an electron-pair acceptor.

electrophilic substitution: a substitution reaction where an electrophile is attracted to an electron-rich atom or part of a molecule and a new covalent bond is formed by the electrophile accepting an electron pair.

elimination reaction: an organic chemical reaction in which one reactant forms two products. Usually a small molecule like water is released.

empirical formula: the simplest whole-number ratio of atoms of each element present in a compound.

enantiomer: an optical isomer.

endothermic: a reaction in which the enthalpy of the products is greater than the enthalpy of the reactants, resulting in heat being taken in from the surroundings (ΔH is positive).

end point: the point in a titration at which there are equal concentrations of the weak acid and conjugate base forms of the indicator. The colour at the end point is midway between the colours of the acid and conjugate base forms.

enthalpy cycle: a pictorial representation showing alternative routes between reactants and products.

enthalpy profile diagram: a diagram of a reaction that allows you to compare the enthalpy of the reactants with the enthalpy of the products.

enthalpy, H: the heat content that is stored in a chemical system.

entropy, S: the quantitative measure of the degree of disorder in a system.

equilibrium law: this law states that for the equilibrium

$$aA + bB \rightleftharpoons cC + dD, K_c = \frac{[C]^c[D]^d}{[A]^a[B]^b}.$$

equivalence point: the point in a titration at which the volume of one solution has reacted exactly with the volume of the second solution. This matches the stoichiometry of the reaction taking place.

equivalent carbon atoms: carbon atoms bonded to the same atom, which therefore experience the same magnetic field in the NMR spectrometer.

equivalent protons: hydrogen atoms bonded to the same atoms that therefore experience the same magnetic field in the NMR spectrometer.

esterification: the chemical reaction which forms an ester.

exothermic: a reaction in which the enthalpy of the products is smaller than the enthalpy of the reactants, resulting in heat loss to the surroundings (ΔH is negative).

E/Z isomerism: a type of stereoisomerism that is caused by the restriction of rotation around the double bond. Two different groups are attached to each carbon atom of the C=C double bond.

first electron affinity: the enthalpy change accompanying the formation of 1 mole of gaseous 1− ions from gaseous atoms.

first ionisation energy: the energy change that accompanies the removal of 1 mole of electrons from 1 mole of gaseous atoms.

free energy change, ΔG: the balance between enthalpy, entropy and temperature for a process: $\Delta G = \Delta H - T\Delta S$. A process can take place spontaneously when $G < 0$.

Friedel–Crafts reaction: a substitution reaction where hydrogen is exchanged for an alkyl, or acyl chain. Friedel–Crafts reactions allow electrophilic substitution to occur on an aromatic ring.

functional group: a group of atoms that is responsible for the characteristic chemical reactions of a compound.

general formula: the simplest algebraic formula for a homologous series.

geometric isomers: molecules that have the same structural formula but a different arrangement in space.

giant ionic lattice: a three-dimensional structure of atoms that are all bonded together by strong covalent bonds.

giant metallic lattice: a three-dimensional structure of positive ions and delocalised electrons, bonded together by strong metallic bonds.

group: a vertical column in the periodic table. Elements in a group have similar chemical properties and their atoms have the same number of outer-shell electrons.

half-life: the time taken for the concentration of a reactant to reduce by half.

Hess' law: states that the enthalpy change in a chemical reaction is independent of the route it takes.

heterogeneous equilibrium: an equilibrium in which species making up the reactants and products are in different physical states.

homogeneous equilibrium: an equilibrium in which all the species making up the reactants and products are in the same physical state.

homologous series: a series of organic compounds that have the same functional group with successive members differing by CH_2.

hydrolysis: a chemical reaction where water is a reactant in a decomposition reaction.

intermediate: a species formed in one step of a multi-step reaction that is used up in a subsequent step, and is not seen as either a reactant or a product of the overall equation.

ion: a positively or negatively charged ion or (covalently bonded) group of atoms (a molecular ion).

ionic bonding: the electrostatic attraction between oppositely charged ions.

ionic product of water, K_w: the ionic product of water is defined as
$K_w = [H^+(aq)][OH^-(aq)]$
At 25 °C, $K_w = 1.00 \times 10^{-14}$ mol² dm⁻⁶.

isoelectric point: the pH value at which the amino acid exists as a zwitterion.

lattice enthalpy, $\Delta_{LE}H^{\ominus}$: the enthalpy change that accompanies the formation of one mole of an ionic lattice from its gaseous ions under standard conditions.

Le Chatelier's principle: states that when a system in dynamic equilibrium is subjected to change, the position of equilibrium will shift to minimise the change.

Lewis acid: an electron-pair acceptor.

ligand: a molecule or ion that can donate a pair of electrons to the transition metal ion to form a coordinate bond.

ligand substitution: a reaction in which one ligand in a complex ion is replaced by another ligand.

lone pair: an outer-shell pair of electrons that is not involved in chemical bonding.

mass (nucleon) number: the number of particles (protons and neutrons) in the nucleus.

mobile phase: the phase that moves in chromatography.

model: a simplified version that allows us to make predictions and understand observations more easily.

molar mass, M: the mass per mole of a substance. The units of molar mass are g mol⁻¹.

mole: the amount of any substance containing as many particles as there are carbon atoms in exactly 12 g of the carbon-12 isotope.

mole fraction: a measure of how much of a given substance is present in a reaction mixture.

molecular formula: shows the numbers and type of the atoms of each element in a compound.

molecular ion, M^+: the positive ion formed in mass spectrometry when a molecule loses an electron.

monomers: small molecules that are used to make polymers.

neutralisation: a chemical reaction in which an acid and a base react together to produce a salt and water.

nitrile: an organic chemical with a –CN functional group.

nucleophile: a species that contains a lone pair of electrons or a negative charge and is attracted to positive areas of a molecule.

optical isomers: molecules which are non-superimposable mirror images of each other. They have the same chemical properties but interact with polarised light differently.

order: with respect to a reactant, the order is the power to which the concentration of the reactant is raised in the rate equation.

overall order: the overall order of a reaction is the sum of the individual orders, $m + n$.

oxidation: loss of electrons, or an increase in oxidation number.

oxidation number: a measure of the number of electrons that an atom uses to bond with atoms of another element. Oxidation numbers are derived from a set of rules.

oxidising agent: the species that is reduced in a reaction and causes another species to be oxidised.

partial pressure: the pressure an individual gaseous substance would exert if it occupied a whole reaction vessel on its own.

phase: a physically distinctive form of a substance.

phenols: a class of aromatic compounds where a hydroxyl group is directly attached to the aromatic ring.

pH $= -\log[H^+(aq)]$. $[H^+(aq)] = 10^{-pH}$

pi (π) bond: sideways overlap of adjacent p-orbitals above and below the bonding C atoms.

polar molecule: a molecule with an overall dipole, having taken into account any dipoles across bonds.

polymer: a macromolecule made from small repeating units.

precipitation reaction: the formation of a solid from a solution during a chemical reaction. Precipitates are often formed when two aqueous solutions are mixed together.

qualitative analysis: an observable change and does not involve observations using numerical values.

rate constant, k: the constant that links the rate of reaction with the concentrations of the reactants raised to the powers of their orders in the rate equation.

rate-determining step: the slowest step in the reaction mechanism of a multi-step reaction.

rate equation: for a reaction A + B ⊠ C, the rate equation is given by rate $= k[A]^m[B]^n$, where m is the order of reaction with respect to A and n is the order of reaction with respect to B.

rate of reaction: the change in concentration of a reactant or a product per unit time.

reaction mechanism: a series of steps that, together, make up the overall reaction; a model with steps to explain and predict a chemical reaction.

recrystallisation: a method for purifying organic compounds.

redox reaction: a reaction in which both reduction and oxidation take place.

reducing agent: the species that is oxidised in a reaction and causes another species to be reduced.

reduction: gain of electrons, or a decrease in oxidation number.

reflux: a technique used to stop reaction mixtures boiling dry.

relative molecular mass: the weighted mean mass of a molecule of a compound compared with one-twelfth of the mass of an atom of carbon-12.

repeat unit: the arrangement of atoms that occurs many times in a polymer.

retention time: the time taken for a component to travel from the inlet to the detector in a gas chromatograph.

retro-synthesis: where the synthetic route is designed by looking at the target molecule and working backwards.

R_f value: a comparison between how far a component has moved compared to the solvent in thin layer chromatography.

salt: any chemical compound formed from an acid when an H^+ ion from the acid has been replaced by a metal ion or another positive ion, often a metal or ammonium ion, NH_4^+.

saturated hydrocarbon: a hydrocarbon with single bonds only.

second electron affinity: the enthalpy change accompanying the formation of 1 mole of gaseous $2-$ ions from 1 mole of gaseous $1-$ ions.

second ionisation energy: the energy change that accompanies the formation of 1 mole of gaseous 2+ ions from 1 mole of gaseous 1+ ions.

shell: a group of atomic orbitals with the same principal quantum number, n. Also known as a main energy level.

skeletal formula: a simplified structural formula drawn by removing hydrogen atoms from alkyl chains.

spectator ions: ions that are present but that play no part in a chemical reaction.

spin–spin coupling: the interaction between spin states of non-equivalent nuclei that results in a group of peaks in an NMR spectrum.

standard conditions: standards set for experimental measurements to allow comparisons to be made between different sets of data. Usually set at 100 kPa and 298 K.

standard electrode potential of a half cell, $E^{\ominus}$: the e.m.f. of a half cell compared with a standard hydrogen half-cell, measured at 298 K with solution concentrations of 1 mol dm^{-3} and a gas pressure of 100 kPa.

standard enthalpy change of atomisation, $\Delta_a H^{\ominus}$: the enthalpy change that accompanies the formation of 1 mole of gaseous atoms from the element in its standard state.

standard enthalpy change of formation, $\Delta_f H^{\ominus}$: the enthalpy change that accompanies the formation of 1 mole of a compound from its elements.

standard enthalpy change of hydration, $\Delta_{hyd} H^{\ominus}$: the enthalpy change that takes place when dissolving 1 mol of gaseous ions in water.

standard enthalpy change of solution, $\Delta_{sol} H^{\ominus}$: the enthalpy change that takes place when one mole of a solute is completely dissolved in water under standard conditions.

standard entropy, $S^{\ominus}$: the entropy content of one mole of a substance under standard conditions.

standard entropy change of reaction, $\Delta S^{\ominus}$: the entropy change that accompanies a reaction in the molar quantities expressed in a chemical equation under standard conditions, all reactants and products being in their standard states.

stationary phase: the phase that does not move in chromatography.

stem: the longest carbon chain present in an organic molecule.

stereoisomers: species with the same structural formula but with a different arrangement of the atoms in space.

stoichiometry: the molar relationship between the relative quantities of substances taking part in a reaction.

strong acid: an acid that completely dissociates in solution.

structural formula: provides the minimum detail to show the arrangement of atoms in a molecule.

structural isomers: compounds with the same molecular formula but different structural formula.

sub-shell: a group of the same type of atomic orbitals (s, p, d or f) within a shell. An 's' sub-shell can hold a maximum of 2 electrons, a 'p' sub-shell can hold a maximum of 6 electrons and a 'd' sub-shell can hold a maximum of 10 electrons.

substitution reaction: when an atom or group of atoms is replaced by a different atom or group of atoms.

synthetic route (or synthetic pathway): a series of reactions that can be used to change a starting chemical into a target molecule.

TMS: an internal standard for both carbon and proton NMR.

transition element (transition metal): a d-block element that has an incomplete d-sub-shell as a stable ion.

unsaturated hydrocarbon: a hydrocarbon containing multiple carbon-to-carbon bonds.

van der Waals' force: a type of intermolecular bonding that includes permanent dipole–dipole bonding and induced dipole–dipole interactions (London forces).

water of crystallisation: water molecules that form an essential part of the crystalline structure of a compound.

weak acid: an acid that partially dissociates in solution.

zwitterion: an internal salt, with no charge, formed by the donation of a proton from a carboxylic acid functional group to the amine functional group in an amino acid.

Group

(1)	(2)												(3)	(4)	(5)	(6)	(7)	(0)
1	2												13	14	15	16	17	18

Key

atomic (proton) number
atomic symbol
name
relative atomic mass

Period																			
1	1 **H** hydrogen 1.0																		2 **He** helium 4.0
2	3 **Li** lithium 6.9	4 **Be** beryllium 9.0											5 **B** boron 10.8	6 **C** carbon 12.0	7 **N** nitrogen 14.0	8 **O** oxygen 16.0	9 **F** fluorine 19.0	10 **Ne** neon 20.2
3	11 **Na** sodium 23.0	12 **Mg** magnesium 24.3	3	4	5	6	7	8	9	10	11	12	13 **Al** aluminium 27.0	14 **Si** silicon 28.1	15 **P** phosphorus 31.0	16 **S** sulfur 32.1	17 **Cl** chlorine 35.5	18 **Ar** argon 39.9
4	19 **K** potassium 39.1	20 **Ca** calcium 40.1	21 **Sc** scandium 45.0	22 **Ti** titanium 47.9	23 **V** vanadium 50.9	24 **Cr** chromium 52.0	25 **Mn** manganese 54.9	26 **Fe** iron 55.8	27 **Co** cobalt 58.9	28 **Ni** nickel 58.7	29 **Cu** copper 63.5	30 **Zn** zinc 65.4	31 **Ga** gallium 69.7	32 **Ge** germanium 72.6	33 **As** arsenic 74.9	34 **Se** selenium 79.0	35 **Br** bromine 79.9	36 **Kr** krypton 83.8
5	37 **Rb** rubidium 85.5	38 **Sr** strontium 87.6	39 **Y** yttrium 88.9	40 **Zr** zirconium 91.2	41 **Nb** niobium 92.9	42 **Mo** molybdenum 95.9	43 **Tc** technetium (98)	44 **Ru** ruthenium 101.1	45 **Rh** rhodium 102.9	46 **Pd** palladium 106.4	47 **Ag** silver 107.9	48 **Cd** cadmium 112.4	49 **In** indium 114.8	50 **Sn** tin 118.7	51 **Sb** antimony 121.8	52 **Te** tellurium 127.6	53 **I** iodine 126.9	54 **Xe** xenon 131.3
6	55 **Cs** caesium 132.9	56 **Ba** barium 137.3	57 **La*** lanthanum 138.9	72 **Hf** hafnium 178.5	73 **Ta** tantalum 180.9	74 **W** tungsten 183.8	75 **Re** rhenium 186.2	76 **Os** osmium 190.2	77 **Ir** iridium 192.2	78 **Pt** platinum 195.1	79 **Au** gold 197.0	80 **Hg** mercury 200.6	81 **Tl** thallium 204.4	82 **Pb** lead 207.2	83 **Bi** bismuth 209.0	84 **Po** polonium (209)	85 **At** astatine (210)	86 **Rn** radon (222)
7	87 **Fr** francium (223)	88 **Ra** radium (226)	89 **Ac*** actinium (227)	104 **Rf** rutherfordium (267)	105 **Db** dubnium (268)	106 **Sg** seaborgium (269)	107 **Bh** bohrium (270)	108 **Hs** hassium (269)	109 **Mt** meitnerium (278)	110 **Ds** darmstadtium (281)	111 **Rg** roentgenium (280)	112 **Cn** copernicium (285)		114 **Fl** flerovium (289)		116 **Lv** livermorium (293)		

58 **Ce** cerium 140.1	59 **Pr** praseodymium 140.9	60 **Nd** neodymium 144.2	61 **Pm** promethium (145)	62 **Sm** samarium 150.4	63 **Eu** europium 152.0	64 **Gd** gadolinium 157.3	65 **Tb** terbium 158.9	66 **Dy** dysprosium 162.5	67 **Ho** holmium 164.9	68 **Er** erbium 167.3	69 **Tm** thulium 168.9	70 **Yb** ytterbium 173.1	71 **Lu** lutetium 175.0
90 **Th** thorium 232.0	91 **Pa** protactinium 231.0	92 **U** uranium 238.0	93 **Np** neptunium (237)	94 **Pu** plutonium (244)	95 **Am** americium (243)	96 **Cm** curium (247)	97 **Bk** berkelium (247)	98 **Cf** californium (251)	99 **Es** einsteinium (252)	100 **Fm** fermium (257)	101 **Md** mendelevium (256)	102 **No** nobelium (259)	103 **Lr** lawrencium (262)

Index